Alfredo Soldati · Cristian Marchioli

Fluid Mechanics
for Mechanical Engineers

 Springer

Alfredo Soldati
Institute of Fluid Mechanics and
Heat Transfer
TU Wien
Wien, Austria

Department of Engineering
and Architecture
University of Udine
Udine, Italy

Cristian Marchioli
Department of Engineering
and Architecture
University of Udine
Udine, Italy

ISBN 978-3-031-53952-7   ISBN 978-3-031-53950-3   (eBook)
https://doi.org/10.1007/978-3-031-53950-3

This Springer imprint is published by the registered company Springer Nature Switzerland AG
The registered company address is: Gewerbestrasse 11, 6330 Cham, Switzerland

Paper in this product is recyclable.

Fluid Mechanics for Mechanical Engineers

Preface

This textbook describes the fundamentals of the phenomena of fluid dynamics in the context of engineering instances. It is designed to replace introductory books and notes on the subject for first-level engineering courses as well as higher-level courses or for professional use. The use of this book requires the basic knowledge of mathematics and physics normally delivered in the early years of undergraduate study. However, the extensive use of examples and solved exercises proposes a parallel intuitive route to understand the necessary mathematical formalisms. It proves that a new fluid dynamics text should not contain new ideas or formalisms, but should present the material in a modern and intuitive way. The approach chosen is primarily practical, so that that readers can practice by solving the proposed problems and examples in order to be prepared to solve the new problems they will encounter in their academic and professional activities. This book serves as a teaching tool for courses in basic fluid dynamics, advanced fluid dynamics, turbulence, and aerodynamics.

Wien, Austria
Udine, Italy

Alfredo Soldati
Cristian Marchioli

Contents

Part I
Fundamental Concepts and Scaling Laws

Chapter 1
Introduction and Fundamentals

1.1 Physical Properties of Fluids

The aim of this textbook is to provide a structured introduction to fluid dynamics to process and mechanical engineers, presenting a unified method of analysis that involves three steps. First, the fundamental principles of fluid dynamics are discussed. Second, the governing equations and the physical models stemming from first principles are presented. Third, equations are solved considering a wide range of relevant engineering examples. It is worthwhile to note that many of the problems that will be analyzed in this book only admit an approximate solution and that the precise identification of the flow field in complex geometries is, in many cases, still an unsolved physical problem. This is why, in the broad area of applied sciences, fluid mechanics is and will continue to be a field of intense research activity. In recent years, research has received a considerable boost from the development of numerical techniques and the availability of increasingly powerful computing infrastructures, which have led to a growing interest in the solution of complex fluid dynamic problems relevant to both science and engineering. Examples include industrial applications such as multiphase flows, combustion, propulsion, fluid machinery and aerodynamics, as well as environmental applications, such as oceanic or atmospheric flows, rain formation and ice crystals growth in clouds.

Fluids can be defined as any material that cannot sustain a tangential, or shearing, force when at rest and that undergoes a continuous change in shape when subjected to such a stress. Fluid mechanics is the branch of physics concerned with the response of fluids subjected to the action of the forces exerted upon them and is based on one fundamental hypothesis: The continuity of the flow field, which implies that the length scale over which the physical properties and velocity of the fluid vary is large with respect to the distance between the molecules of the fluid. This hypothesis ensures that the number of molecules contained in an elementary control volume is extremely high and, therefore, that it is not possible to identify the behavior of the individual molecules. The behavior of fluids is assumed to be the same as they were perfectly continuous in structure, and their physical properties can be regarded as being spread uniformly over the elementary control volume.

Fluids commonly encountered in nature can be classified as either *gases* or *liquids*. Gases tend to fill with a uniform concentration the entire volume of the container into which they are placed. Liquids have their own volume and take on the shape of the container that contains them. In addition, there are fluid systems with special characteristics: (i) Two-phase systems consisting of a gas and a liquid, a gas and a solid or a liquid and a solid; (ii) Multi-phase systems consisting of a gas phase, one or more liquid phases and one or more solid phases.

1.1.1 Density

The density of a fluid is defined as the ratio between the mass and the volume of a fluid element. Usually, the density is denoted by the symbol ρ and is measured in kilograms per cubic meter (kg/m^3) in the SI system, which will be adopted as the reference system hereinafter. As the density of a fluid may vary in space, it is convenient to define it as:

$$\rho = \lim_{\Delta V \to 0} \frac{\Delta m}{\Delta V} \quad , \tag{1.1}$$

where Δm and ΔV are the mass and the volume of the fluid element. The assumption of continuity of the fluid implies that, even in the limit $\Delta V \to 0$ given in Eq. (1.1), the volume of the fluid element is still much larger than the inter-molecular distance.

The density of a fluid is a function of temperature and pressure. This dependence is very limited for liquids and is more pronounced for gases, for which at low pressure or high temperature the ideal gas law applies:

$$\frac{p}{\rho} = \frac{R\,T}{M} \quad , \tag{1.2}$$

where p is the pressure acting on the gas, expressed in Pascal (Pa) or Newton per square meter (N/m^2), T the temperature in Kelvin (K), M the molecular weight of the gas, expressed in kilograms per kilomole (kg/kmol) and R is the universal gas constant, equal to $8, 314$ Joule per kilomole per Kelvin (J/kmolK).

1.1.2 *Viscosity*

Viscosity measures the resistance to flow exerted by a fluid when subject to the action of a shear or tensile stress. Viscosity is usually denoted by the symbol μ and its SI unit is Pa · s.

The definition of viscosity can be obtained from experiments. Let us consider two parallel plates of area A, separated by a layer of fluid having thickness H so that $H \ll \sqrt{A}$. By applying a constant force F to one of the plates, after a short transient, the plate moves with uniform velocity U. It is observed that:

1. for given values of U and H, the force F is proportional to the area A (hence, the ratio F/A is constant);
2. the ratio F/A is a monotonically increasing function of the ratio U/H.

The shear stress, τ, is defined as:

$$\tau = \frac{F}{A} \ . \tag{1.3}$$

We also introduce the strain rate, Γ_s, defined as:

$$\Gamma_s = \frac{U}{H} \ . \tag{1.4}$$

Experimental observations indicate that:

$$\tau = \tau(\Gamma_s) \ , \tag{1.5}$$

and:

$$\frac{d\tau}{d\Gamma_s} > 0 \ . \tag{1.6}$$

For most of the liquids, the experimental curves linking τ with Γ_s are straight lines (for which $d\tau/d\Gamma_s$ is constant). This leads to the following definition of viscosity:

$$\mu = \frac{\tau}{\Gamma_s} \ , \tag{1.7}$$

where the SI units of τ are N/m^2 or Pa and the SI unit of Γ_s is s^{-1}. Fluids with viscosity that does not depend on τ are classified as Newtonian fluids (note that viscosity, however, may still depend on temperature and pressure).

Fluids with viscosity that does depend on the applied strain rate, i.e. Γ_s, are classified as non-Newtonian fluids. In general, the following relation holds:

$$\tau = \mu(\Gamma_s) \cdot \Gamma_s \ . \tag{1.8}$$

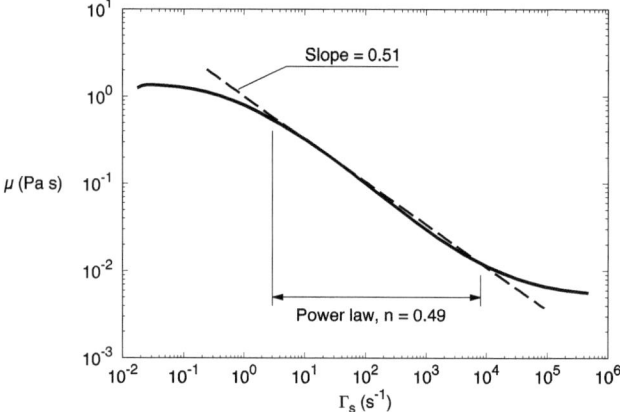

Fig. 1.1 Behavior of viscosity, μ, as a function of the strain rate, Γ_s, for non-Newtonian fluids

Derivation with respect to Γ_s yields:

$$\frac{d\tau}{d\Gamma_s} = \mu(\Gamma_s)\left[1 + \frac{\Gamma_s}{\mu(\Gamma_s)}\frac{d\,\mu(\Gamma_s)}{d\Gamma_s}\right] = \mu(\Gamma_s)\left[1 + \frac{d\,\ln\,\mu(\Gamma_s)}{d\,\ln\Gamma_s}\right]\quad,$$

and, since $d\tau_s/d\Gamma_s > 0$, it follows that:

$$\frac{d\ln\,\mu(\Gamma_s)}{d\ln\,\Gamma_s} > -1\quad,\tag{1.9}$$

which states that, on a logarithmic scale, the viscosity cannot decrease with slope greater than -1 when expressed as a function of Γ_s.

An example of non-Newtonian fluids is provided by the case of fluids whose viscosity follows a power law, as shown in Fig. 1.1. For these fluids, $\mu(\Gamma_s) = K \mid \Gamma_s \mid^{n-1}$, (with K constant and $n > 0$). In other cases, the viscosity increases with Γ_s.

Viscosity can be measured by different methods, one of which is illustrated in the next section. Viscosity can be also calculated using empirical correlations and widely reported theoretical equations, available for both gases and liquids. The reader is referred to the textbooks cited in the list of references for more details on these theoretical equations.

1.1.3 Viscosity Measurement

One simple method to measure viscosity exploits the relative motion between two plates, as for instance the lateral surfaces of two concentric cylinders shown in Fig. 1.2. The difference H between the radii of the two cylinders is much smaller

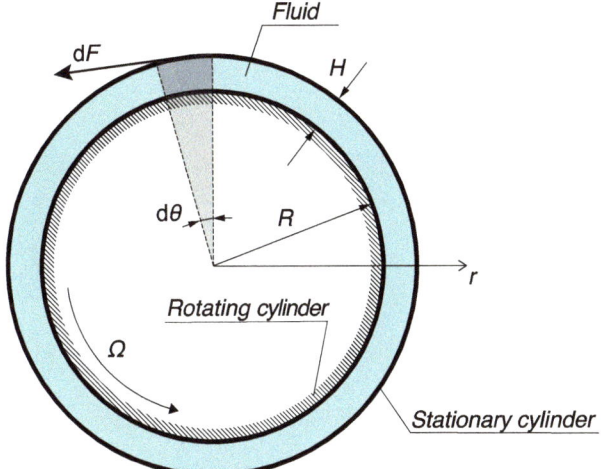

Fig. 1.2 Concentric cylinders for the measurement of viscosity. The distance H between the two cylinders is very small ($H \ll R$). The inner cylinder rotates at constant speed Ω while the outer cylinder is stationary

than both the radius of the inner cylinder, R, and the height of the cylinders, L. The inner cylinder rotates with angular velocity Ω. The tangential velocity is therefore ΩR and $\Gamma_s = R\Omega/H$. The viscous force exerted by the fluid on an elemental surface of area $\mathrm{d}A = LR\mathrm{d}\theta$ on the inner cylinder is given by:

$$\mathrm{d}F = \tau \mathrm{d}A = \mu \Gamma_s \mathrm{d}A = \mu \frac{R\Omega}{H} RL\mathrm{d}\theta \quad , \tag{1.10}$$

and the total torque acting on the cylinder, G, is given by:

$$G = \int_A \mathrm{d}G = \int_0^{2\pi} \mu \frac{R^3 L\Omega}{H} \mathrm{d}\theta \quad , \tag{1.11}$$

and, once G is known (measured), viscosity can be obtained as:

$$\mu = \frac{GH}{2\pi R^3 L\Omega} \quad . \tag{1.12}$$

1.1.4 Surface Tension

At the interface between two fluids, the intermolecular forces are not balanced as they are within the fluid, and a tension usually occurs between the fluids. This tension is

Fig. 1.3 Equilibrium
condition for a gas bubble
(only half bubble is
shown). The bubble is
assumed to be spherical
(with radius R) and
characterized by an
internal pressure p_i. The
bubble is subject to an
external pressure, p_e, and
to surface tension, σ

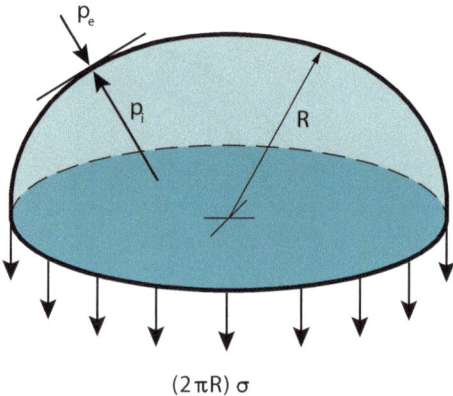

$(2\pi R)\, \sigma$

denoted by the symbol σ and termed surface tension. It is a force per unit length and
its SI units is Newton per meter, N/m. Surface tension is the force that allows soap
bubbles to form or raindrops to maintain a spherical shape, to make a few examples.

To understand the role of surface tension, let us consider a spherical gas bubble
immersed in a liquid. The pressure p_i inside the bubble is larger than the pressure
p_e of the surrounding liquid. If we consider just one half of the bubble, as shown
in Fig. 1.3, the excess pressure acts along the direction normal to the surface of the
bubble, giving rise to a net force only in the direction normal to the one-half cross
section of the bubble (the force component acting along the direction parallel to
such cross section cancels out by symmetry). This net force is equal to $\pi R^2 \Delta p$, with
$\Delta p = p_i - p_e$ and, according to the scheme of Fig. 1.3, is balanced by the surface
tension acting along the perimeter of the cross section, which is a circumference of
radius R. The force balance reads as:

$$\pi R^2 p_i = \pi R^2 p_e + 2\pi R\sigma \quad \rightarrow \quad \pi R^2 \Delta p = 2\pi R\sigma \quad , \tag{1.13}$$

along the direction of action of σ. From Eq. (1.13), we obtain:

$$\Delta p = \frac{2\sigma}{R} \quad . \tag{1.14}$$

Note that the pressure difference Δp is zero in the case of a planar interface, for
which $R \rightarrow \infty$.

1.2   Vector Notation

In a three-dimensional flow field, the position, velocity, and forces acting on a fluid element are vector quantities, the stress terms and the terms describing the convective transport of momentum are described by second-order tensors. It is therefore useful to recall or define the main algebraic and differential operations between vectors and tensors, which will be used in the following to derive the governing equations of fluid dynamics.

1.2.1   Vector and Tensor Algebra

Let us consider a Cartesian frame of reference, having axes denoted by x_1, x_2 and x_3 and the corresponding unit vectors denoted by $\delta_1, \delta_2, \delta_3$.[1] A vector $\mathbf{A}$ can be defined in terms of its three components A_i:

$$\mathbf{A} = \sum_i A_i \delta_i \quad . \tag{1.15}$$

The *scalar product* $\mathbf{A} \cdot \mathbf{B}$ of two vectors $\mathbf{A}$ e $\mathbf{B}$ is given by:

$$\mathbf{A} \cdot \mathbf{B} = \sum_i A_i B_i \quad . \tag{1.16}$$

If φ is the angle between the vectors $\mathbf{A}$ and $\mathbf{B}$, we also have:

$$\mathbf{A} \cdot \mathbf{B} = AB \cos \varphi \quad . \tag{1.17}$$

The *vector product* $\mathbf{A} \times \mathbf{B}$ of two vectors $\mathbf{A}$ and $\mathbf{B}$ can be defined as:

$$\mathbf{A} \times \mathbf{B} = \begin{vmatrix} \delta_1 & \delta_2 & \delta_3 \\ A_1 & A_2 & A_3 \\ B_1 & B_2 & B_3 \end{vmatrix} \quad , \tag{1.18}$$

or, in a more compact notation:

$$\mathbf{A} \times \mathbf{B} = \sum_i \sum_j \sum_k \varepsilon_{ijk} \delta_i A_j B_k \quad , \tag{1.19}$$

---

[1] In the following chapters, for ease of notation, axes and unit vectors will be denoted by x, y and z and $\mathbf{i}$, $\mathbf{j}$ and $\mathbf{k}$, respectively.

where the components ε_{ijk} of the tensor, ε, are defined as:

$$\begin{aligned}
\varepsilon_{ijk} &= 1 \quad \text{if } ijk = 123, 312, 231 \quad, \\
\varepsilon_{ijk} &= -1 \quad \text{if } ijk = 321, 132, 213 \quad, \\
\varepsilon_{ijk} &= 0 \quad \text{if two or more indexes are equal} \quad.
\end{aligned} \tag{1.20}$$

The magnitude of the vector $\mathbf{A} \times \mathbf{B}$ is $AB \sin \varphi$, where the angle φ between the two vectors is measured counterclockwise. The vector $\mathbf{A} \times \mathbf{B}$ is normal to the plane defined by the vectors $\mathbf{A}$ and $\mathbf{B}$ and is directed upward for $\varphi < \pi$. For the vector product, the commutative and associative properties do not apply.

The following vector calculus identities can be easily demonstrated:

$$\mathbf{A} \cdot \mathbf{A} = A^2 \ , \quad A = \left(\sum_i A_i^2 \right)^{1/2} \ , \tag{1.21}$$

$$\mathbf{A} \times \mathbf{A} = 0 \ , \tag{1.22}$$

$$\mathbf{A} \times (\mathbf{B} \times \mathbf{C}) = \mathbf{B}(\mathbf{A} \cdot \mathbf{C}) - \mathbf{C}(\mathbf{A} \cdot \mathbf{B}) \ , \tag{1.23}$$

$$\mathbf{A} \cdot (\mathbf{B} \times \mathbf{C}) = \mathbf{B} \cdot (\mathbf{C} \times \mathbf{A}) \ , \tag{1.24}$$

$$(\mathbf{A} \times \mathbf{B}) \cdot (\mathbf{C} \times \mathbf{D}) = (\mathbf{A} \cdot \mathbf{C})(\mathbf{B} \cdot \mathbf{D}) - (\mathbf{A} \cdot \mathbf{D})(\mathbf{B} \cdot \mathbf{C}) \ . \tag{1.25}$$

The *tensor product* of two vectors is defined as:

$$\mathbf{AB} = \begin{pmatrix} A_1 B_1 & A_1 B_2 & A_1 B_3 \\ A_2 B_1 & A_2 B_2 & A_2 B_3 \\ A_3 B_1 & A_3 B_2 & A_3 B_3 \end{pmatrix} \ . \tag{1.26}$$

The *scalar product* of two tensors, σ and τ, is defined as:

$$\sigma : \tau = \sum_i \sum_j \sigma_{ij} \tau_{ji} \ . \tag{1.27}$$

The *vector product* (or cross product) between a tensor and a vector is given by:

$$\mathbf{A} = \tau \cdot \mathbf{B} = \sum_i \delta_i \left\{ \sum_j \tau_{ij} B_j \right\} \ , \tag{1.28}$$

or:

$$\mathbf{C} = \mathbf{B} \cdot \tau = \sum_i \delta_i \left\{ \sum_j B_j \tau_{ji} \right\} \quad . \tag{1.29}$$

Note that $\mathbf{A} \neq \mathbf{C}$ unless the tensor τ is symmetric.

1.2.2 Differential Operators

The vector differential operator *Del* or *Nabla*, denoted by the symbol ∇, is a mathematical operator used to find higher dimensional derivatives and is defined as:

$$\nabla \equiv \sum_i \delta_i \frac{\partial}{\partial x_i} \quad . \tag{1.30}$$

If *Nabla* is applied to a scalar function $a(x_i)$ the resulting vector is referred to as the gradient of a and is denoted as:

$$\nabla a \equiv \sum_i \delta_i \frac{\partial a}{\partial x_i} \quad . \tag{1.31}$$

If ∇ is applied to a vector $\mathbf{A}$, the quantity $\nabla \cdot \mathbf{A}$ is referred to as divergence of $\mathbf{A}$ and is denoted as:

$$\nabla \cdot \mathbf{A} \equiv \sum_i \frac{\partial A_i}{\partial x_i} \quad . \tag{1.32}$$

Note that $\nabla \cdot A \neq \mathbf{A} \cdot \nabla$, where the differential operator $\mathbf{A} \cdot \nabla$ is defined as:

$$\mathbf{A} \cdot \nabla \equiv \sum_i A_i \frac{\partial}{\partial x_i} \quad . \tag{1.33}$$

For instance:

$$(\mathbf{A} \cdot \nabla)a = \sum_i A_i \frac{\partial a}{\partial x_i} \quad , \tag{1.34}$$

$$(\mathbf{A} \cdot \nabla)\mathbf{B} = \sum_i A_i \frac{\partial \mathbf{B}}{\partial x_i} \quad . \tag{1.35}$$

The scalar product $\nabla \cdot \nabla$, denoted by the symbol ∇^2, is the *Laplacian* or *Laplace operator*, after the French mathematician, physicist and astronomer Pierre-Simon Laplace (1749–1827):

$$\nabla^2 \equiv \sum_i \frac{\partial^2}{\partial x_i^2} \quad . \tag{1.36}$$

The Laplacian ∇^2 can be applied to both scalars and vectors.

The vector product between ∇ and $\mathbf{A}$ yields the *curl* of $\mathbf{A}$:

$$\nabla \times \mathbf{A} \equiv \begin{vmatrix} \delta_1 & \delta_2 & \delta_3 \\ \frac{\partial}{\partial x_1} & \frac{\partial}{\partial x_2} & \frac{\partial}{\partial x_3} \\ A_1 & A_2 & A_3 \end{vmatrix} \quad , \tag{1.37}$$

or, equivalently:

$$\nabla \times \mathbf{A} \equiv \sum_i \sum_j \sum_k \varepsilon_{ijk} \delta_i \frac{\partial A_k}{\partial x_j} \quad . \tag{1.38}$$

For continuous and derivable functions, the following identities apply:

$$\nabla \times \nabla a \equiv 0 \quad , \tag{1.39}$$

$$\nabla \cdot (\nabla \times \mathbf{A}) \equiv 0 \quad , \tag{1.40}$$

since, in both cases, terms of the following type appear:

$$\left(\frac{\partial^2 a}{\partial x_i \partial x_j} - \frac{\partial^2 a}{\partial x_j \partial x_i} \right) \text{ or } \left(\frac{\partial^2 A_i}{\partial x_j \partial x_k} - \frac{\partial^2 A_i}{\partial x_k \partial x_j} \right) \quad .$$

The following vector identities can also be easily demonstrated:

$$\nabla \cdot (a\mathbf{A}) = a(\nabla \cdot \mathbf{A}) + (\mathbf{A} \cdot \nabla)a \quad , \tag{1.41}$$

$$\nabla \times (a\mathbf{A}) = a(\nabla \times \mathbf{A}) + (\nabla a) \times \mathbf{A} \quad , \tag{1.42}$$

$$(\mathbf{A} \cdot \nabla)\mathbf{A} = \nabla \left(\frac{A^2}{2} \right) - \mathbf{A} \times (\nabla \times \mathbf{A}) \quad , \tag{1.43}$$

$$\nabla \times (\mathbf{A} \times \mathbf{B}) = \mathbf{A}(\nabla \cdot \mathbf{B}) + (\mathbf{B} \cdot \nabla)\mathbf{A} - \mathbf{B}(\nabla \cdot \mathbf{A}) - (\mathbf{A} \cdot \nabla)\mathbf{B} \quad , \tag{1.44}$$

$$\nabla \times (\nabla \times \mathbf{A}) = \nabla(\nabla \cdot \mathbf{A}) - \nabla^2 \mathbf{A} \quad . \tag{1.45}$$

Finally, we recall that when the differential operator ∇ is applied to a vector or scalar field bounded by a surface S, Gauss theorem applies:

$$\int_V \nabla a \, dV = \int_S a\mathbf{n} \, dS \quad,$$
(1.46)

$$\int_V \nabla \cdot \mathbf{A} \, dV = \int_S \mathbf{A} \cdot \mathbf{n} \, dS \quad,$$
(1.47)

where dV is the volume element enclosed by the surface S and $\mathbf{n}$ is the unit vector normal to the surface element dS. By convention, $\mathbf{n}$ is positive when pointing towards the unbounded domain (outward orientation).

1.2.3 *Examples*

1 Express the vectors $\mathbf{u}$, $\mathbf{v}$ and $\mathbf{w}$ in terms of their components relative to an orthogonal frame of reference and verify the following vector identity:

$$\mathbf{w} \times (\mathbf{u} \times \mathbf{v}) = \mathbf{u}(\mathbf{w} \cdot \mathbf{v}) - \mathbf{v}(\mathbf{w} \cdot \mathbf{u}) \quad.$$
(1.48)

Solution The product $\mathbf{u} \times \mathbf{v}$ is equal to:

$$\mathbf{u} \times \mathbf{v} = \mathbf{a} = \delta_1(u_2 v_3 - u_3 v_2) + \delta_2(u_3 v_1 - u_1 v_3) + \delta_3(u_1 v_2 - u_2 v_1) =$$

$$= a_1 \delta_1 + a_2 \delta_2 + a_3 \delta_3 \quad,$$
(1.49)

and the left-hand side of Eq. (1.48) becomes:

$$\mathbf{w} \times \mathbf{a} = \delta_1(w_2 a_3 - w_3 a_2) + \delta_2(w_3 a_1 - w_1 a_3) + \delta_3(w_1 a_2 - w_2 a_1) =$$

$$= \delta_1[w_2(u_1 v_2 - u_2 v_1) - w_3(u_3 v_1 - u_1 v_3)] + \delta_2[w_3(u_2 v_3 - u_3 v_2) - w_1(u_1 v_2 - u_2 v_1)] +$$

$$+ \delta_3[w_1(u_3 v_1 - u_1 v_3) - w_2(u_2 v_3 - u_3 v_2)] \quad.$$
(1.50)

The right-hand side of Eq. (1.48) is:

$$\mathbf{u}(\mathbf{w} \cdot \mathbf{v}) - \mathbf{v}(\mathbf{w} \cdot \mathbf{u}) =$$

$$= \mathbf{u}(w_1 v_1 + w_2 v_2 + w_3 v_3) - \mathbf{v}(w_1 u_1 + w_2 u_2 + w_3 u_3) =$$

$$= \sum_{i=1}^{3} \delta_i \left[u_i \left(\sum_{j=1}^{3} w_j v_j \right) - v_i \left(\sum_{j=1}^{3} w_j u_j \right) \right] \quad.$$
(1.51)

Direct term-by-term comparison of the resulting vectors proves the assertion.

1.3 Fluid Statics

1.3.1 Forces Acting on a Quiescent Fluid (Pascal's Law)

A fluid element experiences two different types of forces. First, the surface that separates the fluid element from neighboring fluid elements or from a solid wall can be subject to a force per unit area, denoted here by σ, which in general has both normal and tangential components to the surface. These are referred to as *surface forces*. Second, forces can be also applied to the volume of the fluid. In this case, they are referred to as *volume forces* or *body forces*. An example of a body force is the gravity force. The intensity and direction of the gravity force are given by the product of the fluid parcel mass by the gravity vector **g**.

Surface forces depend both on the *pressure* acting in the direction normal to the surface enclosing the fluid element, and on the relative motion between neighboring fluid elements that cause *viscous stresses*, which can be either normal or tangential. In the absence of motion, the viscous stresses are zero and the only force acting on any surface passing through a given point P within the fluid, is due to pressure.

Since an isotropic fluid has no preferential directions, the pressure acting on point P when the fluid is at rest cannot take different values for different orientations of the surface passing through P. Therefore, at point P, σ must be aligned in the direction as the unit vector **n** normal to the surface and must the same magnitude for any possible orientation of **n**. This result is known as *Pascal's Law*, after the French mathematician, physicist, philosopher and theologian Blaise Pascal (1623–1662).

To show that the magnitude of σ is independent of the orientation of **n**, let us consider a force balance on the fluid element shown in Fig. 1.4. The stresses acting in

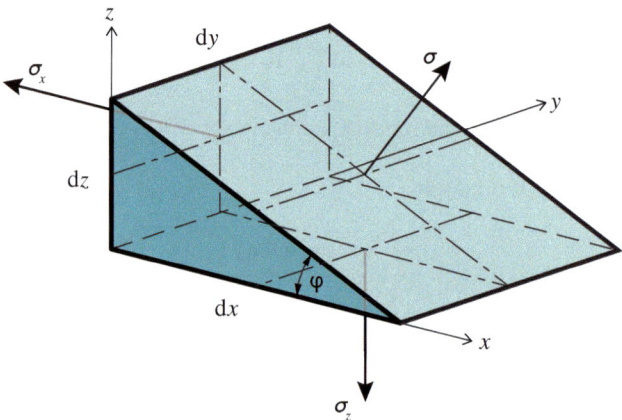

Fig. 1.4 Schematic representation of the stress components σ (forces per unit area) acting on the faces of a fluid element at rest. For clarity of visualization, only some of the stress components are represented

the direction normal to the two faces $dydz$ and $dy(dz/\sin\varphi)$ are denoted respectively by σ_x and $\boldsymbol{\sigma}$. Since the fluid element is at rest, the sum of the forces acting in the x direction must be zero:

$$- \sigma_x dydz + (\sigma \sin\varphi)\left(dy\frac{dz}{\sin\varphi}\right) = 0 \ , \tag{1.52}$$

from which it follows that:

$$(-\sigma_x + \sigma)dydz = 0 \ \rightarrow \ \sigma_x = \sigma \ . \tag{1.53}$$

When considering an infinitesimal element, the normal stresses σ_x and σ are computed at the same point P. In each case, the magnitude of the stress must be the same in each direction since one could have chosen the directions y or z instead of x. Since fluids can only sustain compressive stresses, like those generated by pressure, we identify as pressure p the stress that has the magnitude of σ and direction opposite to $\mathbf{n}$:

$$\boldsymbol{\sigma} = (-p)\mathbf{n} \ . \tag{1.54}$$

Using Eq. (1.54), namely Pascal's law, one can determine the force exerted on a volume V of fluid by the stress acting on the surface S that encloses V. With reference to the schematic of Fig. 1.5, one can integrate the pressure force per unit area, $-p\mathbf{n}$, to compute the pressure force vector:

$$\text{pressure force} = \int_S (-p\mathbf{n})dS \ . \tag{1.55}$$

It is also possible to express this pressure force as a volume integral by means of the Gauss theorem (1.46):

$$\text{pressure force} = \int_V (-\nabla p)dV \ . \tag{1.56}$$

Fig. 1.5 Schematic representation of a fluid volume V enclosed by a surface S

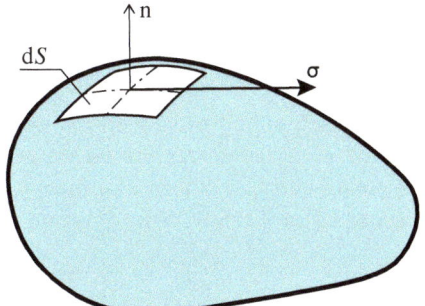

Since the total pressure force exerted on the volume is given by $-\nabla p$ integrated over the volume, the integrand $-\nabla p$ represents a *pressure force per unit volume* at each point in the fluid:

$$\text{pressure force per unit of volume} = -\nabla p \quad . \tag{1.57}$$

Each fluid element having infinitesimal volume $dxdydz$ will experience a pressure force caused by the pressure difference on the faces of the volume element. For example, the pressure difference in the x direction is $(\partial p/\partial x)dx$ and the corresponding force difference is $(\partial p/\partial x)dx(dydz)$, so that the total pressure force is $\nabla p(dxdydz)$.

1.3.2 Pressure Distribution in a Quiescent Fluid

In a quiescent fluid, the sum of the pressure force per unit volume, $-\nabla p$ and the gravity force per unit volume, $\rho \mathbf{g}$, must be zero:

$$-\nabla p + \rho \mathbf{g} = 0 \quad . \tag{1.58}$$

This equation, also known as *Stevin's Law* after the Flemish engineer, physicist and mathematician Simon Stevin of Bruges (1548–1620), expresses the force balance that must hold for each point in a quiescent fluid. The equation shows how the pressure p increases in the direction of $\mathbf{g}$, being $\nabla p = \rho \mathbf{g}$. In Eq. (1.58) the vector $\mathbf{g}$ can be set equal to:

$$\mathbf{g} = \nabla (\mathbf{g} \cdot \mathbf{R}) \quad , \tag{1.59}$$

where $\mathbf{R}$ is the position vector $\mathbf{R} = x\mathbf{i} + y\mathbf{j} + z\mathbf{k}$. Regardless of the orientation of the axes x, y and z, the scalar product $\mathbf{g} \cdot \mathbf{R}$ is equal to the product of the magnitude of $\mathbf{g}$ by the component of $\mathbf{R}$ in the direction of $\mathbf{g}$, denoted here by $-h$ where h represents the vertical distance from some reference elevation. It follows that, for a fluid of uniform density, Eq. (1.58) can be written as follows:

$$p + \rho g h = \text{constant} \quad . \tag{1.60}$$

1.3.3 Pressure Distribution in a Compressible Fluid

The pressure acting on a gas depends on the vertical height z at which it is measured. The dependence of pressure on the height can be easily derived for an ideal gas assuming static conditions. It suffices to combine Eq. (1.2), in which $p = p(z)$, $\rho = \rho(z)$ and $T = T(z)$, with the only non-zero scalar component of Eq. (1.58):

$$-\frac{dp}{dz} + \rho g_z = 0 \quad , \tag{1.61}$$

with $g_z = -g$ assuming that the vertical axis of the frame of reference is oriented upward. Equation (1.2) yields:

$$\rho(z) = \frac{p(z)}{T(z)} \frac{M}{R} \quad , \tag{1.62}$$

Upon substitution of Eq. (1.62) into Eq. (1.61), one finds:

$$\frac{\mathrm{d}p}{p(z)} = -\frac{Mg}{R} \frac{1}{T(z)} \mathrm{d}z \quad . \tag{1.63}$$

This equation can be integrated once the expression of $T(z)$ is known. For atmospheric air, for instance, field measurements show that, up to an altitude of about 10 km, the temperature decreases with increasing altitude according to the following empirical law:

$$T(z) = T_0 - \alpha z \quad , \tag{1.64}$$

where T_0 is the (known) air temperature at the reference altitude z_0. The constant α quantifies the linear decrease of temperature with altitude and is expressed in Kelvin per meter, K/m. Substituting Eq. (1.64) into Eq. (1.63) yields:

$$p(z) = p_0 \left(\frac{T_0 - \alpha z}{T_0 - \alpha z_0} \right)^{\frac{Mg}{R\alpha}} \quad , \tag{1.65}$$

where $p_0 = p(z = z_0)$ is the (known) air pressure at the reference altitude z_0. The equation for density can be obtained combining Eq. (1.65) with Eq. (1.62):

$$\rho(z) = \frac{Mp_0}{RT_0} \left(\frac{T_0 - \alpha z}{T_0 - \alpha z_0} \right)^{\frac{Mg}{R\alpha} - 1} = \rho_0 \left(\frac{T_0 - \alpha z}{T_0 - \alpha z_0} \right)^{\frac{Mg}{R\alpha} - 1} \quad . \tag{1.66}$$

1.3.4 *Pressure Forces on Solid Surfaces*

It is often useful to know the total force $\mathbf{F}$ and the torque $\mathbf{T}$ acting on the walls of a container filled with a quiescent fluid. It is possible to calculate $\mathbf{F}$ and $\mathbf{T}$ by integrating the force $\mathrm{d}\mathbf{F}$ and the torque $\mathrm{d}\mathbf{T}$ exerted by the fluid on a surface element $\mathrm{d}S$ of the walls. With reference to Fig. 1.6, $\mathrm{d}\mathbf{F}$ and $\mathrm{d}\mathbf{T}$ can be expressed as:

$$\mathrm{d}\mathbf{F} = p\mathbf{n}\mathrm{d}S \quad , \tag{1.67}$$

$$\mathrm{d}\mathbf{T} = \mathbf{R} \times p\mathbf{n}\mathrm{d}S \quad . \tag{1.68}$$

Fig. 1.6 Pressure force
per unit surface area
acting on a surface
element dS. Pressure p
acts along the direction
normal to the surface,
indicated by the unit
vector **n**, while **R** is the
position vector

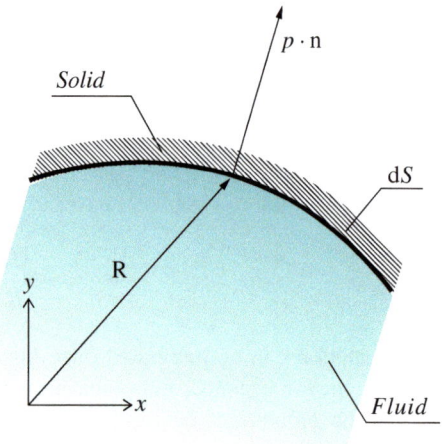

Note that the normal vector **n** points towards the unbounded domain (outward orientation with respect to the surface containing the fluid, inward orientation with respect
to the solid wall). Integration over the surface S, yields:

$$\mathbf{F} = \int_S p\mathbf{n}dS \quad , \tag{1.69}$$

$$\mathbf{T} = \int_S \mathbf{R} \times p\mathbf{n}dS \quad . \tag{1.70}$$

In general, these integrals do not apply to an enclosed volume but, rather, to a
portion of the wall over which the fluid forces and torques are exerted. The pressure
forces acting on the surface can be replaced with a single force **F** applied at a specific
point $\mathbf{R}_{cp}$, called *center of pressure*, such that the same torque is produced:

$$\mathbf{R}_{cp} \times \mathbf{F} = \mathbf{T} \quad . \tag{1.71}$$

Note that the torque of the pressure forces is zero when computed with respect to the
center of pressure.

When calculating pressure forces on walls that are completely or partially surrounded by a gas at atmospheric pressure, the absolute pressure p can be replaced
with the relative pressure, $p - p_{atm}$, since the total pressure force acting on the wall
is independent of the magnitude of atmospheric pressure:

$$\mathbf{F} = \int_S p\mathbf{n}\mathrm{d}S = \int_S (p - p_{atm})\mathbf{n}\mathrm{d}S + \int_S p_{atm}\mathbf{n}\mathrm{d}S =$$

$$= \int_S (p - p_{atm})\mathbf{n}\mathrm{d}S + \int_V \nabla p_{atm}\mathrm{d}V = \int_S (p - p_{atm})\mathbf{n}\mathrm{d}S \quad . \qquad (1.72)$$

In this derivation, the Gauss theorem has been used to derive the volume integral, which is zero being $\nabla p_{atm} = 0$. It follows that *a uniform pressure applied to the surface of a solid wall produces no net force or moment on the wall.*

1.3.5 Archimedes' Principle

Consider a body completely enclosed by a solid surface and immersed in a fluid, like the one shown in Fig. 1.7, which we assume to have a spherical shape for ease of discussion. According to Pascal's law, the distribution of the hydrostatic pressure p around the body generates a force per unit area σ normal to the surface of the body, in the direction of the unit vector $\mathbf{n}$. This force per unit area is proportional to p, according to Eq. (1.54). Because of the finite size of the body, the pressure distribution is not isotropic, as it is for an infinitesimal fluid element (which can be approximated to a material point). In particular, it is not isotropic along the vertical direction, with the consequence that p increases linearly with depth, Eq. (1.58), as shown in the right-hand plot in Fig. 1.7. The body will therefore be subject to a pressure force that changes with depth: the pressure force exerted by the fluid on the lower part of the immersed body will be larger than the pressure force exerted on the upper part and will generate a non-zero hydrostatic upward thrust. In the horizontal direction, on the other hand, the depth does not change and therefore the pressure distribution always results in a zero net force.

The total pressure force acting on the body, termed *buoyancy force* and denoted by $\mathbf{F}_g$, can be easily calculated using Eq. (1.69). Indeed, the integral of the pressure force with respect to the surface S of the body can be rewritten as a volume integral with respect to the volume of fluid V displaced by the body:

$$\mathbf{F}_g = \int_S p\mathbf{n}\mathrm{d}S = -\int_V \nabla p\mathrm{d}V = -\int_V \rho\mathbf{g}\mathrm{d}V = -\rho\mathbf{g}V \quad . \qquad (1.73)$$

In this derivation, we have used the Gauss theorem and Eq. (1.58) to replace the pressure gradient with the gravity force per unit volume. Equation (1.73) states that *the pressure force (or buoyancy force) for a body immersed in a fluid is equal in magnitude and opposite in direction to the gravity force acting on the displaced fluid.* This is known as Archimedes' principle. The position of the center of buoyancy of a body partially or fully immersed in a fluid, $\mathbf{R}_g$, is defined as the center of mass of the displaced fluid:

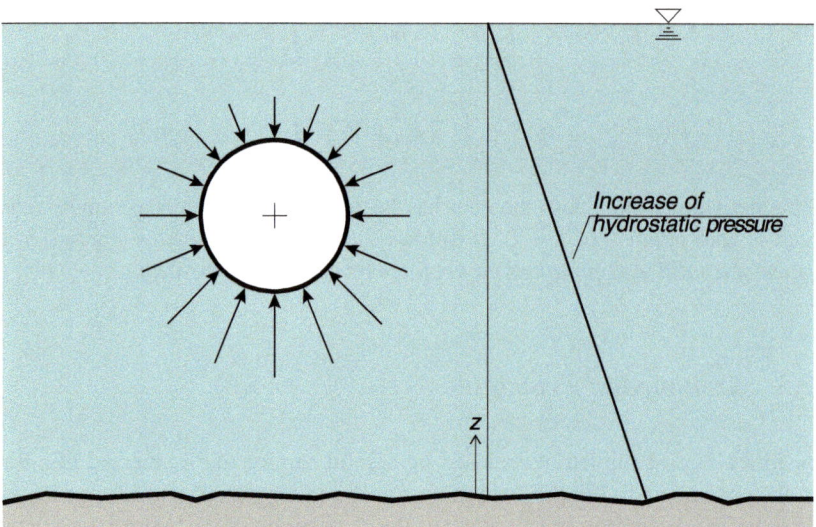

Fig. 1.7 Pressure force distribution around a body of spherical shape. The body is at rest and completely immersed in the fluid. Because of the finite size of the body and the hydrostatic distribution of pressure around it, the latter is not isotropic along the vertical direction

$$\mathbf{R}_g = \frac{1}{V} \int_V \mathbf{R} \mathrm{d}V \quad .$$ (1.74)

When a body floats at the interface between two fluids (e.g. a boat on water), both fluids will contribute to the total buoyancy force according to a quantity $\rho g V$, equal to the gravitational force acting on the displaced volume of each fluid. However, for a body floating on water, the density of air is so small compared to that of water that its contribution to the total buoyancy force exerted on the body is negligible: The gravity force acting on the displaced volume of water is predominant. This is always the case for gases with respect to liquids.

1.3.6   Hydraulic Transmission of Forces

Pressurized fluids can be exploited to exert forces on pistons that are used to drive mechanical components. Since pressure can be transmitted through long pipes, the force can be applied at a point far away from the source of the pressurized fluid. For example, the hydraulic braking system of a car is designed to activate the wheel brakes when a force is exerted by a pushrod on the piston(s) in a master cylinder, causing fluid contained in the brake reservoir to flow into a pressure chamber through a compensating port. This results in an increase in the pressure of the entire hydraulic

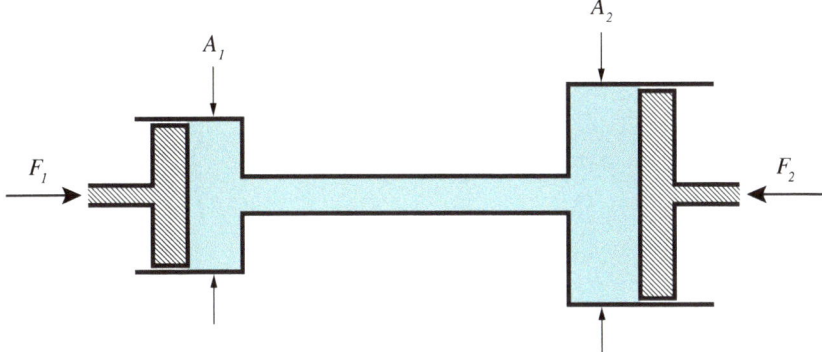

Fig. 1.8 Schematic of hydraulic transmission system and force amplification

system. Since the fluid transmits the same pressure within the whole system, just as the electric potential is held constant within a conducting wire, an identical braking force will be applied to each wheel, thus stopping the vehicle. In such a system, the pressure p is much greater than both the product of ρg and the difference in height between any two given points in the braking system, so that the pressure within the system is essentially constant.

One property of hydraulic force transmission systems is that they can amplify the intensity of the actuating force, as if a mechanical lever was used. This principle is illustrated in Fig. 1.8, which shows two piston-cylinder pairs connected by a pipe filled with oil. The smaller piston, having area A_1, is subjected to a force F_1 that pressurizes the oil in the cylinder at a pressure $p_1 = F_1/A_1$. If there is no flow, the larger piston, having area A_2, must apply a resisting force F_2 so that $p_2 = p_1$. Hence:

$$\frac{F_1}{A_1} = p_1 = p_2 = \frac{F_2}{A_2} \rightarrow F_2 = \left(\frac{A_2}{A_1}\right) F_1 \ . \tag{1.75}$$

If $A_2 \gg A_1$, the applied force F_1 induces a much greater force F_2 on the mechanism to be operated.

1.3.7 Pressure Measurement

The standard device for measuring atmospheric pressure is the mercury barometer. As can be seen in Fig. 1.9, this tool consists of a glass tube approximately one meter in length and closed at one end. After being filled with mercury, the tube is turned upside down and the open end is placed below the free surface of a tank containing mercury. The mercury in the column falls below the top end of the tube, leaving a space containing only mercury at a pressure, denoted by p_1, equal to the vapor pressure of mercury at room temperature, which can be set to zero for practical

Fig. 1.9 Sketch of a
mercury barometer

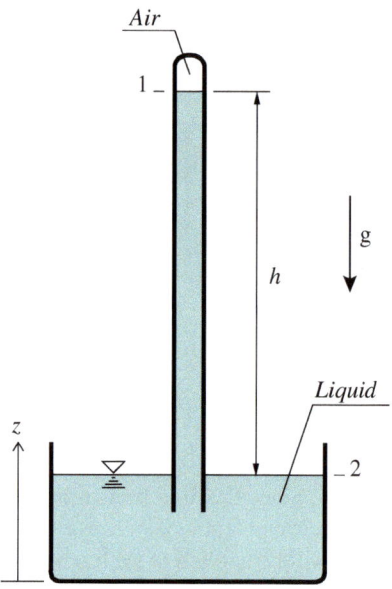

purposes. The atmospheric pressure at the level of the free surface in the tank is
calculated by measuring the vertical distance h between the height of the mercury in
the column (Sect. 1.1) and the free surface in the tank (Sect. 1.2), as shown in Fig. 1.9.
Using the equation of hydrostatic equilibrium, Eq. (1.60):

$$p_1 + \rho g h_1 = p_2 + \rho g h_2 \quad , \tag{1.76}$$

from which we obtain:

$$p_2 = p_1 + \rho g (h_1 - h_2) = \rho g h \quad , \tag{1.77}$$

where we impose $p_1 = 0$. The density of mercury at 0 °C is $1.36 \cdot 10^4$ kg/m^3,
and the product ρg is equal to $1.3337 \cdot 10^5$ Pa/m, using for g the standard value
$g = 9.8066$ m/s^2. If the atmospheric pressure is equal to the standard one, 101325 Pa,
Eq. (1.77) yields $h = 0.760$ m $= 760$ mm.

The operating principle of the barometer can be also used to measure pressure
in enclosed spaces using a manometer, Fig. 1.10. A U-shaped glass tube is partially
filled with a liquid, typically water or mercury. One end of the tube is open to the
atmosphere, while the other end is connected to the container within which the
pressure is to be measured. Applying Eq. (1.60) to the manometric fluid (of density
ρ_m) between sections 1 and 2, we have:

$$p_2 + \rho_m g h_2 = p_1 + \rho_m g h_1 \quad , \tag{1.78}$$

Fig. 1.10 Schematic of a U-shaped differential manometer for measuring pressure within containers

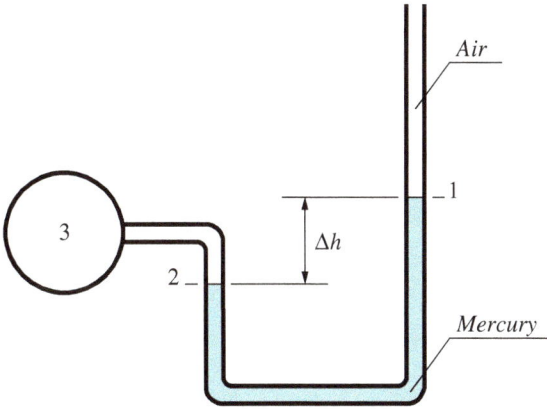

from which we obtain:

$$p_2 = p_1 + \rho_m g(h_1 - h_2) = p_{atm} + \rho_m g(h_1 - h_2) \ , \tag{1.79}$$

where we assume p_1 equal to the atmospheric pressure. Pressure p_2 is not necessarily equal to pressure p_3 at the center of the container if the fluid is a liquid. To account for this difference, we apply Eq. (1.60) to the fluid within the container (density ρ_c) between points 3 and 2:

$$p_3 + \rho_c g h_3 = p_2 + \rho_c g h_2 \ . \tag{1.80}$$

From Eqs. (1.79) and (1.80), we obtain:

$$p_3 = p_2 - \rho_c g(h_3 - h_2) = p_{atm} + \rho_m g(h_1 - h_2) - \rho_c g(h_3 - h_2) \ . \tag{1.81}$$

If the fluid in the container is a gas, the density ρ_c is much less than that of the manometric fluid, ρ_m, and hence p_2 and p_3 are essentially equal. On the other hand, if the fluid is a liquid, the pressure difference $p_2 - p_3 = \rho_c g(h_3 - h_2)$ may contribute significantly in determining the pressure in the container and cannot be neglected.

1.3.8 Examples

$\boxed{1}$ The barrage dam shown in Fig. 1.11 bounds a stream of water having width W and height H. Determine the force and torque due to water pressure on the dam.

Solution To calculate the pressure force, we determine the pressure acting on the vertical wall of the dam at a height h from the bottom, using the condition of hydrostatic equilibrium:

Fig. 1.11 Schematic of a barrage dam. The dam experiences a torque that tends to rotate it about the z axis but also a torque that tends to rotate it about the y axis

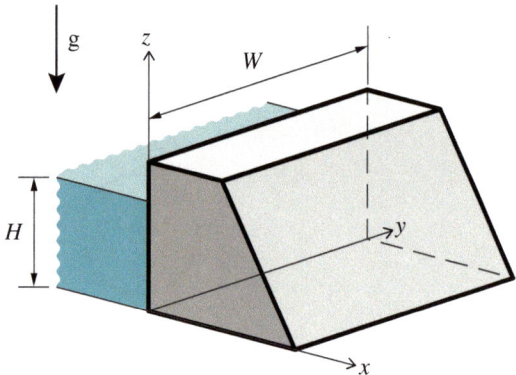

$$p(h) = p_1 + \rho g h_1 - \rho g h = \rho g H - \rho g h \quad , \tag{1.82}$$

where position 1 is on the free surface of water so that $p_1 = 0$ and $h_1 = H$. Substituting Eq. (1.82) into Eq. (1.69), the force $\mathbf{F}$ can be determined as:

$$\mathbf{F} = \int_S p\mathbf{n}\,dS = \int_0^W \int_0^H \rho g(H - h)\mathbf{i}\,dy\,dz = \rho g \left(\frac{WH^2}{2} \right) \mathbf{i} \quad . \tag{1.83}$$

Note that the average pressure exerted on the dam is $\rho g H/2$. The torque $\mathbf{T}$ is calculated by substituting Eq. (1.83) into Eq. (1.70):

$$\mathbf{T} = \int_S (\mathbf{R} \times p\mathbf{n})\,dS = \int_0^W \int_0^H (x\mathbf{i} + y\mathbf{j} + z\mathbf{k}) \times \rho g(H - h)\mathbf{i}\,dy\,dz =$$

$$= \rho g \int_0^W \int_0^H (z\mathbf{j} - y\mathbf{k})(H - h)\,dy\,dz = \rho g \left(\frac{WH^3}{6} \right) \mathbf{j} - \rho g \left(\frac{W^2 H^2}{4} \right) \mathbf{k} \quad . \tag{1.84}$$

2 The wall of a tank containing water has a hemispherical bump of diameter D, as shown in Fig. 1.12. The center of the bump is located at a distance h below the water level. Calculate the force acting on the bump.

Solution To solve this problem, it is possible to integrate the Eq. (1.69) over the surface of the hemispherical bump, thus obtaining the horizontal component, F_o, and the vertical component, F_v, of the force exerted by the water.

Alternatively, it can be assumed that the given surface bounds a closed hemispheric volume that contains water and is surrounded by water. This volume must be in equilibrium with the pressure forces acting on its surface. In the horizontal direction, the pressure force acting on the hemispherical portion of the surface must be balanced

Fig. 1.12 Schematic
sketch to compute the
pressure force on a
hemispherical bump

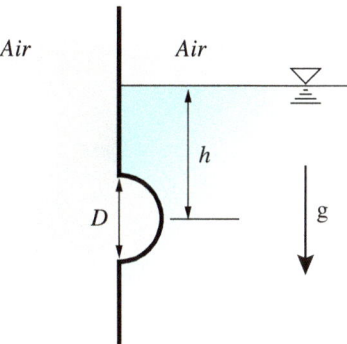

by the force acting on the horizontal projection of the surface, which is a vertical
flat circle of area $\pi D^2/4$. The pressure in the center of the circle is $p_c = \rho g h$, which
yields:

$$F_o = \rho g h \left(\frac{\pi D^2}{4} \right) \quad . \tag{1.85}$$

In the vertical direction, the pressure force exerted on the flat part of the surface is
zero and, therefore, one is left only with the upward pressure force, F_v, acting on the
hemispherical surface. This force must balance the gravitational pull exerted by the
liquid within the volume, which is equal to $\pi D^3/12$, and yields:

$$F_v = \frac{\pi \rho g D^3}{12} \quad . \tag{1.86}$$

3 A rectangular tank, partially filled with water, is inclined by an angle φ with
respect to the horizontal direction, as shown in Fig. 1.13. To calculate the stresses
in the tilted tank, it is necessary to know the pressure distribution within the fluid
as a function of the distances x and y measured from the bottom left corner of the
container. Derive an expression for $p(x, y)$.

Solution The components of the acceleration of gravity, **g**, along x and y are:

$$\mathbf{g} = g \sin \varphi \mathbf{i} - g \cos \varphi \mathbf{j} \quad . \tag{1.87}$$

Equation (1.60) can be rewritten as follows:

$$p - \rho \mathbf{g} \cdot \mathbf{R} = \text{constant} \quad , \tag{1.88}$$

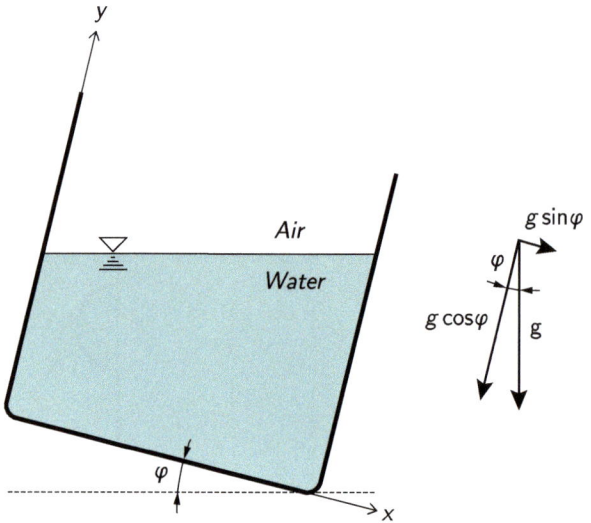

Fig. 1.13 Sketch of a tilted tank

where $\mathbf{R}$ is the position vector. The invariant $p - \rho\mathbf{g} \cdot \mathbf{R}$ becomes:

$$p - \rho\mathbf{g} \cdot \mathbf{R} = p - \rho(g \sin \varphi \mathbf{i} - g \cos \varphi \mathbf{j}) \cdot (x\mathbf{i} + y\mathbf{j} + z\mathbf{k}) =$$

$$= p - \rho g(x \sin \varphi - y \cos \varphi) \quad . \tag{1.89}$$

The resulting pressure distribution is:

$$p(x, y) = p_1 - \rho g(x_1 \sin \varphi - y_1 \cos \varphi) + \rho g(x \sin \varphi - y \cos \varphi) \quad . \tag{1.90}$$

For this pressure distribution, the horizontal lines, $y = x \tan \varphi + \text{constant}$, are isobars (lines of constant pressure).

4   A piece of soap floats on water surface, with its bottom part submerged for a height equal to D. The piece of soap has width W, thickness H and length L. Determine the density of soap.

Solution The density of soap, ρ_s, can be expressed as $\rho_r \rho$, where ρ is the density of water and ρ_r is the density of soap relative to that of water, $\rho_r = \rho_s/\rho$. The weight of the piece of soap is:

$$F_p = \rho_s g V = \rho_r \rho g W H L \quad . \tag{1.91}$$

According to Archimedes' principle, F_p must be balanced by buoyancy, which is equal to the gravitational force acting on the volume of fluid displaced by the soap:

$$F_g = \rho g W D L \quad . \tag{1.92}$$

Equating the two expressions, we have:

$$\rho_r \rho g W H L = \rho g W D L \quad , \tag{1.93}$$

and:

$$\rho_r = \frac{D}{H} \;\; \rightarrow \;\; \rho_s = \rho \frac{D}{H} \quad . \tag{1.94}$$

5   A large boat was capsized and sunk, and is now lying at a depth of 30 m below water level. The boat is 30 m long, 10 m deep and 10 m wide, with a total mass, m_b, of 300 tons. To recover the boat from the bottom, a marine crane is used, which has a lifting capacity, C_s, of 100 tons. It is necessary to remove water from inside the boat by pumping air until the force required to lift the boat is equal to the lifting capacity of the crane. Calculate the volume of water that must be removed so that the lifting force available is sufficient for recovery. Calculate the relative air pressure in the capsized boat.

Solution To calculate the volume of water to be removed, V_0, it is sufficient to calculate the Archimedes' upward thrust, F_g, that is created by removing that volume from inside the boat. Under static conditions, the following equilibrium of forces in the vertical direction applies:

$$F_s + F_p + F_g = 0 \quad . \tag{1.95}$$

where $F_s = C_s g$ is the magnitude of the lifting force of the crane, which is directed upward, $F_p = m_b g$ is the magnitude of the weight of the boat, which is directed downward, and $F_g = \rho g V_0$, with ρ density of water. Substitution of these expressions into Eq. (1.95) gives:

$$m_b g = C_s g + \rho g V_0 \quad , \tag{1.96}$$

and, therefore:

$$V_0 = \frac{m_b - C_s}{\rho} = 200 \, \text{m}^3 \quad . \tag{1.97}$$

To calculate the relative air pressure inside the capsized boat, we consider that, below sea level, the pressure inside the boat and outside is the same at a fixed level of depth:

$$p_{air} = p_{atm} + \rho g (30 - h) \quad . \tag{1.98}$$

Being:

$$h = \frac{V - V_0}{30 \cdot 10} = 9.3 \, \text{m} \quad , \tag{1.99}$$

we obtain:

$$p_{air} = 3.04 \cdot 10^5 \, \text{Pa} \quad , \tag{1.100}$$

and:

$$\Delta p_{rel} = p_{air} - p_{atm} = 2.04 \cdot 10^5 \, \text{Pa} \quad . \tag{1.101}$$

6 | Moving away from the Earth's surface, the temperature varies as a function of height (z) as $T(z) = T_0 - \alpha z$, with $T_0 = 293$ K at $z_0 = 0$ m above the sea-level and $\alpha = 5.0 \cdot 10^{-3}$ K/m. A small weather balloon of spherical shape, radius $R_0 = 1$ m and mass $m_p = 2.5$ kg is filled with a mass $m_g = 0.8$ kg of light gas. Assuming that the radius of the balloon remains constant, determine the height $\bar{z}$ at which the balloon stops.

Solution The total weight of the balloon is given by $p = (m_p + m_g)\mathbf{g}$, while the buoyancy force is $\mathbf{F}_g = -\rho V_s g$, where $V_s = 4\pi R_0^3/3$ is the volume of the balloon and ρ is the density of air (which is a function of height in this problem). Combining Eq. (1.66) with the equilibrium condition:

$$m_p + m_g = rho(\bar{z}) \frac{4}{3} \pi R_0^3 \quad , \tag{1.102}$$

we obtain:

$$\rho(\bar{z}) = \frac{3}{4} \frac{m_p + m_g}{\pi R_0^3} = \frac{M p_0}{R T_0} \left(\frac{T_0 - \alpha \bar{z}}{T_0} \right)^{\frac{Mg}{R\alpha} - 1} . \tag{1.103}$$

Solving Eq. (1.103) for $\bar{z}$, we find:

$$\bar{z} = \frac{T_0}{\alpha} \left[1 - \left(\frac{3}{4} \frac{m_p + m_g}{\pi R_0^3} \frac{R T_0}{M p_0} \right)^{\frac{R\alpha}{Mg - R\alpha}} \right] = 3943.5 \, \text{m} \quad . \tag{1.104}$$

1.3.9   Problems

a | An oil well, 4000 m deep, terminates in an oil reservoir that completely fills the well up to the surface. The oil pressure at the surface, p_1, is equal to $3 \cdot 10^6$ Pa.

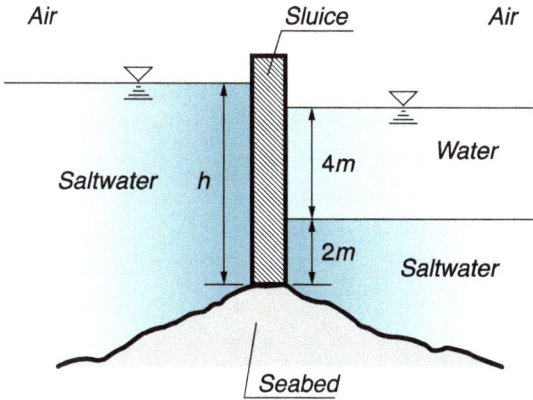

Fig. 1.14 Sketch of a sluice gate

1. Calculate the oil pressure at the bottom of the well, if the oil density is $\rho_o = 750 \text{ kg/m}^3$.
2. Drilling mud is a a heavy, viscous fluid mixture that is used in oil and gas drilling operations. The density ρ_m of drilling mud can be tuned and, therefore, set equal to an arbitrary value greater than that of water. To continue drilling beyond 4000 m, drilling mud is introduced within the wellbore to displace the oil. Calculate the density of the mud that is required to reduce the pressure at the top of the well to zero once the well is completely filled with mud.

$\boxed{b}$ A sluice gate separates a river basin from sea water. The river basin consists of an upper layer of fresh water with density $\rho_d = 1 \cdot 10^3 \text{ kg/m}^3$, 4 m deep, and a lower layer of saltwater with density $\rho_s = 1.03 \cdot 10^3 \text{ kg/m}^3$. The interface between fresh water and saltwater is located at an height of 2 m from the bottom of the sluice gate, as shown in Fig. 1.14. On the seawater side, the water level, h, also measured from the bottom of the sluice gate, can vary over time depending on the tide.

1. Calculate the level h at which the force exerted horizontally by the seawater on the sluice gate is exactly balanced by the force exerted by the two water layers in the basin.
2. Calculate the level h at which the moment exerted horizontally by the seawater on the sluice gate is exactly balanced by the moment exerted by the two water layers in the basin. Compute the moments considering the bottom of the sluice gate as pivot point.
3. Using the value of h computed at point 1., and assuming there is a crack at the bottom of the sluice, determine whether the water flows to or from the basin.

$\boxed{c}$ Determine the pressure on the surface of a sphere of radius R placed at a depth H below the free surface of a fluid of density ρ. Calculate the resulting surface force and verify that it is equal to the weight of the volume of fluid displaced by the sphere.

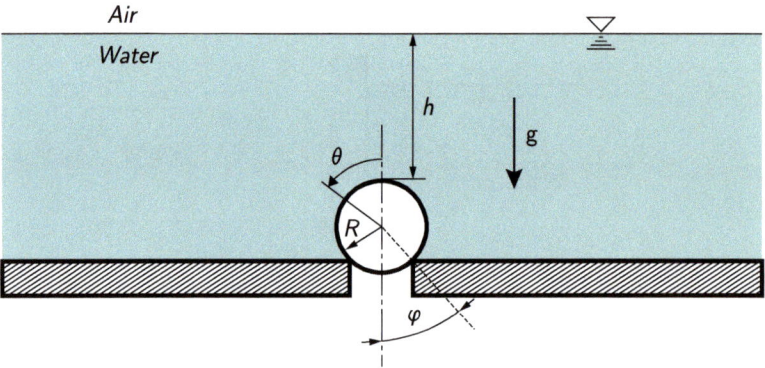

Fig. 1.15 Schematic of a sanitary drain

$\boxed{d}$ The foot valve of a sanitary drain consists of a sphere of radius R and negligible mass that plugs a circular opening at the bottom of a tank containing water, as shown in Fig. 1.15. The contact line between the sphere and the opening is located at an angle φ measured with respect to the vertical direction. The air/water interface at the top of the tank is located at a height h above the valve. The water pressure at the top of the sphere holds the sphere in place if the opening is too small, namely when the angle φ is very close to π.

1. Derive an expression for the relative pressure $p(\theta)$ on the surface of the sphere, being θ the angle measured with respect to to the vertical axis, as a function of the parameters R, h and the density ρ of water.
2. Derive an expression for the net force, F, directed downward due to water pressure as a function of the angle φ.
3. Calculate the minimum value of the ratio h/R for which $F > 0$ when $\varphi = 3\pi/4$.

$\boxed{e}$ In the North Sea, characterized by strong tidal fluctuations, a cylindrical tidal drum (with diameter $D = 0.4$ m and height $h = 0.8$ m) is used as a marker buoy. Under equilibrium conditions, the drum is submerged to an height $h_{sub} = 0.6$ h and is anchored to the seabed bottom with a rope of length $L = 9.5$ m.

1. Determine the mass of the tidal drum.
2. Due to the effect of the tide, the water level rises and at certain times of the day the bin is completely submerged and the top is under tension. The sea level varies over time t with the law:

$$H(t) = H_M + \Delta H \sin(\omega t)$$

with $H_M = 10$ m, $\Delta H = 2$ m, $\omega = 2\pi/T$ e $T = 12$ h. Determine how the buoyancy force acting on the tidal drum varies over time.

3. To reduce the tension acting on the rope, a hole is drilled at the bottom of the drum. Calculate how much water enters when the drum is free to float (equilibrium conditions) assuming that the air in the drum cannot escape and is initially at atmospheric pressure.

Chapter 2
Physical Models for Friction Forces

2.1 Dimensional Analysis

2.1.1 Introduction

The complexity of the equations of fluid dynamics has directed the research activity, since the origins of this scientific field, to combine theoretical analysis with careful experiments, which have often been the main source of new methods and models that can be used for the design and for the analysis of complex flow fields.

In fluid mechanics, the complexity of the flow field is primarily related to the occurrence of *turbulent flow* conditions. Through experimental observations, Northern Irish physicist and engineer Osborne Reynolds (1842–1912) was the first to note how the motion of a high viscosity fluid is characterized by the relative motion of adjacent fluid elements, referred to as *laminar flow* conditions. As the velocity of the fluid increases, the interaction between adjacent fluid elements becomes more pronounced and takes on the characteristics of chaotic mixing. The transition between laminar and turbulent flow conditions occurs quite sharply and is discussed in this chapter as well as in Chap. 8. Turbulence can be considered as a form of flow instability for which the velocity and pressure at each point vary over time, while maintaining, in the case of *steady-state conditions*, their average values constant.[1]

Changes in the flow field over time induce additional stresses than the viscous shear stresses described in the previous chapter and make the solution of the problem particularly difficult, even in the case of simple geometries. For example, the stationary motion of a fluid in a cylindrical duct can be described, from the mathematical point of view, rather simply in the case of laminar flow using a linear second-order ordinary differential equation with constant coefficients. Once the laminar/turbulent transition occurs, however, the system cannot be considered stationary anymore and

---

[1] Broadly speaking, the time-averaged values of the field variables are constant when the flow is at steady-state. A more rigorous definition will be given in Chap. 8.

A. Soldati and C. Marchioli, *Fluid Mechanics for Mechanical Engineers*,
https://doi.org/10.1007/978-3-031-53950-3_2

is described by a system of four partial differential equations in space and time. The solution of these equations, which are highly non-linear, is not feasible analytically and has become possible numerically only in recent years.

A second case that will be discussed in this chapter is the motion of a fluid around a spherical particle. In this case, the increased complexity associated to the particle geometry makes it difficult to solve for the equations of motion even in the case of laminar flow, with the exception very viscous fluids, a limiting case that will be discussed in Chap. 5.

The examples given in this chapter are all cases of practical interest, widely studied in the past, and analyzed using the principles of dimensional analysis, which represents a very powerful tool for the design of an experiment and the subsequent analysis of the results.

2.1.2 Buckingham's Theorem

The dimensional analysis of an experiment is based on identifying the physical and geometrical parameters that may influence the outcome of the experiment and on the consideration that any mathematical model describing the phenomenon to be investigated can be formulated in a dimensionless form, thus allowing for the generalization of the results.

To define the parameters of an experiment, it is first necessary to identify the fundamental physical quantities that characterize the experiment. In fluid mechanics, the fundamental quantities are mass, M, length, L, and time, T. Any physical parameter that affects the experiment can be expressed with these quantities.

The result of an experiment that depends on n parameters, p_i, with $i = 1, ..., n$, can be formulated as follows:

$$p_1 = f(p_2, p_3 \ldots p_n) \quad . \tag{2.1}$$

In this equation, the function $f(p_i, i \neq 1)$ is one of the possible forms in which the functional dependency among the n parameters that determine the outcome of the experiment can be expressed. The functional dependence (2.1) can be written in dimensionless form. This operation is made easy by the application of *Buckingham's Theorem*, which makes it possible to disregard the analytic form that the function $f(p_i, i \neq 1)$ takes. Buckingham's Theorem can be formulated as follows: *If an experiment depends on n parameters and involves m fundamental physical quantities, the results of the experiment can be described in terms of n − m independent dimensionless groups.*

This theorem, the proof of which is omitted, allows us to replace the functional dependence (2.1) with the following equation:

$$\pi_1 = g(\pi_i; i = 2, n - m) \quad , \tag{2.2}$$

where the terms π_i are the dimensionless groups that can be used to express the results of the experiment. It is clear that, when evaluating the results of an experiment, the use of the relation (2.2) instead of the relation (2.1) makes it possible to reduce the number of variables to be considered from n to $n - m$.

2.2 Friction Forces for Flows in Pipelines

2.2.1 Flow in Smooth Pipes

Assume that an incompressible Newtonian fluid flows in a horizontal pipe with length L and diameter D. Assume further that the inner wall of the pipe, which is in contact with the fluid, is perfectly smooth. Let ρ and μ be the density and viscosity of the fluid, respectively. The flowrate is defined as the mass or volume of fluid flowing through any given section of the pipe per unit time. The flowrate can be expressed as a function of the *mean velocity*, v, defined as the mean value of the fluid velocity over the pipe section:

$$\text{Volumetric flowrate:} \quad Q = v\frac{\pi D^2}{4} \tag{2.3}$$

$$\text{Mass flowrate:} \quad w = \rho v\frac{\pi D^2}{4} \quad . \tag{2.4}$$

Sometimes, the mass flow rate is also denoted as $\dot{m}$.

Let us now assume that, for the considered pipe flow, the average velocity is assigned. The fluid flows due to an imposed pressure difference Δp between the two ends of the pipe. Six variables and three dimensions are involved in the process, as indicated in Table 2.1.

Based on Buckingham's Theorem, three independent dimensionless groups can be introduced. Among the different dimensionless groups that can be obtained combining the variables reported in Table 2.1, it is convenient to use the following:

Table 2.1 Flow in a smooth pipe: Variables and physical quantities

Variable		Physical quantity
D	length	L
L	length	L
Δp	force/surface	$ML^{-1}T^{-2}$
v	length/time	LT^{-1}
ρ	mass/volume	ML^{-3}
μ	(force/area) × time	$ML^{-1}T^{-1}$

$$\frac{\Delta p}{\rho v^2} , \quad \frac{\rho v D}{\mu} , \quad \frac{L}{D} . \tag{2.5}$$

Using these groups, we can write the following functional dependency:

$$\frac{\Delta p}{\rho v^2} = g\left(\frac{\rho v D}{\mu}, \frac{L}{D}\right) , \tag{2.6}$$

which contains all the information that the dimensional analysis can provide. Since it is clear that, except for an initial portion of pipe that is needed for flow development, the pressure drop increases proportionally to the pipe length, Eq. (2.6) can be written as:

$$\frac{\Delta p}{\rho v^2} = \frac{L}{D} \cdot g\left(\frac{\rho v D}{\mu}\right) , \tag{2.7}$$

where the dimensionless group $\rho v D/\mu$ is the *Reynolds number*, Re. Hence, we have:

$$\frac{\Delta p}{\rho v^2} \frac{D}{L} = g(Re) . \tag{2.8}$$

The dimensionless group $\Delta p D/\rho v^2 L$ is proportional to the friction factor, which is defined as:

$$f = \frac{2\tau_w}{\rho v^2} , \tag{2.9}$$

where τ_w is the wall shear stress. This proportionality can be derived by considering that in a horizontal pipe: (a) the pressure forces acting at the two ends of the pipe are equal to $p_1(\pi D^2/4)$ and $p_2(\pi D^2/4)$, respectively, being p_1 and p_2 the pressures at the two ends, and therefore the net pressure force exerted on the fluid is $F_{\Delta p} = (p_1 - p_2)\pi D^2/4$; (b) the friction forces generated by the fluid in contact with the pipe wall is $F_\tau = \tau_w \pi D L$. Since the two forces must balance, it is found that $\tau_w = \Delta p D/4L$ and:

$$\frac{1}{2}\frac{\Delta p}{\rho v^2}\frac{D}{L} = f(Re) . \tag{2.10}$$

Equation (2.10), also known as Darcy-Weisbach equation, is of great importance because it allows to calculate the pressure difference that is required for the transport of any fluid in a horizontal pipe of any diameter and length. Furthermore, to determine Eq. (2.10), it is enough to perform a single experiment in which the physical properties as well as the flow geometry are assigned and constant and only the average fluid velocity is changed. The validity of relation (2.10) has been proven by experimental measurements, which also made it possible to determine the form of the function $f(Re)$. These measurements, presented in graphical form in Fig. 2.1, confirm that

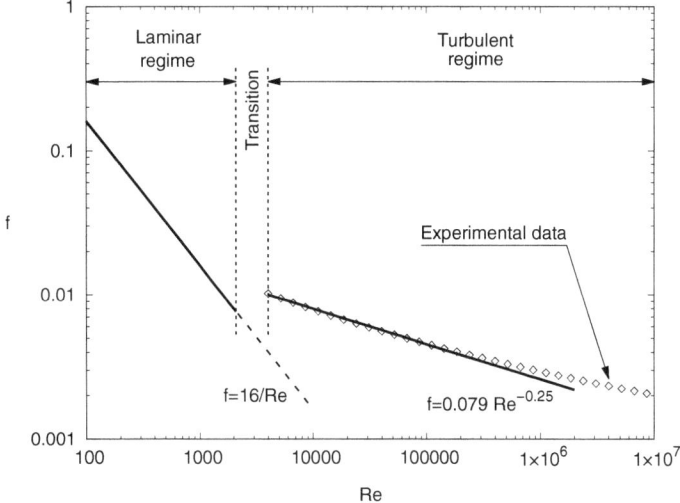

Fig. 2.1 Friction factor for flow in smooth pipes as a function of the Reynolds number

two main flow regimes can be distinguished. A laminar regime when $Re \leq 2100$, for which:

$$f = \frac{16}{Re} \quad , \tag{2.11}$$

and a turbulent regime when $Re \geq 4000$. In this case, the friction factor is given by the *Blasius* correlation, proposed by the German physicist and mathematician Paul Richard Heinrich Blasius (1883–1970):

$$f = 0.079 Re^{-1/4} \quad . \tag{2.12}$$

This correlation is applicable to smooth pipes, as it does not take into account the effects (in terms of generated friction) due to the surface roughness of the inner pipe wall.

In the range $2100 < Re < 4000$, there is a transition between the laminar regime and the turbulent regime. In this region, the flow is characterized by the occurrence of local instabilities that are amplified as the Reynolds number increases, resulting in a significant increase of the friction factor. In this transition zone, neither the Eq. (2.11) nor the Eq. (2.12) provide an accurate value for the friction coefficient, which can therefore be determined only by experimental measurements or numerical simulations.

A more complex equation for the calculation of the friction factor in the turbulent regime, which has the advantage of being valid over a wider range of Re compared to the Blasius correlation, is the von Kármán-Nikuradse equation, proposed by the

Hungarian engineer, physicist and mathematician Theodore von Kármán (1881–1963) and the Georgian engineer and physicist Johann Nikuradse (1894–1979):

$$\frac{1}{\sqrt{f}} = 1.7\ln(Re\sqrt{f}) - 0.4 \quad . \tag{2.13}$$

If we substitute Eqs. (2.11) and (2.3) in Eq. (2.8), we have:

$$Q = \frac{\pi}{128}\frac{\Delta p D^4}{L\mu} \quad . \tag{2.14}$$

This equation, valid for laminar flow, is known as the Hagen-Poiseuille equation and will be derived theoretically in Chap. 4. It can be used to measure the viscosity of a fluid.

2.2.2 Physical Meaning of the Reynolds Number

Consider a small mass m of fluid moving with velocity v and acceleration a along a given trajectory. The Reynolds number represents the ratio between the force associated with the acceleration of the mass of fluid (called inertial force) and the force arising from the action of shear stresses in the moving viscous fluid (called viscous force).

By defining acceleration a as the ratio of the change in velocity Δv of the mass m over an infinitesimal time interval Δt, it is possible to express the component of the inertial force along the trajectory as:

$$F_I = m \cdot a = m\frac{\Delta v}{\Delta t} = m\frac{\Delta v^2}{\Delta l} \quad , \tag{2.15}$$

where $\Delta t = \Delta l/\Delta v$ with Δl length covered by the mass in the interval Δt.

On the other hand, the viscous force is associated with the action of a shear stress $\tau = \mu \cdot \Gamma_s$, where the strain rate Γ_s is given by the ratio $\Delta v/\Delta l$. Therefore, we have:

$$F_v = \tau \cdot A = \mu\frac{\Delta v}{\Delta l} \cdot A \quad , \tag{2.16}$$

where A is the area of the surface over which the shear stress τ acts. If we assume $A \approx \Delta l^2$ and $m \approx \rho\Delta l^3$, the ratio between F_I and F_v is given by:

$$\frac{F_I}{F_v} = \frac{\rho\Delta v\Delta l}{\mu} \tag{2.17}$$

This expression is analogous to the definition of the Reynolds number Re that appears in the relation (2.7), with Δv replacing v and Δl replacing D.

Also the friction factor, f, has a physical meaning. It represents the ratio between the force applied to the fluid element, $\Delta p D^2$, and the inertial force, $\rho v^2 D^2$.

$$\frac{\Delta p \, D^2}{F_I} \approx f\left(\frac{F_I}{F_v}\right) \quad . \tag{2.18}$$

In the laminar regime, inertial forces are negligible and so the function f on the right-hand side of the functional dependence (2.18) must be independent of F_I. This implies:

$$\frac{\Delta p \, D^2}{F_I} \approx \frac{F_v}{F_I} = Re^{-1} \quad ,$$

and, in turn:

$$f \approx Re^{-1} \quad . \tag{2.19}$$

This last relation has been verified experimentally.

2.2.3 Power and Dissipation

Since the net pressure force acting on a fluid flowing in a horizontal pipe is equal to $F_{\Delta p} = (p_1 - p_2)\pi D^2/4$, the work J necessary to displace the fluid by a distance Δl between sections 1 and 2 over a time Δt, is $J = F\Delta l$. The corresponding power, P, equal to the work of the pressure force per unit time, is:

$$P = \frac{J}{\Delta t} = (p_1 - p_2)\frac{\pi D^2}{4}v = \Delta p Q \quad , \tag{2.20}$$

where Q is the volumetric flowrate. In an adiabatic system, the power is dissipated, resulting in an increase of the temperature of the fluid:

$$P = \Delta p Q = \rho c_v Q \Delta T \quad , \tag{2.21}$$

where ΔT is the temperature jump and c_v the specific heat.

2.2.4 Flows in Commercial Pipes

The internal wall surface of commercial pipes is not smooth but, rather, character-ized by a certain roughness, k, which measures the irregularity or unevenness of the surface. The SI unit of roughness is m. It is clear that k must be taken into account in the dimensional analysis of the process. As a consequence, an additional dimensionless group must be introduced, which is the ratio $\varepsilon = k/D$, referred to as *relative roughness*. Therefore, for the friction factor, we have:

$$f = f\left(Re, \frac{k}{D}\right) \quad . \tag{2.22}$$

Experimental measurements have demonstrated that f is independent of k/D in the laminar flow regime. In the turbulent flow regime, i.e. at high Reynolds numbers, f follows very closely the behavior characteristic of smooth pipes for low values of k/D. When both Re and k/D are sufficiently high, f becomes independent of Re because viscous forces are negligible and only depends on k/D. The most common equation for the calculation of f in commercial pipes is the *Colebrook equation*, proposed by C.F. Colebrook in 1939. This equation reads as:

$$\frac{1}{\sqrt{f}} = -1.7 \ln\left(\frac{k}{D} + \frac{4.67}{Re\sqrt{f}}\right) + 2.28 \quad , \tag{2.23}$$

and is graphically sketched in Fig. 2.2, where the so-called Moody diagram is shown. The Moody diagram is used extensively in calculating head loss and pressure drops in pipes in combination with the Darcy-Weisbach equation, Eq. (2.10). Figure 2.2 shows clearly that the friction factor in the turbulent regime increases, even by several orders of magnitude, as the relative roughness k/D increases.

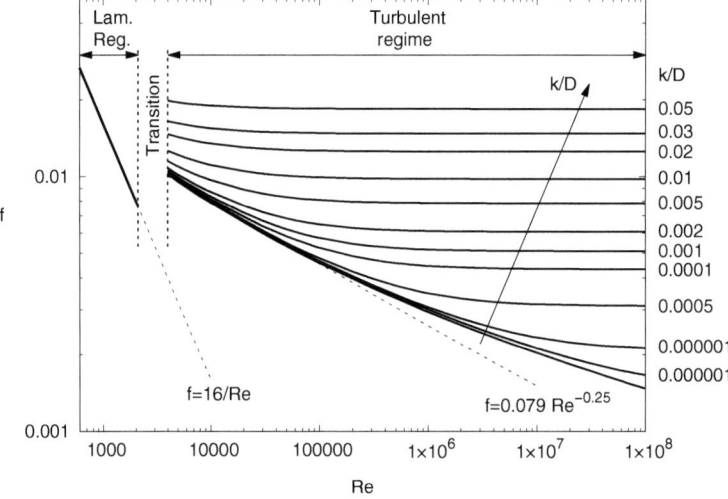

Fig. 2.2 Friction factor for flow in commercial pipes (Moody diagram)

2.2.5 *Flow in Pipes of Non-Circular Cross-Section*

To calculate the pressure loss in pipes or ducts of non-circular cross-section, it is useful to introduce the concept of *hydraulic diameter*, D_H, which represents the characteristic cross-sectional dimension of the pipe or duct. The hydraulic diameter is defined as:

$$D_H = \frac{4A_w}{\mathcal{P}_w} \quad , \tag{2.24}$$

where A_w is the wetted cross-sectional area of the pipe or duct and $\mathcal{P}_w$ is the wetted perimeter. An equivalent definition is:

$$D_H = \frac{4V_w}{S_w} \quad , \tag{2.25}$$

where V_w is the total wetted volume and S_W is the total wetted surface area. Both definitions yield $D_H = D$ for a smooth pipe with circular cross section.

For a duct with rectangular cross-section $a \times b$, b being the largest dimension, the hydraulic diameter is:

$$D_H = \frac{4ab}{2(a+b)} = \frac{2ab}{a+b} \quad . \tag{2.26}$$

In the limit $a \ll b$ (i.e., $a + b \simeq b$), the hydraulic diameter can be approximated as $D_H \simeq 2a$, i.e., the characteristic dimension of the duct is represented by the smaller dimension of the cross-section.

In general, the calculation of the friction factor for pipes or ducts with non-circular cross section, is carried out by replacing D with D_H in the correlations available for the calculation of the friction factor in pipes of circular cross section.

2.2.6 *Examples*

1 Consider the flow of water at 93 °C in a smooth pipe with diameter $D = 100$ mm. Knowing that the flowrate is 100 m³/h, compute the pressure loss per unit length along the pipe.

Solution The properties of water at $93^o C$ are $\rho = 960$ kg/m³ and $\mu = 0.31 \cdot 10^{-3}$ Pa · s. For the given flowrate and pipe diameter, the flow Reynolds number is:

$$Re = \frac{\rho v D}{\mu} = 1.096 \cdot 10^6 \quad . \tag{2.27}$$

with $v = 4Q/\pi D^2 = 3.54$ m/s. This value of Re indicates that the flow inside the pipe is within the turbulent regime. Therefore, the friction factor is:

$$f = 0.079\, Re^{-0.25} = 2.44 \cdot 10^{-3} \quad , \tag{2.28}$$

and the resulting pressure loss is:

$$\frac{|\Delta p|}{L} = \frac{2\rho v^2 f}{D} = \frac{32 w^2 f}{\pi^2 \rho D^5} = 586.1 \text{ Pa/m} \quad . \tag{2.29}$$

2 A smooth pipe with diameter $D = 50$ mm is used to transport a flowrate $Q = 0.01$ m^3/s of an oil with density 980 kg/m^3 and viscosity $\mu = 0.1$ Pa $\cdot$ s. Calculate the friction losses per unit length along the pipe.

Solution From Eq. (2.10), it follows that the friction losses per unit length, l_v/L, can be expressed as:

$$\frac{l_v}{L} = 2f\frac{v^2}{D} = \frac{32 f Q^2}{\pi^2 D^5} \quad , \tag{2.30}$$

where $l_v = \Delta p/\rho$. The Reynolds number is:

$$Re = \frac{\rho v D}{\mu} = \frac{4\rho Q}{\mu \pi D} = \frac{4 \cdot 980 \cdot 0.01}{0.1 \cdot 3.14 \cdot 0.05} = 2.50 \cdot 10^3 \quad . \tag{2.31}$$

Using Eq. (2.12), the friction factor is $f = 0.079 Re^{-0.25} = 0.011$. Therefore:

$$\frac{l_v}{L} = \frac{32 \cdot 0.011 \cdot 10^{-4}}{3.14^2 \cdot 0.05^5} = 11.6 \text{ m/s}^2 \quad . \tag{2.32}$$

The power losses per unit length are:

$$\frac{\Delta p \cdot Q}{L} = \frac{\rho l_v \cdot Q}{L} = 11.6 \cdot 980 \cdot 0.01 = 113.6 \text{ W/m} \quad . \tag{2.33}$$

3 A flowrate $Q = 0.01$ m^3/s of oil (density 980 kg/m^3 and viscosity $\mu = 0.01$ Pa $\cdot$ s) is transported through a commercial pipe ($k = 0.05$ mm) over a horizontal distance $L = 5$ km. Calculate the diameter of the pipe knowing that the maximum head that the pump can supply is $F = 10^3$ J/kg.

Solution The maximum head, F, is defined as:

$$F = 2f\frac{v^2}{D}L = f\frac{32 Q^2}{\pi^2 D^5}L = 1000 \text{ J/kg} \quad . \tag{2.34}$$

From this equation, the diameter of the pipe can be obtained as:

$$D = \left(f \frac{32 Q^2}{\pi^2 F} L \right)^{1/5} = 0.277 f^{1/5} \quad .$$ (2.35)

Likewise, we can express the Reynolds number as:

$$Re = \frac{1248}{D} \quad .$$ (2.36)

From Eq. (2.35), using an arbitrary tentative value for the friction factor, e.g. $f = 0.007$, we obtain $D \simeq 0.1$ m and, from Eq. (2.36), $Re = 12480$. Substituting the values of f, Re and $k/D = 5 \cdot 10^{-4}$ in the Colebrook equation, we obtain $f = 0.0076$, which results in $D \simeq 0.1$ m.

2.3 Friction Forces for Flow Past a Sphere

2.3.1 Steady Flow Past a Sphere

In many cases of industrial and environmental interest, a common problem is the motion of a spherical particle in still fluid or, alternatively, the flow of a fluid past a stationary sphere. The variables that must be taken into account to describe this problem are: The drag force, F_D, acting on the sphere; the velocity, v_p, at which, after an initial transient, the sphere moves with respect to the fluid; the diameter, D_p, of the sphere and the density and viscosity of the fluid, ρ and μ, respectively. These five variables involve three physical quantities (mass, length, time). Therefore, according to Buckingham's theorem, the problem can be fully described in terms of two dimensionless groups. These are the Reynolds number of the sphere:

$$Re_p = \frac{D_p v_p \rho}{\mu} \quad ,$$ (2.37)

and the *drag coefficient*, C_D, defined as:

$$C_D = \frac{F_D}{\frac{1}{2} \rho v_p^2 A_p} \quad ,$$ (2.38)

where $A_p = \pi D_p^2 / 4$ is the projection of the area of the sphere in the direction of v_p. In most of the situations of practical interest for mechanical and process engineers, the drag coefficient is a function of the Reynolds number only, as shown in Fig. 2.3. For $Re_p < 1$ the flow around the sphere is purely viscous and is referred to as *Stokes flow regime* (from George Gabriel Stokes, British physicist and mathematician, 1819–1903). In this flow regime, C_D is given by the following expression:

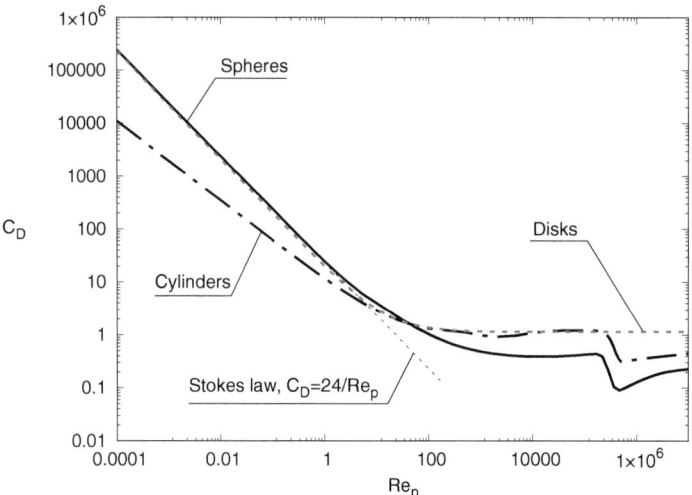

Fig. 2.3 Drag coefficient for spheres, disks and cylinders as a function of the Reynolds number, Re_p. The dashed line represents the values of C_D provided by the Stokes law

$$C_D = \frac{24}{Re_p} \quad . \tag{2.39}$$

For $1 < Re_p < 10^3$, C_D can be approximated by the following expression:

$$C_D = 18Re_p^{-0.6} \quad . \tag{2.40}$$

For $10^3 < Re_p < 2 \cdot 10^5$, the drag coefficient is $C_D \cong 0.44$ and the flow regime is referred to as *Newton regime*.

A sphere settling in a still fluid reaches, after an initial transient, a constant velocity referred to as *terminal velocity*. The forces acting on the sphere in the direction of motion are those shown in Fig. 2.4:

$$\text{Gravity force:} \quad \longrightarrow \quad F_G = -\rho_p \frac{\pi D_p^3}{6} g \quad , \tag{2.41}$$

$$\text{Buoyancy force:} \quad \longrightarrow \quad F_B = \rho \frac{\pi D_p^3}{6} g \quad , \tag{2.42}$$

$$\text{Drag force:} \quad \longrightarrow \quad F_D = \frac{1}{2} \rho v_p^2 A_p C_D \quad , \tag{2.43}$$

where ρ_p is the density of the sphere. Gravity and buoyancy forces are volume forces and therefore act on the sphere even in the absence of motion (static conditions). The

Fig. 2.4 Forces acting on
a sphere settling in a
viscous fluid

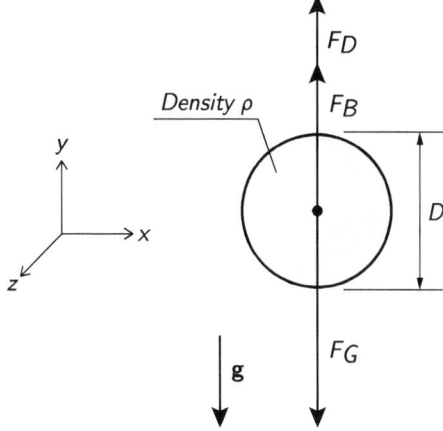

drag force is a surface force and acts on the sphere only when there is a relative motion between the sphere and the surrounding fluid. When the sphere moves at constant velocity, the gravity, buoyancy and drag forces balance out:

$$g(\rho - \rho_p)\frac{\pi D_p^3}{6} + \frac{\pi}{8}\rho D_p^2 v_p^2 C_D = 0 \quad . \tag{2.44}$$

From this force balance, the expression of v_p can be computed once the flow regime is known, namely the expression of C_D is known. For instance, in the Stokes regime, the terminal velocity of the sphere reads as:

$$v_p = \frac{g D_p^2(\rho_p - \rho)}{18\mu} \quad . \tag{2.45}$$

This expression is used in the *falling sphere viscosimeter* for viscosity measurements.

2.3.2 Unsteady Flow Past a Sphere

A sphere immersed in a three-dimensional flow field experiences the forces shown in Fig. 2.4: gravity force, $\mathbf{F}_G$, drag force, $\mathbf{F}_D$, and buoyancy force, $\mathbf{F}_B$. When $\rho_p \gg \rho$, the drag force can be expressed as follows:

$$\mathbf{F}_D = \frac{1}{2}\rho(\mathbf{v} - \mathbf{v}_p)|\mathbf{v} - \mathbf{v}_p|\frac{\pi D_p^2}{4}C_D \quad , \tag{2.46}$$

where $\mathbf{v}$ is the velocity of the fluid surrounding the sphere. For the drag coefficient C_D, the same expressions that apply to the case of a sphere settling in still fluid can be used, with the only difference that the Reynolds number must be now expressed as $Re = \rho|\mathbf{v} - \mathbf{v}_p|D_p/\mu$. From Eq. (2.46), it can be concluded that $\mathbf{F}_D$ has the same direction as the difference between the fluid velocity and the sphere velocity, $(\mathbf{v} - \mathbf{v}_p)$. In the case of unsteady flow, the sum of the forces acting on the sphere is equal to the rate of change of momentum of the sphere:

$$\frac{d(m_p\mathbf{v}_p)}{dt} = \frac{1}{2}\rho(\mathbf{v} - \mathbf{v}_p)|\mathbf{v} - \mathbf{v}_p|\frac{\pi D_p^2}{4}C_D + (\rho - \rho_p)V_p\mathbf{g} \ , \qquad (2.47)$$

where m_p is the mass of the sphere and V_p its volume.

2.3.2.1 Sphere with Constant Mass

When the mass of the sphere remains constant during its motion, Eq. (2.47) can be simplified as:

$$m_p\frac{d\mathbf{v}_p}{dt} = \frac{1}{2}\rho(\mathbf{v} - \mathbf{v}_p)|\mathbf{v} - \mathbf{v}_p|\frac{\pi D_p^2}{4}C_D + (\rho - \rho_p)V_p\mathbf{g} \ . \qquad (2.48)$$

To further examine the equation of motion of the sphere, it is convenient to assume that the motion occurs in the Stokes regime. In this regime, Eq. (2.48) becomes:

$$\frac{d\mathbf{v}_p}{dt} = \frac{\mathbf{v} - \mathbf{v}_p}{\tau_p} + \hat{\rho}\mathbf{g} \ , \qquad (2.49)$$

where $\hat{\rho} = (\rho - \rho_p)/\rho_p$ and τ_p is the relaxation time of the sphere, defined as:

$$\tau_p = \frac{\rho_p D_p^2}{18\mu} \ . \qquad (2.50)$$

The relaxation time is the time required for the sphere to adjust its velocity to a new condition of forces and, hence, quantifies the ability of the sphere to respond to changes in the flow field around it.

The solution of Eq. (2.49), assuming constant fluid velocity, is:

$$\mathbf{v}_p(t) = \mathbf{v}_0 e^{-t/\tau_p} + \mathbf{v}(1 - e^{-t/\tau_p}) + \hat{\rho}\mathbf{g}\tau_p(1 - e^{-t/\tau_p}) \ , \qquad (2.51)$$

where $\mathbf{v}_0 = \mathbf{v}_p(t = 0)$ is the initial velocity of the sphere. This equation shows that the velocity of the sphere results from the superposition of three contributions: one due to $\mathbf{v}_0$, which decays exponentially with time until it cancels out in the limit $t \to \infty$; one due to the entrainment exerted by the fluid on the sphere and one due to gravity. The last two contributions are initially zero and grow exponentially, such that $\mathbf{v}_p(t \to \infty) = \mathbf{v} + \hat{\rho}\mathbf{g}\tau_p$.

The components of $\mathbf{v}_p(t)$ in the direction perpendicular to gravity and in the direction parallel to it are, respectively:

$$v_{p,x}(t) = v_{0,x}e^{-t/\tau_p} + v_x(1 - e^{-t/\tau_p}) \quad , \tag{2.52}$$

and:

$$v_{p,y}(t) = v_{0,y}e^{-t/\tau_p} + v_y(1 - e^{-t/\tau_p}) + \hat{\rho}g\tau_p(1 - e^{-t/\tau_p}) \quad , \tag{2.53}$$

being $g_x = 0$ and $g_y = g$. Note that $\hat{\rho}g\tau_p$ is the terminal velocity in the Stokes regime, given by Eq. (2.45).

The trajectory, $\mathbf{x}_p(t)$, of the sphere can be obtained upon integration of Eq. (2.51) over time, since $\mathbf{v}_p(t) = d\mathbf{x}_p(t)/dt$ by definition. Integration yields:

$$\mathbf{x}_p(t) = \mathbf{v}_0\tau_p\left(1 - e^{-t/\tau_p}\right) + \left(\mathbf{v}\tau_p + \hat{\rho}\mathbf{g}\tau_p^2\right)\left[t - \tau_p(1 - e^{-t/\tau_p})\right] \quad . \tag{2.54}$$

The components of $\mathbf{x}_p(t)$ in the direction perpendicular to gravity and in the direction parallel to it are:

$$x_p(t) = v_{0,x}\tau_p\left(1 - e^{-t/\tau_p}\right) + v_x\tau_p\left[t - \tau_p(1 - e^{-t/\tau_p})\right] \quad . \tag{2.55}$$

and:

$$y_p(t) = v_{0,y}\tau_p\left(1 - e^{-t/\tau_p}\right) + \left(v_y\tau_p + \hat{\rho}g\tau_p^2\right)\left[t - \tau_p(1 - e^{-t/\tau_p})\right] \quad . \tag{2.56}$$

This equation shows that also the trajectory results from the superposition of the contribution due to $\mathbf{v}_0$, which decays exponentially with time, and the contribution due to $\mathbf{v}$ and $\mathbf{g}$, which instead increase over time such that $\mathbf{x}_p(t) \to \infty$ for $t \to \infty$.

By way of illustration, let us consider the case of a spherical particle with zero initial velocity ($v_{0,x} = v_{0,y} = 0$), immersed in a flow characterized by constant and uniform velocity U in the horizontal direction ($v_x = U$, $v_y = 0$) and subject to gravity in the vertical direction. The horizontal velocity component of the particle is:

$$v_{p,x}(t) = U(1 - e^{-t/\tau_p}) \quad \to \quad v_{p,x}(t)/U = 1 - e^{-t/\tau_p} \quad . \tag{2.57}$$

The behavior of $v_{p,x}/U$ is shown in Fig. 2.5. After a time $t = \tau_p$, $1 - e^{-t/\tau_p} = 1 - 1/e$ and the velocity is approximately equal to 63% of the asymptotic value, which is the terminal velocity. From a mathematical point of view, Eq. (2.57) dictates that $v_{p,x} = U$ only in the limit $t \to \infty$. From a practical point of view, it is safe to assume that the sphere reaches its terminal velocity for $t/\tau_p \simeq \mathcal{O}(10)$.

The vertical velocity component reads as:

$$v_{p,y}(t) = \hat{\rho}g\tau_p(1 - e^{-t/\tau_p}) \quad \to \quad v_{p,y}(t)/\hat{\rho}g\tau_p = 1 - e^{-t/\tau_p} \quad . \tag{2.58}$$

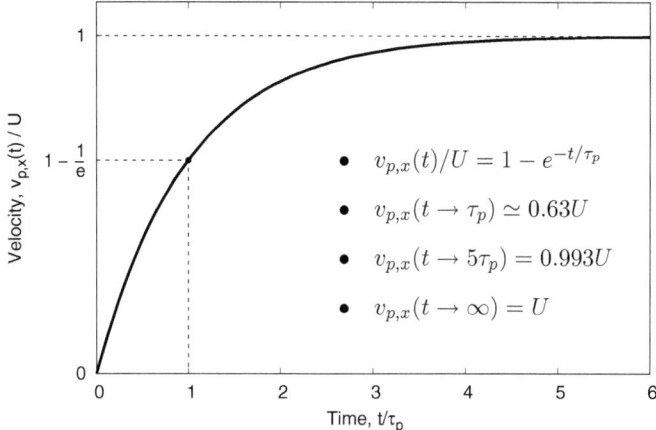

Fig. 2.5 Time behavior of the horizontal velocity component, $v_{p,x}(t)/U$, of a sphere immersed in uniform flow with velocity $v_x = U$ and subject to the action of gravity. Time t is normalized with respect to τ_p on the horizontal axis

Comparing Eqs. (2.57) and (2.58), it can be seen that the velocity components $v_{p,x}$ and $v_{p,y}$ differ only by the constant in front of the term $(1 - e^{-t/\tau_p})$, and that their evolution over time is analogous to that of the electric potential during the charging process of a capacitor.

The scalar components of the trajectory, obtained upon integration of Eqs. (2.57) and (2.58) respectively, read as:

$$x_p(t) = U\left[t - \tau_p(1 - e^{-t/\tau_p})\right] \quad \rightarrow \quad x_p(t)/U\tau_p = t/\tau_p - (1 - e^{-t/\tau_p}) \tag{2.59}$$

and:

$$y_p(t) = \hat{\rho} g \tau_p\left[t - \tau_p(1 - e^{-t/\tau_p})\right] \quad \rightarrow \quad y_p(t)/\hat{\rho} g \tau_p^2 = t/\tau_p - (1 - e^{-t/\tau_p}) \quad . \tag{2.60}$$

Also the components of the trajectory exhibit the same qualitative time behavior. As an example, the time evolution of $x_p(t)/U$ is shown in Fig. 2.6. Starting at time $t = \tau_p$, the trajectory grows linearly with time such that $x_p(t \rightarrow \infty) \rightarrow \infty$.

2.3.2.2 Sphere with Variable Mass

When the mass of the sphere changes during the motion, a situation encountered in processes like combustion, evaporation or condensation, Eq. (2.47) becomes:

$$m_p \frac{d\mathbf{v}_p}{dt} + \mathbf{v}_p \frac{dm_p}{dt} = \frac{1}{2}\rho(\mathbf{v} - \mathbf{v}_p)|\mathbf{v} - \mathbf{v}_p|\frac{\pi D_p^2}{4}C_D + (\rho - \rho_p)V_p\mathbf{g} \quad . \tag{2.61}$$

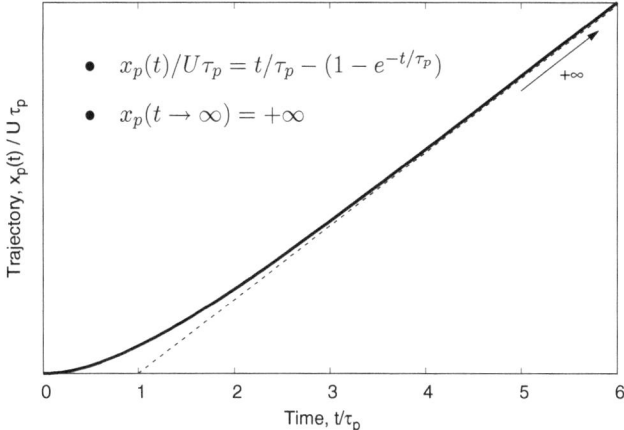

Fig. 2.6 Time behavior of the horizontal trajectory component, $x_p(t)/U\tau_p$, of a sphere immersed in uniform flow with velocity $v_x = U$ and subject to the action of gravity. Time t is normalized with respect to τ_p on the horizontal axis

To integrate this equation, it is first necessary to express dm_p/dt as a function of time, t. For this purpose, the conservation law can be invoked and applied to the mass of the sphere:

$$\frac{dm_p}{dt} = \dot{m}_{p,in} - \dot{m}_{p,out} \quad , \tag{2.62}$$

where dm_p/dt is the rate of change of mass and $\dot{m}_{p,in} - \dot{m}_{p,out}$ is the net mass flux across the surface of the sphere, all terms in Eq. (2.62) being expressed in kg/s. Note that $\dot{m}_{p,in} = 0$ for combustion or evaporation, as it can be assumed that there is no inward mass flux entering the surface of the sphere, and $\dot{m}_{p,out} = 0$ for condensation, as it can be assumed that there is no outward mass flux leaving the surface of the sphere. In general, it is always possible to express mass fluxes as a function of the diameter of the sphere. For example, $\dot{m}_p = K\pi\rho_p D_p$, where K is expressed in m²/s and $D_p = D_p(t)$.

Since $m_p = \rho_p \pi D_p^3/6$, taking the time derivative yields:

$$\frac{dm_p}{dt} = \rho_p \frac{\pi D_p^2}{2} \frac{dD_p}{dt} \quad . \tag{2.63}$$

Substituting Eq. (2.63) into Eq. (2.62) gives:

$$\frac{dm_p}{dt} = \rho_p \frac{\pi D_p^2}{2} \frac{dD_p}{dt} = \rho_p(K_{in} - K_{out})\pi D_p \quad , \tag{2.64}$$

with $\rho_p K_{in}$ specific rate of mass increase and $\rho_p K_{out}$ specific rate of mass decrease. Solving for D_p, leads to the following expression:

$$\frac{dD_p}{dt} = \frac{2(K_{in} - K_{out})}{D_p} \quad \rightarrow \quad D_p(t) = \sqrt{D_0^2 + 4(K_{in} - K_{out})t} \quad , \qquad (2.65)$$

under the assumption that both K_{in} and K_{out} are constant. In Eq. (2.65) $D_0 = D_p(t = 0)$ is the initial diameter of the sphere.

By using Eqs. (2.65) and (2.64), Eq. (2.61) can be rewritten in the Stokes regime as:

$$\frac{d\mathbf{v}_p}{dt} + \mathbf{v}_p \frac{6(K_{in} - K_{out})}{D_p^2} = \frac{18\mu}{\rho_p D_p^2}(\mathbf{v} - \mathbf{v}_p) + \hat{\rho}\mathbf{g} \quad . \qquad (2.66)$$

This equation shows clearly that the relaxation time of the sphere is not constant as it is inversely proportional to time t.

2.3.3 Examples

1 A hollow steel sphere with diameter $D_p = 5$ mm and mass of $5 \cdot 10^{-5}$ kg, is released into a liquid of density $\rho = 900$ kg/m^3. The terminal velocity of the sphere is equal to 5 mm/s. Estimate the viscosity of the liquid.

Solution The apparent density of the sphere is:

$$\rho_p = \frac{6m_p}{\pi D_p^3} = 763.9 \text{ kg} \cdot \text{m}^{-3} \quad . \qquad (2.67)$$

Being $\rho_p < \rho$, the buoyancy force, F_B, overcomes the gravity force, F_G, and therefore the drag force, F_D, is directed downwards. When the sphere settles with its terminal velocity, the force balance on the sphere reads as:

$$F_B - F_D - F_G = 0 \quad \rightarrow \quad \rho \frac{\pi D_p^3 g}{6} - \rho_p \frac{\pi D_p^3 g}{6} - \frac{\pi}{8}\rho v_p^2 D_p^2 C_D = 0 \quad , \qquad (2.68)$$

Let us now assume that *Stokes regime* conditions apply. The validity of this assumption shall be verified a posteriori by checking that the condition $Re < 1$ is satisfied. Using Eq. (2.68), the following expression for the terminal velocity of the sphere can be derived:

$$v_p = \frac{g D_p^2 (\rho - \rho_p)}{18\mu} \quad , \qquad (2.69)$$

with $\rho_p < \rho$. Solving for viscosity yields $\mu = 0.370$ Pa · s and, in turn:

$$Re = \frac{\rho v_p D_p}{\mu} = 60.8 \cdot 10^{-3} < 1 \quad . \tag{2.70}$$

This verifies the validity of the Stokes regime assumption.

2 A falling-sphere viscometer is a device used to measure the viscosity of a fluid by measuring the time required for a sphere with known density and diameter to cover a certain distance under gravity inside a glass tube filled with the fluid whose viscosity is to be determined. Calculate the viscosity of a fluid with density $\rho = 800\,\text{kg/m}^3$ given that the sphere has diameter $D_p = 5\,\text{mm}$, density $\rho_p = 2700\,\text{kg/m}^3$ and terminal velocity equal to 0.07 m/s.

Solution Assuming that Stokes regime conditions ($Re_p < 1$) apply when the sphere falls through the tube, the terminal velocity can be expressed as:

$$v_p = \frac{g D_p^2 (\rho - \rho_p)}{18\mu} = -0.07\,\text{m/s} \quad , \tag{2.71}$$

which yields:

$$\mu = \frac{g D_p^2 (\rho - \rho_p)}{18 v_p} \quad \longrightarrow \quad \mu = 0.37\,\text{Pa} \cdot \text{s} \quad . \tag{2.72}$$

For the $Re_p < 1$ condition to be verified, it must be:

$$\mu > \rho v_p D_p \quad \longrightarrow \quad \mu > 0.28\,\text{Pa} \cdot \text{s} \quad . \tag{2.73}$$

The value calculated using Eq. (2.72) satisfies this condition and is, therefore, acceptable.

3 A glass sphere of apparent density $\rho_p = 2.62\,\text{kg/m}^3$ falls in a fluid of density $\rho = 1.59\,\text{kg/m}^3$ and viscosity $\mu = 9.58 \cdot 10^{-5}\text{Pa} \cdot \text{s}$. What is the diameter of the sphere if its terminal velocity is $v_{term} = 1.22\,\text{m/s}$?

Solution To determine the diameter, it is necessary to solve the following equation:

$$F_G + F_B + F_D = 0 \quad , \tag{2.74}$$

with respect to D_p. This yields:

$$D_p = \frac{3}{4} \left(\frac{\rho}{\rho_p - \rho} \right) \frac{v_{term}^2 C_D}{g} \quad . \tag{2.75}$$

To proceed, it is now necessary to know the (unknown) diameter on which the drag coefficient C_D depends. To derive such diameter, an iterative procedure can be used. First, an initial guess for C_D must be assumed, e.g. $C_D = C_D^0 = 0.44$. Using this initial guess, Eq. (2.75) yields $D_p = 0.0773$ m, which corresponds to a Reynolds number of the sphere $Re_p = 1.57 \cdot 10^3$. Based on this value, a new value C_D' of the drag coefficient can be determined, using in particular the diagram of Fig. 2.3 since $Re_p = 1.57 \cdot 10^3$ falls in the Newton regime of motion, for which no simple relation linking C_D to Re_p exists. The iterative process stops if the difference between C_D^0 and C_D' is negligible (ideally zero, practically below an arbitrarily small threshold value, ϵ). Otherwise the process continues, using the value C_D^n obtained at the end of the n-th iteration to compute a new value for D_p from Eq. (2.75) and then repeating the steps just described until convergence ($C_D^{n+1} - C_D^n < \epsilon$) is reached.

Another possible way to proceed is the following. From Eq. (2.47), evaluated at steady state, i.e. when $d(m_p \mathbf{v}_p)/dt = 0$, it is possible to express C_D as:

$$C_D = \frac{4}{3} \frac{(\rho_p - \rho)}{\rho} \frac{g D_p}{(\mathbf{v} - \mathbf{v}_p)|\mathbf{v} - \mathbf{v}_p|} \quad . \tag{2.76}$$

Recalling that $Re_p = \rho |\mathbf{v} - \mathbf{v}_p| D_p / \mu$ and expressing the vectorial quantities by their magnitude, Eq. (2.76) can be rewritten as:

$$\frac{C_D}{Re_p} = -\frac{4}{3} \frac{\mu g}{\rho (v - v_p)|v - v_p|^2} \left(\frac{\rho_p - \rho}{\rho} \right) = \frac{4}{3} \frac{\mu g}{\rho (v_p - v)^3} \left(\frac{\rho_p - \rho}{\rho} \right) \tag{2.77}$$

It is apparent that C_D/Re is independent of D_p. The value of the right-hand side of Eq. (2.77) can be calculated from the available data, and assuming zero fluid velocity, v. This value will be denoted as C in the following and is equal to $2.8 \cdot 10^{-4}$.

The diameter can be computed by solving the system:

$$\begin{cases} C_D = C \cdot Re_p \\ C_D = C_D(Re_p) \end{cases} ,$$

where $C_D = C_D(Re_p)$ is the curve providing the value of C_D for a sphere, in the C_D vs Re_p diagram shown in Fig. 2.3. In this diagram, the equation $C_D = C \cdot Re_p$ is a straight line of slope equal to 1 in logarithmic scale, which crosses the curve $C_D = C_D(Re)$ at $Re_p = \rho v_p D_p / \mu = 1.59 \cdot 10^3$. Hence, the diameter of the sphere can be calculated as:

$$D_p = \frac{\mu Re_p}{\rho v_p} = \frac{(2.4 \cdot 10^4)(9.58 \cdot 10^{-5})}{1.59 \cdot 12.2} \simeq 0.079 \text{ m} \quad . \tag{2.78}$$

4 An electrostatic precipitator consists of two vertical parallel plates, placed at a distance $h = 0.1$ m. A potential difference ΔV is applied between the plates. A stream of air with average velocity $v_f = 1\ m/s$, laden with charged dust particles of diameter D_p and electric charge q_p, flows between the plates. Calculate the minimum length of the plates that is required to remove all the particles from the air stream.

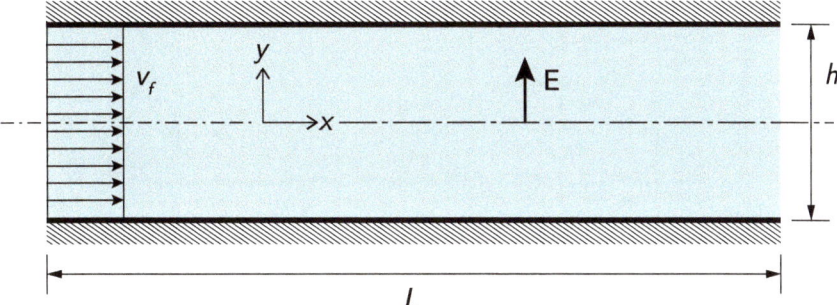

Fig. 2.7 Schematic of an electrostatic precipitator

Solution In an electrostatic precipitator, particles suspended in a gas stream can be separated using the force of an induced electrostatic charge. Imposing an electric field **E** between the plates, as shown in Fig. 2.7, allows to generate a uniform and constant electrostatic force, $\mathbf{F}_E = q_p\mathbf{E}$, which acts to push the particles toward the collector plate. The electrostatic force is a volume force and, in a precipitator, it is much larger than the gravity force.

Considering the schematic of an electrostatic precipitator shown in Fig. 2.7 and assuming the particle relaxation time to be sufficiently small to assume a steady motion condition for the particles, their equation of motion reads as:

$$0 = \mathbf{F}_D + \mathbf{F}_E = \frac{1}{2}\rho(\mathbf{v} - \mathbf{v}_p)\,|\,\mathbf{v} - \mathbf{v}_p\,|\,\frac{\pi D_p^2}{4}C_D + q_p\mathbf{E}\ ,\qquad (2.79)$$

where q_p is the electric charge of the particle and the gravity and buoyancy force have been neglected. It is also assumed that particles move in the Stokes regime ($Re_p < 1$, $C_D = 24/Re_p$) with velocity equal to that of the air stream in the x direction and velocity equal to v_E in the y direction, where v_E can be obtained from the y-component of Eq. (2.79).

$$3\pi\mu D_p v_E = q_p E \quad \rightarrow \quad v_E = \frac{q_p E}{3\pi D_p \mu}\ .\qquad (2.80)$$

A particle that enters the precipitator near the non-collector plate will deposit on the collector plate after a time $\tau = h/v_E$. This time must be long enough to ensure that $L_{min} = v_f\tau$. Therefore, the minimum length of the plates that ensures deposition of all particles within the precipitator is:

$$L_{min} = v_f\tau = v_f\frac{h}{v_E} = \frac{3\pi D_p \mu v_f h}{q_p E}\ .\qquad (2.81)$$

5 Gravity settling chambers are industrial separators whose mode of operation is schematically shown in Fig. 2.8: A gas stream laden with suspended particles flows

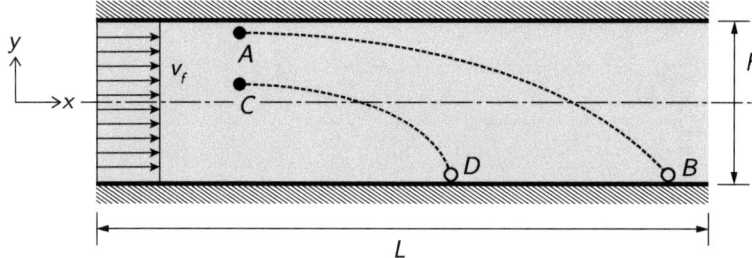

Fig. 2.8 Schematic of a gravity settling chamber

through the chamber and the particles of larger size get separated by gravitational settling, eventually depositing at the bottom wall of the chamber. Settling chambers can be sized by imposing that particles with diameter larger than or equal to a given threshold value are deposited before exiting the chamber, regardless of their position at the inlet section. With reference to the figure, for instance, the length L must be such that any particle at position A at the inlet will be deposited at position B just before the outlet. Assuming that the particle moves with the velocity of the fluid in the horizontal direction, the time required to cover the distance L is:

$$t_L = \frac{L}{v_f} \ , \tag{2.82}$$

where v_f is the average velocity of the fluid, equal to:

$$v_f = \frac{Q}{Wh} \ , \tag{2.83}$$

with Q the flowrate of the gas stream and W the depth of the chamber (measured along the z axis). The velocity v_f should not exceed a value of about 3 m/s, above which turbulent motions may be established within the chamber: These motions are undesirable as they could prevent particle deposition by keeping the particles into suspension.

Assuming that the particle moves vertically with its terminal velocity v_p, the time required to cover the distance h is:

$$t_h = \frac{h}{v_p} \ . \tag{2.84}$$

Imposing $t_L = t_h$, and considering Stokes regime conditions, the following expression for L can be obtained:

$$L = \frac{Q}{W} \frac{1}{v_p} = \frac{Q}{W} \frac{18\mu}{g D_p^2 (\rho_p - \rho)} \ . \tag{2.85}$$

This equation yields the minimum length L required to deposit any particle of diameter D_p.

6 The flue gas ($\rho = 1.4\,\text{kg/m}^3$ $\mu = 1.8 \cdot 10^{-5}\,\text{Pa} \cdot \text{s}$) exiting from a combustion chamber is laden with ash particles ($\rho_p = 800\,\text{kg/m}^3$) that have diameters $D_{p,min} = 0.5\,\mu\text{m}$ and $D_{p,max} = 50\,\mu\text{m}$. To separate the ash particles from the gas stream, a gravity settling chamber is used to separate the larger $D_{p,max}$ particles, followed by an electrostatic precipitator, used to separate the smaller $D_{p,min}$ particles. The flowrate of the flue gas is $Q = 0.6\,\text{m}^3/\text{s}$. Assume that the particles move in the Stokes regime and steady-state conditions.

1. Determine the length L of the gravity settling chamber knowing that the width of the chamber is $W = 2$ m and its height $H = 1.5$ m.
2. Determine the potential, V, to be applied to the plates of the electrostatic precipitator for the smaller particles to be collected, knowing that the distance between the plates is $d = 0.2\ m$ and that the height and length of the plates are $H = 1.5$ m and $L = 3$ m, respectively. Assume that the charge of the particles is $q_p = 1.6 \cdot 10^{-15}\ C$.

Solution The motion of the ash particles is described by this force balance:

$$m_p \frac{d\mathbf{v}_p}{dt} = \mathbf{F}_D + \mathbf{F}_B + \mathbf{F}_G + \mathbf{F}_E \quad . \tag{2.86}$$

At steady state, $d\mathbf{v}_p/dt = 0$. In the Stokes regime, the drag force is:

$$\mathbf{F}_D = 3\pi\mu D_p(\mathbf{v} - \mathbf{v}_p) \quad . \tag{2.87}$$

The sum of gravity and buoyancy forces is:

$$\mathbf{F}_G + \mathbf{F}_B = \mathbf{g}\frac{\pi D_p^3}{6}(\rho - \rho_p) \quad , \tag{2.88}$$

while the electrostatic force is:

$$\mathbf{F}_E = q_p\mathbf{E} = q_p\frac{V}{d} \quad , \tag{2.89}$$

with d distance between the plates.

For the gravity settling chamber, the relevant directions for particle motion are the flow direction (x) and the vertical direction (z). The force balance in the x direction yields $F_{D,x} = 0$, from which $v_{p,x} = v = Q/WH$. In the vertical direction, with the z-axis pointing upward, the force balance simplifies to:

$$3\pi\mu(v_z - v_{p,z}) + g\frac{\pi D_p^3}{6}(\rho - \rho_p) = 0 \quad , \tag{2.90}$$

and, being $v_z = 0$:

$$v_{p,z} = -\frac{g D_p^2 (\rho_p - \rho)}{18\mu} < 0 \quad . \tag{2.91}$$

The resulting velocity is negative as the particle moves in the direction opposite to that of the z-axis. To design the gravity settling chamber, the condition to be met is that the time t_H required for the particle to cover the vertical distance H must be at least equal to the time t_L required to cover the length L of the chamber. Therefore:

$$t_H = t_L \quad \longrightarrow \quad \frac{H}{v_{p,z}} = \frac{L}{v} \quad , \tag{2.92}$$

from which:

$$L = \frac{18\mu Q}{g W D_p^2 (\rho_p - \rho)} = 4.95 \, \text{m} \quad . \tag{2.93}$$

For the electrostatic precipitator, located downstream of the chamber, the relevant directions for particle motion are the flow direction (x) and the transverse direction (y). The force balance in the x direction yields $F_{D,x} = 0$, hence $v_{p,x} = v = Q/Hd$. In the transverse direction, the electrostatic force is only balanced by the drag force:

$$\frac{q_p V}{d} - 3\pi\mu D_p v_{p,y} = 0 \quad , \tag{2.94}$$

and:

$$v_{p,y} = \frac{q_p V}{3\pi\mu D_p d} \quad . \tag{2.95}$$

To design the precipitator, the condition to be met is that the particles must cover the distance between the plates in a time t_d at least equal to the time t_L. Therefore:

$$t_d = t_L \quad \longrightarrow \quad \frac{d}{v_{p,y}} = \frac{L}{v} \quad , \tag{2.96}$$

and:

$$V = \frac{3\mu\pi D_p Q d}{L H q_p} = 1413.7 \, V \quad . \tag{2.97}$$

$\boxed{7}$ Consider the laminar flow of a fluid (with viscosity μ) between two vertical cylinders with inner radius R_1 and outer radius R_2, respectively. The inner cylinder is stationary, the outer cylinder rotates with angular velocity Ω. In this configuration, the only non-zero fluid velocity component is the tangential one, v_θ, which can be expressed as follows:

$$v_\theta(r) = \Omega \frac{R_2^2\, r}{R_2^2 - R_1^2}\left(1 - \frac{R_1^2}{r^2}\right) \quad . \tag{2.98}$$

1. A particle of diameter D_p and density ρ_p is initially placed at the position r_i, θ_i inside the fluid. Derive an expression for the radial component of the particle velocity, assuming that the particle moves tangentially with the same velocity as the fluid and that the Stokes' law applies.
2. Calculate the arc $\Delta\theta$ subtended by the trajectory of the particle when the particle hits the outer cylinder of radius R_2.

Solution

1. During its motion, the particle experiences a centrifugal force F_c that generates the following radial component of the particle terminal velocity:

$$v_r = \frac{F_c}{3\pi\,\mu D_p} \quad , \tag{2.99}$$

with:

$$F_c = m_p \frac{v_\theta^2}{r} \quad . \tag{2.100}$$

Substituting the expression of v_θ, yields:

$$v_r = \frac{\rho_p D_p^2}{18\mu} \frac{\Omega^2 R_2^4}{(R_2^2 - R_1^2)^2} r \left[1 - \left(\frac{R_1}{r}\right)^2\right]^2 \quad . \tag{2.101}$$

2. An expression for $\theta = \theta(r)$ is sought. Since $v_\theta = r d\theta/dt$ and $v_r = dr/dt$:

$$r\frac{d\theta}{dr} = \frac{v_\theta}{v_r} \quad , \tag{2.102}$$

and, therefore:

$$\frac{d\theta}{dr} = \frac{1}{r}\frac{v_\theta}{v_r} = \frac{18\mu}{\rho_p D_p^2}\frac{R_2^2 - R_1^2}{\Omega R_2^2}\frac{r}{r^2 - R_1^2} \quad . \tag{2.103}$$

Upon integration of Eq. (2.103) between the initial position of the particle r_i and the outer cylinder R_2, the expression of the arc can be obtained as:

$$\Delta\theta = \frac{18\mu}{\rho_p D_p^2} \frac{R_2^2 - R_1^2}{\Omega R_2^2} \int_{R_i}^{R_2} \frac{r}{r^2 - R_1^2} dr =$$

$$= \frac{18\mu}{\rho_p D_p^2} \frac{R_2^2 - R_1^2}{\Omega R_2^2} \frac{1}{2} \ln\left(\frac{R_2^2 - R_1^2}{R_i^2 - R_1^2}\right) \ . \tag{2.104}$$

8 A sphere of diameter $D_p = 1$ mm and density $\rho_p = 10^3 \text{kg/m}^3$ moves in a flow of air ($\rho = 1.38 \text{ kg/m}^3$, $\mu = 21.5 \cdot 10^{-5}\text{Pa} \cdot \text{s}$) with velocity $v = 30$ m/s. The sphere is initially placed at position $x = 0$ with zero velocity. It can be assumed that the sphere moves in the Newton regime ($C_D = 0.44$). Calculate the time required for the particle to reach 80% of the maximum velocity and the distance covered by the particle in the two following cases:

1. the velocity vector $\mathbf{v}$ is parallel to the horizontal plane (neglect the effect of gravity),
2. the velocity vector $\mathbf{v}$ is anti-aligned with gravity in the vertical direction.

Solution

1. Recalling Eq. (2.48) and dividing all terms in this equation by the mass m_p of the sphere, the following equation can be obtained:

$$\frac{d\mathbf{v}_p}{dt} = \frac{3}{4} C_D \frac{\rho}{\rho_p D_p} (\mathbf{v} - \mathbf{v}_p)|\mathbf{v} - \mathbf{v}_p| + \frac{\rho - \rho_p}{\rho_p} \mathbf{g} \ . \tag{2.105}$$

The scalar component of this equation in the x direction reads as:

$$\frac{d\dot{x}}{dt} = C_D \frac{3}{4} \frac{\rho}{\rho_p D_p} (v - \dot{x})^2 \ \longrightarrow \ \ddot{x} = C_D \frac{3}{4} \frac{\rho}{\rho_p D_p} (v - \dot{x})^2 \ . \tag{2.106}$$

In the Newton regime, $C_D = 0.44$ and the following constant can be introduced: $\lambda = (3C_D\rho)/(4\rho_p D_p) = 0.455 \text{ m}^{-1}$. Equation (2.106) becomes:

$$\ddot{x} = \lambda(v - \dot{x})^2 \ \longrightarrow \ \dot{\xi} = -\lambda\xi^2 \ , \tag{2.107}$$

with $\xi = v - \dot{x}$. The solution of this equation is:

$$\frac{1}{\xi} = \lambda t + C \ \longrightarrow \ \xi = \frac{1}{\lambda t + C} \ , \tag{2.108}$$

and, in turn:

$$\dot{x} = v - \frac{1}{\lambda t + C} \ . \tag{2.109}$$

At $t = 0$, the velocity of the sphere is zero. This initial condition yields $C = 1/v$, and therefore:

$$\dot{x} = v \left(1 - \frac{1}{1 + v\lambda t}\right) \quad . \tag{2.110}$$

The velocity of the sphere in the x direction tends asymptotically to v. The time required to reach 80% of this velocity is obtained as:

$$t_{80\%} = \frac{\dot{x}/v}{1 - \dot{x}/v} \frac{1}{\lambda v} = 0.292 \, s \quad , \tag{2.111}$$

with $\dot{x}/v = 0.8$. Upon integration of Eq. (2.110), with the initial condition $x(0) = 0$, the following expression for the horizontal distance covered by the particle can be obtained:

$$x = vt - \frac{1}{\lambda} \ln(\lambda vt + 1) \quad . \tag{2.112}$$

Using this expression, the distance covered at time $t_{80\%}$ can be calculated as:

$$x_{80\%} = x(t_{80\%}) = 5.23 \, m \quad . \tag{2.113}$$

2. When the velocity vector is anti-aligned with gravity, namely the drag force anti-aligned with the gravity force, the x-component of the momentum equation is:

$$\ddot{x} = \lambda(v - \dot{x})^2 - \hat{\rho}g \quad , \tag{2.114}$$

with $\hat{\rho} = (\rho_p - \rho)/\rho_p \simeq 1$. By defining $\xi = v - \dot{x}$, such that $d\xi/dt = \ddot{x}$, Eq. (2.114) can be rewritten as:

$$\frac{d\xi}{\hat{\rho}g - \lambda\xi^2} = dt \quad . \tag{2.115}$$

Integration of this equation yields:

$$\ln \frac{\left|\xi - \sqrt{(\hat{\rho}g)/\lambda}\right|}{\xi + \sqrt{(\hat{\rho}g)/\lambda}} = -2\sqrt{\hat{\rho}g\lambda}\, t + C' \quad . \tag{2.116}$$

The constant terms in this expression are: $\sqrt{\hat{\rho}g/\lambda} \equiv \kappa = 4.64 \, m/s$ and $2\sqrt{\hat{\rho}g\lambda} \equiv \alpha = 4.225 \, s^{-1}$. Recalling that $\xi = v - \dot{x}$, the following expression for $\dot{x}$ is obtained:

$$\dot{x} = v - \kappa \frac{1 + Ce^{-\alpha t}}{1 - Ce^{-\alpha t}} \quad , \tag{2.117}$$

with $C = (v - \kappa)/(v + \kappa) = 0.732$ by virtue of the initial condition of zero particle velocity. The *flight time* of the particle is then calculated using the expression:

$$t = -\frac{1}{\alpha} \ln \left(\frac{1}{C} \frac{v - \dot{x} - \kappa}{v - \dot{x} + \kappa} \right) \quad , \tag{2.118}$$

which yields:

$$t_{80\%} = 0.35 \, \text{s} \quad . \tag{2.119}$$

Integration of Eq. (2.117) provides the expression for the displacement of the
particle, which is:

$$x = vt + \frac{\kappa}{\alpha} \ln \left(\frac{e^{\alpha t}}{e^{\alpha t} - C} \right) - \frac{\kappa}{\alpha} \ln(e^{\alpha t} - C) + C_2 \quad . \tag{2.120}$$

The constant C_2 can be calculated from the initial condition $x(0) = 0$ and is
equal to $C_2 = \kappa/\alpha \ln[(1 - C)^2] = -2.89$. Therefore, the distance covered by the
particle at time $t_{80\%}$ is:

$$x_{80\%} = 6.386 \, \text{m} \quad . \tag{2.121}$$

---

9   A burner injects fuel droplets into a combustion chamber. Droplets have an
initial diameter $D_p(t = 0) = D_0$ and initial velocity $v_p(t = 0) = v_0$ in the horizontal
direction x (the initial velocity in the vertical direction is zero). Assuming that the
droplets move in the Stokes regime and neglecting the influence of gravity:

1. determine the *stopping distance* of the droplets, namely the maximum distance
   from the injection point that the droplets can reach.
2. During their motion, the fuel droplets burn at a rate $c = K\pi\rho_p D_p$, expressed in
   kg/s, where D_p is the diameter of the droplets and K a constant measured in
   m^2/s. Derive the balance equations for the motion of the droplets in this case and
   determine their new stopping distance.

Solution

1. Neglecting gravity, the balance of the forces acting on the droplets is simply $\mathbf{F}_I =
   \mathbf{F}_D$ being $\mathbf{F}_B = \mathbf{F}_G = 0$. Since the velocity of the fluid in which the droplets move
   is zero ($\mathbf{v} = 0$) and the particle velocity, $\mathbf{v}_p$, has only one non-zero component
   ($v_{p,x}$), the balance of forces in the Stokes regime becomes:

$$\rho_p \frac{\pi D_p^3}{6} \frac{dv_{p,x}}{dt} = -3\pi \mu D_p v_{p,x} \quad , \tag{2.122}$$

where it is assumed that the mass of the droplets, m_p, is constant.

This equation is a differential equation that can be solved by separation of variables, which yields:

$$\frac{dv_{p,x}}{v_{p,x}} = -\frac{dt}{\tau_p} \quad , \tag{2.123}$$

and, upon integration:

$$v_{p,x}(t) = C e^{-t/\tau_p} \quad . \tag{2.124}$$

The initial condition $v_{p,x}(t = 0) = v_0$ imposes $C = v_0$. By integrating further with respect to time, the horizontal distance covered by the fuel droplets can be obtained:

$$x_p(t) = C - v_0 \tau_p e^{-t/\tau_p} \quad . \tag{2.125}$$

The initial condition $x_p(t = 0) = x_0$ imposes $C = v_0 \tau_p + x_0$. Assuming for simplicity $x_0 = 0$, the expression for the horizontal component of the droplet trajectory reads as:

$$x(t) = v_0 \tau_p \left(1 - e^{-t/\tau_p}\right) \quad . \tag{2.126}$$

The stopping distance, covered by the droplets in the horizontal direction, will be reached when $v_{p,x} = 0$. Eq. (2.124) allows to conclude that $v_{p,x} = 0$ for $t \to \infty$. Therefore, the stopping distance will be:

$$x_{p,max} = \lim_{t \to +\infty} x_p(t) = v_0 \tau_p \quad . \tag{2.127}$$

2. The force balance $\mathbf{F}_I = \mathbf{F}_D$ must now be solved considering that the mass of the droplets reduces over time due to combustion. The stopping distance can be determined from Eq. (2.66), written in scalar form and simplified for the case of horizontal motion in still fluid with $K_{in} = 0$ and $K_{out} = K$. This yields:

$$\frac{dv_{p,x}}{dt} - v_{p,x}\frac{6\,K}{D_p^2} = -\frac{18\mu}{\rho_p D_p^2} v_{p,x} \quad , \tag{2.128}$$

with $D_p(t)$ given by Eq. (2.65). Equation (2.128) can be recast as:

$$\frac{dv_{p,x}}{v_{p,x}} = \left(6\,K\frac{1}{D_p^2} - \frac{18\mu}{\rho_p}\frac{1}{D_p^2}\right)dt = \underbrace{\left(6\,K - \frac{18\mu}{\rho_p}\right)}_{C=\text{const.}}\frac{1}{D_p^2}\,dt \quad , \tag{2.129}$$

Equation (2.129) can be integrated recalling from relation (2.65) that $dt = -D_p dD_p/2K$, thus obtaining:

$$v_{p,x} = C \, D_p^{-C/2K} \quad , \tag{2.130}$$

where constant C can be obtained by imposing the initial conditions $v_{p,x}(t = 0) = v_0$ and $D_p(t = 0) = D_0$, which yield $C = v_0/D_0^{-C/2K}$ and:

$$v_{p,x} = v_0 \left(\frac{D_p}{D_0}\right)^{-C/2K} \quad . \tag{2.131}$$

From Eqs. (2.65) and (2.131), it can be concluded that the velocity $v_{p,x}$ of the droplets vanishes when their diameter D_p goes to zero. This happens at time:

$$\hat{t} = \frac{D_0^2}{4K} \quad . \tag{2.132}$$

The droplet trajectory $x_p(t)$ can be obtained by time integration of Eq. (2.131), recalling that $dx_p/dt = v_{p,x}$ and imposing the initial condition $x_p(t = 0) = x_0$:

$$x_p = x_0 + \frac{v_0 D_0^{C/2K}}{C - 4K} \left[(D_0^2 - 4\,Kt)^{1-C/4K} - (D_0^2)^{1-C/4K}\right] \quad . \tag{2.133}$$

By combining Eqs. (2.133) and (2.132), it is possible to obtain the expression for the new stopping distance simply by imposing $t = \hat{t}$.

[10] During the pandemic of COVID-19 it was made public that a safe distance between people was two meters. Since it is clear that the virus is transported by small droplets, it is safely assumed that this distance is related to the maximum distance at which a droplet, ejected by an infected person's mouth with a certain initial velocity, can travel. This 2-meters rule was devised after the Spanish flu pandemic, when it was apparent that airborne transmission was crucial to spreading virus contagion: researchers produced several fundamental works like the experiments of Duguid (1946)[2] and the model of Wells (1934).[3] These seminal works have been used also to establish current guidelines published by health organizations. In the following, we aim to clarify how these guidelines have been obtained.

When a person sneezes, many respiratory droplets are released. Suppose that the initial average size of the droplets is equal to $D_0 = 100 \ \mu m$, their density is $\rho_p = 10^3 \ kg/m^3$ and that they are released in air ($\rho = 1.2 \ kg/m^3$, $\mu = 2 \cdot 10^{-5} \ Pa \cdot s$) with an initial velocity of $v_0 = 17$ m/s.

---

[2] J. P. Duguid, The size and the duration of air-carriage of respiratory droplets and droplet-nuclei. J. Hyg. (Lond.) 44, 471–479 (1946).

[3] W. F. Wells, On air-borne infection. Study II. Droplets and droplet nuclei. Am. J. Hyg. 20, 611–618 (1934).

1. Determine the stopping distance of the droplets assuming that their mass is constant during motion.
2. Determine the stopping distance in the case droplets undergo evaporation during motion. Assume an evaporation rate $c = K\pi\rho_p D_p$, where D_p is the diameter of the droplets and $K = 10^{-8}$ m^2/s the evaporation constant.

Solution

1. The stopping distance can be obtained by assuming that the motion of the droplets occurs only in the horizontal direction x (the influence of gravity is neglected) and in the Stokes regime. With these assumptions, which match those adopted by the WHO during the Covid-19 pandemic, the time behavior of the velocity and position of the droplet is governed by Eqs. (2.124) and (2.126). Therefore, the stopping distance is given by Eq. (2.127): $x_{p,max} = v_0$, $\tau_p = 0.472$ m, being $\tau_p = \rho_p D_0^2/18\mu = 0.0278$ s.
2. Due to evaporation, the mass of the droplet decreases over time according to the following mass conservation relation:

$$\frac{dm_p}{dt} = -\dot{m}_{p,out} = -c \quad . \tag{2.134}$$

Recalling Eq. (2.63):

$$\rho_p \frac{\pi D_p^2}{2}\frac{dD_p}{dt} = -c = -K\pi\rho_p D_p \qquad \longrightarrow \qquad D_p dD_p = -2\,K dt \tag{2.135}$$

which is a simplified form of Eq. (2.65). By integrating this differential equation using the technique of separating variables, it is possible to derive an expression for the variation of the diameter over time:

$$D_p(t) = \sqrt{D_0^2 - 4\,K t} \quad . \tag{2.136}$$

To obtain the velocity $v_{p,x}$, it suffices to combine Eq. (2.134) with the x-component of Eq. (2.61), obtaining:

$$m_p \frac{dv_{p,x}}{dt} + v_{p,x}\frac{dm_p}{dt} = -3\pi\mu D_p v_{p,x} \quad , \tag{2.137}$$

with $D_p = D_p(t)$ given by Eq. (2.136). Dividing by the mass m_p and separating the variables, Eq. Eq. (2.137) can be recast as:

$$\frac{dv_{p,x}}{v_{p,x}} = \underbrace{\left[6K - \frac{18\mu}{\rho_p}\right]}_{C\,\equiv\,cost.}\frac{dt}{D_p^2} \quad , \tag{2.138}$$

with $C = -3 \cdot 10^{-7}$ m^2/s. From relation (2.135), $dt = -D_p dD_p/2K$. Plugging this expression into Eq. (2.138) and integrating its right-hand side terms with respect to D_p yields:

$$v_{p,x} = \mathcal{C} \cdot D_p^{-C/2K} \quad , \tag{2.139}$$

where constant $\mathcal{C} = v_0/D_0^{-C/2K}$ is obtained by imposing the initial conditions $v_{p,x}(t = 0) = v_0$ and $D_p(t = 0) = D_0$. Therefore, the droplet velocity is:

$$v_{p,x} = v_0 \left(\frac{D_p}{D_0} \right)^{-C/2K} . \tag{2.140}$$

This velocity vanishes when the droplet diameter D_p becomes zero, and this happens at time:

$$\hat{t} = \frac{D_0^2}{4K} = 0.25 \text{ s} \quad . \tag{2.141}$$

The trajectory of the droplets can be obtained by integrating Eq. (2.140), and imposing the initial condition $x_p(t = 0) = x_0$, obtaining:

$$x_p = x_0 + \frac{v_0 D_0^{C/2K}}{C - 4K} \left[(D_0^2 - 4Kt)^{1-C/4K} - (D_0^2)^{1-C/4K} \right] \quad . \tag{2.142}$$

The time evolution of the droplet diameter, velocity and position is shown in Fig. 2.9. Assuming $x_0 = 0$, the stopping distance is equal to:

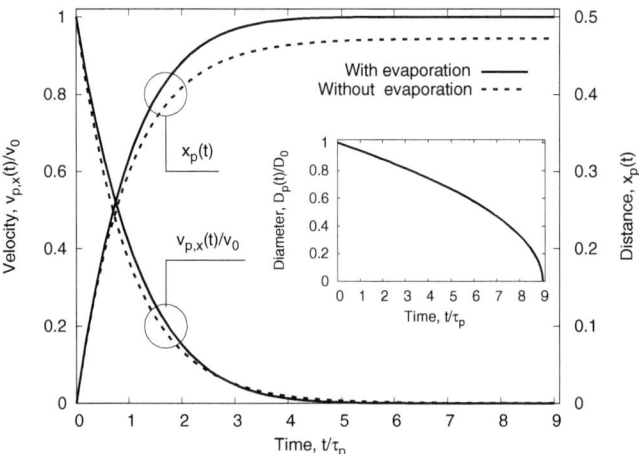

Fig. 2.9 Time evolution of velocity, trajectory and diameter of an evaporating droplet. Velocity, time and diameter are normalized with respect to v_0, τ_p and D_0, respectively

$$x_{p,max} = x_p(t = \hat{t}) = -\frac{v_0 D_0^2}{C - 4K} = 0.5\,\text{m} \quad . \tag{2.143}$$

Note how this value is larger than that obtained in the absence of evaporation. In general, the distance $x_p(t)$ covered by the droplets is always larger in the presence of evaporation, as shown in Fig. 2.9. This is due to a slower decrease of the droplet velocity during the motion.

2.4 Flow in Porous Beds

Porous beds consist of a collection of beads of various sizes and shapes through which a fluid flows at velocities small enough to avoid any appreciable particle entrainment effect. A porous bed is mainly characterized by its porosity ε, which represents the volume fraction of the bed not occupied by the beads; its specific surface area, S, defined as the surface area exposed to the flow per unit volume of the bed; and the specific surface area of the beads in the bed, S_B, which is the surface area exposed per unit of volume of the beads. The three parameters ε, S and S_B obey the relation:

$$S = (1 - \varepsilon)S_B \quad . \tag{2.144}$$

For a bed consisting of spherical beads of diameter D_p, the following expressions can be used:

$$S_B = \frac{\pi D_p^2}{\pi D_p^3/6} = \frac{6}{D_p} \quad , \quad \varepsilon = \frac{D_p^3 - \pi D_p^3/6}{D_p^3} = 1 - \frac{\pi}{6} \quad , \quad S = \frac{\pi}{D_p} \quad .\tag{2.145}$$

To determine the pressure drop, Δp, through the porous bed, the following assumptions can be made:

1. The pressure drop is proportional to the bed height, L, so the relevant quantity to be considered is the pressure gradient $\Delta p/L$;
2. The pressure gradient $\Delta p/L$ is dependent on the specific velocity $v_S = Q/A$, where Q is the volumetric flowrate through the bed and A the cross-sectional area of the bed.

In analogy with the flow in a pipe or channel, and on the basis of a dimensional analysis, it can be concluded that the friction factor, f_p, associated with the fluid flow through the porous bed, which is defined as:

$$f_p = \frac{D_C}{\rho v_S^2} \frac{\Delta p}{L} \quad , \tag{2.146}$$

where D_C is the characteristic size of the bed (e.g. D_p for a bed of spheres with equal diameter), is a function of just two dimensionless groups: ε and $Re_p = \rho v_S D_C / \mu$.

To define the form of Eq. (2.146), we assume that the bed can be modeled as a series of channels of diameter D through which the fluid flows with velocity v, such that Eq. (2.146) become independent of ε:

$$\frac{D}{\rho v^2} \frac{\Delta p}{L} = f \left(\frac{D v \rho}{\mu} \right) \quad . \tag{2.147}$$

It is now necessary to express D and v as a function of known parameters of the problem. In each section of the bed, the average cross-sectional area for the fluid is εA and the flowrate can be expressed as:

$$Q = \varepsilon A v \quad , \tag{2.148}$$

which, in turn, leads to:

$$v = \frac{v_S}{\varepsilon} \quad . \tag{2.149}$$

The diameter of the channels, D, can be defined in terms of hydraulic diameter, already defined in Sect. 2.2.5. Recalling Eq. (2.24), D can be therefore expressed as:

$$D = 4 \frac{\varepsilon}{S} = \frac{4\varepsilon}{(1 - \varepsilon) S_B} \quad . \tag{2.150}$$

For a bed of spheres:

$$D = \frac{2}{3} \left(\frac{\varepsilon}{1 - \varepsilon} \right) D_p \quad . \tag{2.151}$$

By substituting Eqs. (2.149) and (2.150) into Eq. (2.147), the following functional dependence is obtained:

$$\frac{4\varepsilon^3}{\rho v_S^2 (1 - \varepsilon) S_B} \frac{\Delta p}{L} = f_p \left[\frac{6\rho v_S}{S_B (1 - \varepsilon) \mu} \right] \quad . \tag{2.152}$$

The function f_p is known from experimental measurements, which have been formulated as a function of the Reynolds number Re_p in the so-called Ergun equation:

$$f_p = \frac{150}{Re_p} + 1.75 \quad . \tag{2.153}$$

This equation can be applied over a wide range of fluid velocities in both laminar and turbulent flow conditions. In Eq. (2.153), f_p is defined as:

$$f_p = \frac{\varepsilon^3}{\rho v_S^2(1 - \varepsilon)} D_p \frac{\Delta p}{L} \quad . \tag{2.154}$$

Problems

$\boxed{a}$ A steel pipe with diameter $D = 0.15$ m is used to transport water at a flowrate $Q = 7.5$ m^3 per minute.

1. Demonstrate that the flow is turbulent.
2. Calculate the pressure drop Δp over a pipe length $L = 1$ km.
3. Calculate the pumping power required to realize the transport.

$\boxed{b}$ Describe a way to represent graphically the relationship between the friction factor and the Reynolds number so that problems like:

- calculation of the flowrate for a given pressure drop and a given pipe diameter,
- calculation of the pipe diameter required to transport a given mass flowrate by means of an imposed pressure drop,

can be solved via a direct (non-iterative) procedure.

$\boxed{c}$ The use of the diagram representing the drag coefficient as a function of the Reynolds number for solving the problem of a falling sphere requires an iterative process for the calculation of the terminal velocity or of the diameter of the sphere. Derive an expression to represent the values of the drag coefficient and the Reynolds number so that problems like:

- the calculation of the particle diameter for a given value of the terminal velocity,
- the calculation of the terminal velocity for a given value of the particle diameter,

can be solved via a direct (non-iterative) procedure, given the physical properties of both the fluid and the sphere.

$\boxed{d}$ A metal screen is used to support a catalytic bed of height 0.3 m and radius 0.1 m. The metal screen offers negligible resistance to flow. The catalyst particles have an average diameter of 5 mm and density equal to 1500 kg/m^3. Calculate the power required to establish an upward flow through the bed at a flowrate equal to 0.03 m^3/s.

Part II
Conservation Equations

Chapter 3
Differential Form of Conservation Equations

3.1 Conservation Law

The detailed structure of a flow field and its evolution over time are described by a system of partial differential equations that stem from the conservation of scalar quantities such as mass and energy, or vector quantities such as momentum. Let Γ be some field variable, generally defined as a function of space and time, that can be associated to the fluid and V be a control volume that encloses some finite region in space occupied by the fluid at a given instant of time. In general, the conservation law applied to Γ in the control volume V states that: *The rate of change of Γ over time is equal to the net flux of Γ across the surface of the control volume, which is given by the difference between the inward and outward fluxes of Γ through the surface, referred to as $\dot{\Gamma}_{in}$ and $\dot{\Gamma}_{out}$ respectively.* In mathematical terms:

$$\frac{d\Gamma}{dt} = \dot{\Gamma}_{in} - \dot{\Gamma}_{out} \quad . \tag{3.1}$$

In some cases, the rate of change may also depend on the presence of source or sink terms for Γ within the control volume.

3.2 Mass Conservation and Continuity Equation

With reference to Fig. 3.1, let us consider as control volume ΔV a parallelepiped with dimensions Δx, Δy and Δz. The mass of fluid contained in ΔV is equal to $\rho \, \Delta x \, \Delta y \, \Delta z$ and its time variation (referred to as accumulation) is expressed as:

$$\text{Accumulation} = \frac{\partial \rho}{\partial t} \, \Delta x \, \Delta y \, \Delta z \quad . \tag{3.2}$$

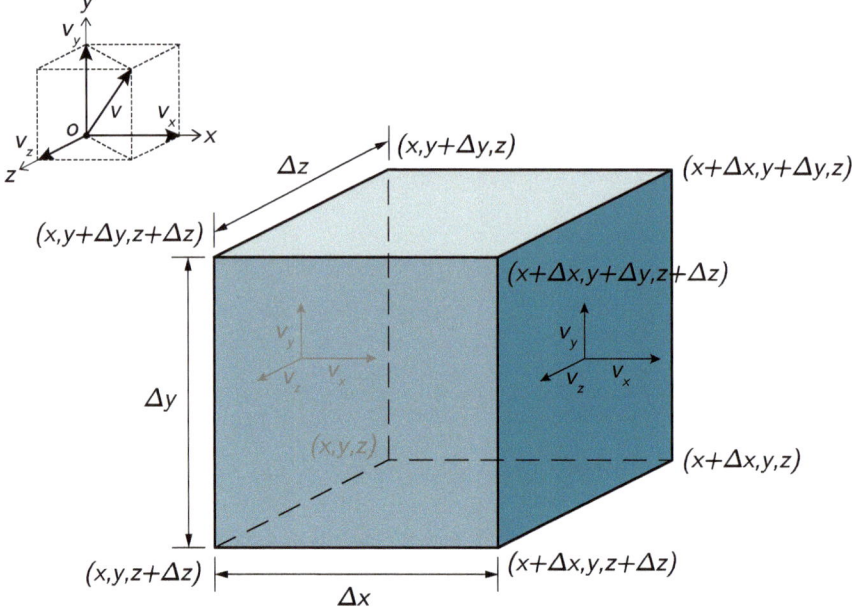

Fig. 3.1 Control volume used for deriving the mass conservation law in Cartesian coordinates

The mass fluxes across the faces of the volume along the x direction can be expressed as:

$$\text{Inward flux at position } x = \rho v_x \mid_x \Delta y \Delta z,$$
$$\text{Outward flux at position } x + \Delta x = \rho v_x \mid_{x+\Delta x} \Delta y \Delta z \quad . \tag{3.3}$$

By proceeding in a similar fashion for the mass fluxes across the other faces along the y and z directions, the overall mass balance can be written as:

$$\frac{\partial \rho}{\partial t} \Delta x \, \Delta y \, \Delta z = (\rho v_x \mid_x - \rho v_x \mid_{x+\Delta x}) \Delta y \Delta z +$$

$$+ (\rho v_y \mid_y - \rho v_y \mid_{y+\Delta y}) \Delta x \Delta z + (\rho v_z \mid_z - \rho v_z \mid_{z+\Delta z}) \Delta y \Delta x \quad . \tag{3.4}$$

In this equation, each term on the right-hand side can be expressed using a Taylor series expansion truncated at the first order. For instance, the term $\rho v_x \mid_{x+\Delta x}$ providing the mass flux in the x direction at position $x + \Delta x$, can be expressed as:

$$\rho v_x \mid_{x+\Delta x} = \rho v_x \mid_x + \left. \frac{\partial \rho v_x}{\partial x} \right|_x \Delta x \quad . \tag{3.5}$$

This approximation is accurate in the limit $\Delta x \to 0$. Adopting the same procedure for the other terms, again in the limit $\Delta x, \Delta y, \Delta z \to 0$, Eq. (3.4) becomes:

$$\frac{\partial \rho}{\partial t} + \frac{\partial (\rho v_x)}{\partial x} + \frac{\partial (\rho v_y)}{\partial y} + \frac{\partial (\rho v_z)}{\partial z} = 0 \quad . \tag{3.6}$$

Equation (3.6) is referred to as *continuity equation*. In vector notation, it reads as:

$$\frac{\partial \rho}{\partial t} + \nabla \cdot (\rho \mathbf{v}) = 0 \quad . \tag{3.7}$$

The same equation can be derived also considering a mass balance with respect to a finite volume V bounded by a surface S within the flow field. The time variation of the mass of fluid contained in V is given by:

$$\frac{\partial}{\partial t} \int_V \rho \mathrm{d}V.$$

The net mass flux is given by:

$$- \int_S \rho \mathbf{v} \cdot \mathbf{n} \mathrm{d}S,$$

where the negative sign is due to the fact that $\mathbf{n}$ is an outward-pointing unit vector. Equating the two terms yields:

$$\frac{\partial}{\partial t} \int_V \rho \mathrm{d}V = \int_V \frac{\partial \rho}{\partial t} \mathrm{d}V = - \int_S \rho \mathbf{v} \cdot \mathbf{n} \mathrm{d}S \quad . \tag{3.8}$$

Invoking the Gauss theorem, the term on the right-hand side of Eq. (3.8) can be rewritten as a volume integral:

$$\int_S \rho \mathbf{v} \cdot \mathbf{n} \mathrm{d}S = \int_V \nabla \cdot (\rho \mathbf{v}) \mathrm{d}V \quad , \tag{3.9}$$

and, therefore:

$$\int_V \left[\frac{\partial \rho}{\partial t} + \nabla \cdot (\rho \mathbf{v}) \right] \mathrm{d}V = 0 \quad . \tag{3.10}$$

This equation is satisfied if the integrand is zero, a condition that leads exactly to Eq. (3.7). In the remainder of this chapter, for simplicity, the momentum and energy balances will be derived only in differential form. However, it is always possible to derive an integral form of these balances.

3.3 Material Derivative (or Lagrangian)

The differential operator:

$$\frac{D}{Dt} = \frac{\partial}{\partial t} + v_x \frac{\partial}{\partial x} + v_y \frac{\partial}{\partial y} + v_z \frac{\partial}{\partial z} \tag{3.11}$$

is defined as the *material derivative* (also known as *Lagrangian derivative*) of a function. Using this operator, it is possible to rewrite Eq. (3.6) as:

$$\frac{D\rho}{Dt} = -\rho \left(\frac{\partial v_x}{\partial x} + \frac{\partial v_y}{\partial y} + \frac{\partial v_z}{\partial z} \right) = -\rho \nabla \cdot \mathbf{v} \quad . \tag{3.12}$$

To understand the meaning of material derivative, let us consider a fluid element located at position x, y, z at time t. At time $t + \Delta t$, the element will be at position:

$$x(t + \Delta t) = x(t) + \frac{dx}{dt} \Delta t = x(t) + v_x \Delta t \quad , \tag{3.13}$$

and similarly for the other two coordinates. Consider now a generic property ξ of the fluid that may vary with position and time. The value of ξ at time $t + \Delta t$ will be related to the value of ξ at time t as follows:

$$\xi(t + \Delta t) = \xi(t) + \frac{\partial \xi}{\partial t} \Delta t + \frac{\partial \xi}{\partial x} \Delta x + \frac{\partial \xi}{\partial y} \Delta y + \frac{\partial \xi}{\partial z} \Delta z \quad , \tag{3.14}$$

and, being $\Delta x = v_x \Delta t$, $\Delta y = v_y \Delta t$, and $\Delta z = v_z \Delta z$:

$$\xi(t + \Delta t) = \xi(t) + \left(\frac{\partial \xi}{\partial t} + v_x \frac{\partial \xi}{\partial x} + v_y \frac{\partial \xi}{\partial y} + v_z \frac{\partial \xi}{\partial z} \right) \Delta t \quad , \tag{3.15}$$

or, equivalently:

$$\frac{D\xi}{Dt} = \lim_{\Delta t \to 0} \frac{\xi(t + \Delta t) - \xi(t)}{\Delta t} = \frac{\partial \xi}{\partial t} + v_x \frac{\partial \xi}{\partial y} + v_y \frac{\partial \xi}{\partial y} + v_z \frac{\partial \xi}{\partial z} \quad . \tag{3.16}$$

Therefore, the material derivative D/Dt can be interpreted as the rate of change of a physical quantity (like velocity or temperature) in time, experienced by an observer (referred to as Lagrangian observer) that moves with the velocity of a fluid element. The derivative $\partial/\partial t$, on the other hand, represents the rate of change experienced by an observer (referred to as Eulerian observer) that is stationary in a fixed or inertial frame of reference.

3.3.1 Examples

1 The velocity of a fluid is given by:

$$\mathbf{v} = u_0 e^{-at} \left(\mathbf{i}\, b\, x + \mathbf{j}\, c\, y^2 \right) \quad . \tag{3.17}$$

Derive the expression for the acceleration $D\mathbf{v}/Dt$ of the fluid as a function of a, b, c and u_0.

Solution For two-dimensional problems, the material derivative is defined as:

$$\frac{D\mathbf{v}}{Dt} = \frac{\partial \mathbf{v}}{\partial t} + v_x \frac{\partial \mathbf{v}}{\partial x} + v_y \frac{\partial \mathbf{v}}{\partial y} \quad . \tag{3.18}$$

Equation (3.17) yields:

$$\frac{\partial \mathbf{v}}{\partial t} = -a\, u_0 e^{-at} \left(\mathbf{i}\, b\, x + \mathbf{j}\, c\, y^2 \right) \quad , \tag{3.19}$$

and:

$$v_x \frac{\partial \mathbf{v}}{\partial x} = u_0^2 e^{-2at} b^2 x \mathbf{i} \quad , \qquad v_y \frac{\partial \mathbf{v}}{\partial y} = 2 u_0^2 e^{-2at} c^2 y^3 \mathbf{j} \quad . \tag{3.20}$$

3.4 Momentum Conservation and Navier–Stokes Equations

3.4.1 Eulerian Derivation

Momentum is a vector quantity that is conserved in all three physical directions at the same time. For each direction, the conservation law states that *the rate of accumulation of momentum in a given control volume is equal to the net rate of momentum entering the control volume (momentum flux) plus the sum of all external forces acting on the volume.* Conservation of momentum can thus be formulated in terms of a vector balance equation, namely three scalar equations. The schematic of the control volume used for deriving the momentum balance equation is shown in Fig. 3.2.

The momentum of the fluid in the control volume along the x direction is equal to $\rho v_x\, \Delta x\, \Delta y\, \Delta z$. The rate of change of momentum in time is:

$$\text{Accumulation of momentum} = \frac{\partial (\rho v_x)}{\partial t} \Delta x\, \Delta y\, \Delta z \quad . \tag{3.21}$$

The flowrate entering the control volume, e.g. through the face normal to the x axis, is $\rho v_x\, \Delta y\, \Delta z$. The momentum in the x direction associated with this flowrate is

$\rho v_x \, \Delta y \, \Delta z \, \cdot v_x \, |_x$. Similarly, the momentum in the x direction associated with the flowrate entering the control volume through the y and z faces is $\rho v_y \, \Delta x \, \Delta z \, \cdot v_x \, |_y$ and $\rho v_z \, \Delta x \, \Delta y \, \cdot v_x \, |_z$, respectively. When the momenta associated with the outgoing flowrates are also considered, the momentum balance equation reads as:

$$\frac{\partial(\rho v_x)}{\partial t} \, \Delta x \, \Delta y \, \Delta z \; = (\rho v_x v_x \, |_x - \rho v_x v_x \, |_{x+\Delta x}) \, \Delta y \, \Delta z \; +$$

$$+ \, (\rho v_x v_y \, |_y - \rho v_x v_y \, |_{y+\Delta y}) \, \Delta x \, \Delta z \; + (\rho v_x v_z \, |_z - \rho v_x v_z \, |_{z+\Delta z}) \, \Delta x \, \Delta y \; +$$

$$+ \text{ Sum of the forces acting along } x \quad . \tag{3.22}$$

The forces acting on the control volume are of two types: Body forces and surface forces. The main body force is the one due to gravity and its component in the x direction is $\rho g_x \, \Delta x \, \Delta y \, \Delta z$. The surface force contributions are due to the presence of *normal stresses* and *tangential stresses* acting on the faces of the control volume. The surface force exerted by the surrounding fluid on the faces of the control volume that are normal to the x direction can be expressed as $\sigma_x \, \Delta y \, \Delta z$, where:

$$\boldsymbol{\sigma}_x = \sigma_{xx}\mathbf{i} + \tau_{xy}\mathbf{j} + \tau_{xz}\mathbf{k} \quad , \tag{3.23}$$

σ_{xx} being the normal stress acting on the face of the control volume, and τ_{xy} and τ_{xz} the tangential stresses (see Fig. 3.2). Note that, in each stress component, the first subscript identifies the face on which the stress acts by the direction normal to the face while the second subscript identifies the direction along which the stress acts.

Regarding the sign of the stresses, different conventions exist. The following convention is adopted in this textbook: a stress acting on a surface with normal $\mathbf{n}$ has positive sign when exerted *by* the fluid on the side of the surface to which the normal points, *on* the fluid from which the normal points. As a consequence, tensile normal stresses have positive sign, while compressive normal stresses have negative sign.

With reference to Fig. 3.2, the normal and tangential stresses acting on the faces of the control volume in the x direction, are σ_{xx} (at x and $x + \Delta x$), τ_{yx} (at y and $y + \Delta y$) and τ_{zx} (at z and $z + \Delta z$). Hence:

$$\text{Shear stress along } x \; =$$

$$= (\sigma_{xx} \, |_{x+\Delta x} - \sigma_{xx} \, |_x) \, \Delta y \, \Delta z \; + (\tau_{yx} \, |_{y+\Delta y} - \tau_{yx} \, |_y) \, \Delta x \, \Delta z$$

$$+ \, (\tau_{zx} \, |_{z+\Delta z} - \tau_{zx} \, |_z) \Delta y \Delta x \quad . \tag{3.24}$$

Plugging the terms in Eq. (3.24) into Eq. (3.22), the momentum balance in the x direction becomes:

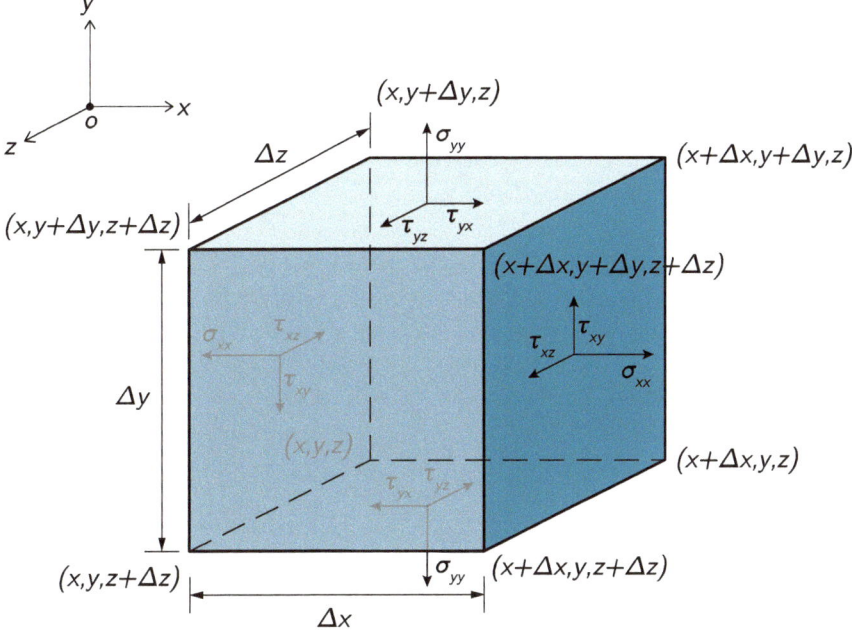

Fig. 3.2 Control volume for the momentum balance in Cartesian coordinates

$$\frac{\partial(\rho v_x)}{\partial t} + \frac{\partial(\rho v_x v_x)}{\partial x} + \frac{\partial(\rho v_x v_y)}{\partial y} + \frac{\partial(\rho v_x v_z)}{\partial z} =$$

$$= \frac{\partial \sigma_{xx}}{\partial x} + \frac{\partial \tau_{yx}}{\partial y} + \frac{\partial \tau_{zx}}{\partial z} + \rho g_x \quad , \tag{3.25}$$

which, using Eq. (3.6), can be rewritten as:

$$\rho\frac{Dv_x}{Dt} = \rho\frac{\partial v_x}{\partial t} + \rho v_x\frac{\partial v_x}{\partial x} + \rho v_y\frac{\partial v_x}{\partial y} + \rho v_z\frac{\partial v_x}{\partial z}$$

$$= \frac{\partial \sigma_{xx}}{\partial x} + \frac{\partial \tau_{yx}}{\partial y} + \frac{\partial \tau_{zx}}{\partial z} + \rho g_x \quad . \tag{3.26}$$

Similar scalar equations can be written for the momentum balance along the y and z directions. In vector notation, the momentum conservation equation can be expressed as:

$$\rho\frac{D\mathbf{v}}{Dt} = \nabla \cdot \boldsymbol{\sigma} + \rho\mathbf{g} \quad , \tag{3.27}$$

where $\boldsymbol{\sigma}$ is referred to as *stress tensor*, and the product $\nabla \cdot \boldsymbol{\sigma}$ is a vector. In general, the kth component of $\nabla \cdot \boldsymbol{\sigma}$ is expressed as $\sum_i \partial \tau_{ik}/\partial x_i$ using index notation.

Equation (3.27) is known as the Cauchy equation, from the French mathematician and engineer Augustin-Louis Cauchy (1789–1857), and applies to any continuous medium.

3.4.2   Lagrangian Derivation

The Cauchy equation can also be derived by applying Newton's second law to a mass of fluid $\Delta m = \rho \Delta V$ moving with acceleration $\mathbf{a} = D\mathbf{v}/Dt$. In this case, Newton's second law states that the change in momentum, given by the product of mass and acceleration in the case of constant mass, is equal to the sum of the external forces acting on the fluid inside the volume ΔV, these forces being the volume (body) and surface forces introduced in Chap. 1:

$$\Delta m \mathbf{a} = \sum_i \mathbf{F}_i = \mathbf{F}_{vol} + \mathbf{F}_{sur} \quad , \tag{3.28}$$

with:

$$\Delta m \mathbf{a} = \int_{\Delta V} \rho \mathbf{a} dV,$$

$$\mathbf{F}_{vol} = \int_{\Delta V} \rho \mathbf{g} dV,$$

$$\mathbf{F}_{sur} = \int_{\Delta S} \boldsymbol{\sigma} \cdot \mathbf{n} dS \quad . \tag{3.29}$$

with ΔS the surface enclosing the volume ΔV, $\mathbf{n}$ the unit vector normal to the surface element dS, and $\boldsymbol{\sigma} \cdot \mathbf{n}$ the product between the tensor $\boldsymbol{\sigma}$ and the unit vector $\mathbf{n}$. By applying Gauss' divergence theorem (1.47), the surface force can be expressed as:

$$\mathbf{F}_{sur} = \int_{\Delta V} \nabla \cdot \boldsymbol{\sigma} dV \quad . \tag{3.30}$$

Substituting Eqs. (3.29) and (3.30) into Eq. (3.28) gives:

$$\int_{\Delta V} \rho \mathbf{a} dV = \int_{\Delta V} \rho \mathbf{g} dV + \int_{\Delta V} \nabla \cdot \boldsymbol{\sigma} dV \quad , \tag{3.31}$$

and, removing the integrals:

$$\rho \frac{D\mathbf{v}}{Dt} = \underbrace{\rho \mathbf{g}}_{\mathbf{f}_{vol}} + \underbrace{\nabla \cdot \boldsymbol{\sigma}}_{\mathbf{f}_{sup}} \quad , \tag{3.32}$$

where $\mathbf{f}_{vol} = \mathbf{F}_{vol}/\Delta V$ and $\mathbf{f}_{sur} = \mathbf{F}_{sur}/\Delta V$.

3.4.3 Stress Tensor

The stress tensor $\boldsymbol{\sigma}$, also called Cauchy stress tensor, can be represented in matrix form as follows:

$$\boldsymbol{\sigma} = \begin{bmatrix} \boldsymbol{\sigma}_x \\ \boldsymbol{\sigma}_y \\ \boldsymbol{\sigma}_z \end{bmatrix} = \begin{bmatrix} \sigma_{xx} & \tau_{xy} & \tau_{xz} \\ \tau_{yx} & \sigma_{yy} & \tau_{yz} \\ \tau_{zx} & \tau_{zy} & \sigma_{zz} \end{bmatrix} \quad . \tag{3.33}$$

The tensor components are symmetric ($\tau_{xy} = \tau_{yx}$, $\tau_{xz} = \tau_{zx}$ and $\tau_{yz} = \tau_{zy}$) due to *conservation of angular momentum*. Let us consider the torque exerted by the forces that induce rotation of a given fluid element about the z-axis:

$$\text{Torque around } z =$$

$$= \frac{1}{2}(\tau_{xy}\mid_x + \tau_{xy}\mid_{x+\Delta x} - \tau_{yx}\mid_y - \tau_{yx}\mid_{y+\Delta y})\Delta x\,\Delta y\,\Delta z \quad . \tag{3.34}$$

The angular momentum of the element can be expressed as:

$$\text{Angular momentum} =$$

$$= \int_{\Delta V} \rho r^2 \Omega \mathrm{d}x\mathrm{d}y\mathrm{d}z = \rho \Omega r_g^2\,\Delta x\,\Delta y\,\Delta z \quad , \tag{3.35}$$

where the integral is extended to the control volume ΔV, r is the distance from the z axis, Ω is the angular velocity about the z axis, and $r_g^2 = \Delta x \Delta y/6$. Conservation of angular momentum yields:

$$\lim_{\substack{\Delta x \to 0 \\ \Delta y \to 0}} \left(\frac{\rho}{6} \frac{\mathrm{d}\Omega}{\mathrm{d}t} \Delta x\,\Delta y \right) = \tau_{xy} - \tau_{yx} \quad . \tag{3.36}$$

Since $\mathrm{d}\Omega/\mathrm{d}t$ must be finite, it follows that $\tau_{xy} - \tau_{yx}$ must become zero in the limit and, therefore, $\tau_{xy} = \tau_{yx}$. Similar relations can be obtained for the other off-diagonal components of $\boldsymbol{\sigma}$.

3.4.4 Navier–Stokes Equations for Newtonian Fluids

To solve the momentum conservation equation, it is necessary to derive constitutive relations between the stress tensor and the fluid velocity field. In the case of Newtonian fluids, the shear stresses are directly proportional to the first derivatives of the velocity field. Consider, for instance, a fluid motion in the x direction only, between two surfaces at distance Δy from each other. The following relation holds:

$$\tau_{yx} = \mu \frac{dv_x}{dy} = \tau_{xy} \quad . \tag{3.37}$$

If the frame of reference is rotated by $90°$, so that the motion is in the y direction only, the following relation holds:

$$\tau_{xy} = \mu \frac{dv_y}{dx} = \tau_{yx} \quad . \tag{3.38}$$

Since, in general, the motion may occur in both directions simultaneously, the previous equations can be generalized as follows:

$$\tau_{xy} = \tau_{yx} = \mu \left(\frac{\partial v_x}{\partial y} + \frac{\partial v_y}{\partial x} \right) \quad . \tag{3.39}$$

Similar relations can be found for the other components of σ.

To derive the expression of the normal stresses, a rigorous analytical treatment is required. The derivation is based on the following assumptions, which are valid for Newtonian fluids:

1. The stresses are symmetrical.
2. The stress at a given position depends only on the instantaneous value of the velocity gradient at that position.
3. The stresses are linear functions of the velocity gradient.
4. The stresses are isotropic in the absence of motion.

Based on these properties of Newtonian fluids, the following expressions for the normal stresses can be derived:

$$\sigma_{xx} = -p + 2\mu \frac{\partial v_x}{\partial x} - \frac{2}{3}\mu \nabla \cdot \mathbf{v} \quad , \tag{3.40}$$

$$\sigma_{yy} = -p + 2\mu \frac{\partial v_y}{\partial y} - \frac{2}{3}\mu \nabla \cdot \mathbf{v} \quad , \tag{3.41}$$

$$\sigma_{zz} = -p + 2\mu \frac{\partial v_z}{\partial z} - \frac{2}{3}\mu \nabla \cdot \mathbf{v} \quad . \tag{3.42}$$

Summing up the normal stresses yields:

$$\sigma_{xx} + \sigma_{yy} + \sigma_{zz} = -3p + 2\mu \left(\frac{\partial v_x}{\partial x} + \frac{\partial v_y}{\partial y} + \frac{\partial v_z}{\partial z} \right) - 2\mu \nabla \cdot \mathbf{v} =$$

$$= -3p + 2\mu \nabla \cdot \mathbf{v} - 2\mu \nabla \cdot \mathbf{v} = -3p \quad , \tag{3.43}$$

and thus:

$$p = -\frac{1}{3}(\sigma_{xx} + \sigma_{yy} + \sigma_{zz}) \quad . \tag{3.44}$$

This relation shows that pressure can be interpreted as a compressive stress equal to the average value of the normal stresses acting on the control volume, namely to the isotropic part of the stress tensor The normal stresses can be decomposed into two components. One component is the hydrostatic or dilatational stress that acts to change the volume of the material only; the other is the deviatoric stress that acts to change the shape only:

$$\tau_{xx} = p + \sigma_{xx} = 2\mu\frac{\partial v_x}{\partial x} - \frac{2}{3}\mu\nabla \cdot \mathbf{v} \quad . \tag{3.45}$$

Similar equations can be derived for the other stress components. Therefore, for a Newtonian fluid whose viscosity does not depend on position, the momentum conservation equations, known as the *Navier–Stokes equations* (*N–S*) after the French engineer and scientist Claude-Louis Navier (1785–1836) and G. G. Stokes, can be written in the following form (*x*-component only):

$$\rho\frac{Dv_x}{Dt} = -\frac{\partial p}{\partial x} + \mu\left(\frac{\partial^2 v_x}{\partial x^2} + \frac{\partial^2 v_x}{\partial y^2} + \frac{\partial^2 v_x}{\partial z^2}\right) + \frac{1}{3}\mu\frac{\partial}{\partial x}(\nabla \cdot \mathbf{v}) + \rho g_x \quad .\tag{3.46}$$

The *N–S* equations in vector form read as:

$$\rho\frac{D\mathbf{v}}{Dt} = -\nabla p + \mu\nabla^2\mathbf{v} + \frac{1}{3}\mu\nabla(\nabla \cdot \mathbf{v}) + \rho\mathbf{g} \quad . \tag{3.47}$$

3.4.5 Navier–Stokes Equations for Incompressible Fluids

For an incompressible fluid, it is possible to express the *N–S* equations in a more compact form by introducing the equivalent pressure, $\mathcal{P}$, defined as:

$$\mathcal{P} = p + \rho gh \quad , \tag{3.48}$$

where ρgh is the potential energy of the fluid. Using this definition, and recalling that $\nabla \cdot \mathbf{v} = 0$, the *N–S* equations take the form:

$$\rho\frac{D\mathbf{v}}{Dt} = \rho\left[\frac{\partial \mathbf{v}}{\partial t} + (\mathbf{v} \cdot \nabla)\mathbf{v}\right] = -\nabla\mathcal{P} + \mu\nabla^2\mathbf{v} \quad . \tag{3.49}$$

In Eq. (3.49), the terms $\rho(\mathbf{v} \cdot \nabla)\mathbf{v}$ represent the convective transport of momentum in the control volume, while the terms $\mu\nabla^2\mathbf{v}$ represent the diffusive transport of momentum in the control volume. *Convective transport* is associated with the macroscopic bulk motion of the fluid molecules within the control volume. Diffusive transport is instead associated with the microscopic random-like motion of individual fluid molecules and the interactions that occur between the molecules. Therefore, diffusive transport is present even in the absence of bulk motion of the fluid molecules. Convective transport and diffusive transport may coexist in a fluid and may not necessarily occur in the same direction, but they will always be characterized by different transport rates, the convective one being much faster, and by different timescales, the diffusive one being much longer.

3.5   Energy Conservation

The energy per unit mass of a fluid element is given by the sum of the internal energy, e, the kinetic energy, $v^2/2$, and the potential energy, gh. The conservation of energy for a flowing system can be interpreted as an extension of the first law of thermodynamics and can be formulated as follows: *The rate of change of the fluid energy in the control volume is equal to the net energy flux through the control volume plus the net heat transfer into the control volume plus the net work done on the control volume.*

When applied to a fluid element of volume $\Delta x\ \Delta y\ \Delta z$, this law leads to the identification of the following terms:

$$\text{Accumulation of energy} = \frac{\partial}{\partial t}\left[\rho\left(e + \frac{1}{2}v^2 + gh\right)\right]\Delta x\ \Delta y\ \Delta z,$$

$$\text{Net convective flux (input - output)} =$$

$$= \Delta y\ \Delta z\left[v_x\rho\left(e + \frac{1}{2}v^2 + gh\right)\Big|_x - v_x\rho\left(e + \frac{1}{2}v^2 + gh\right)\Big|_{x+\Delta x}\right] +$$

$$+ \Delta x\ \Delta z\left[v_y\rho\left(e + \frac{1}{2}v^2 + gh\right)\Big|_y - v_y\rho\left(e + \frac{1}{2}v^2 + gh\right)\Big|_{y+\Delta y}\right] +$$

$$+ \Delta x\ \Delta y\left[v_z\rho\left(e + \frac{1}{2}v^2 + gh\right)\Big|_z - v_z\rho\left(e + \frac{1}{2}v^2 + gh\right)\Big|_{z+\Delta z}\right],$$

$$\text{Heat flux due to thermal conduction} =$$

$$= \Delta y\ \Delta z(q_x\ |_x - q_x\ |_{x+\Delta x}) + \Delta x\ \Delta z(q_y\ |_y - q_y\ |_{y+\Delta y}) + \Delta x\ \Delta y(q_z\ |_z - q_z\ |_{z+\Delta z}),$$

where q_x, q_y and q_z are the components of the vector $\mathbf{q}$, which represents conductive heat flux. For case in which the Fourier law for thermal conduction applies, $\mathbf{q}$ can be expressed as a function of the temperature gradient as:

$$\mathbf{q} = -k\nabla T \quad , \tag{3.50}$$

where k is the thermal conductivity of the fluid and T is its temperature.

The work done per unit time on the control volume is:

$$\text{Work per unit time } =$$

$$= \Delta y\,\Delta z\,\left\{(\sigma_{xx}v_x + \tau_{xy}v_y + \tau_{xz}v_z)\,|_{x+\Delta x} -(\sigma_{xx}v_x + \tau_{xy}v_y + \tau_{xz}v_z)\,|_x\right\} +$$

$$+ \Delta x\,\Delta z\,\left\{(\tau_{yx}v_x + \sigma_{yy}v_y + \tau_{yz}v_z)\,|_{y+\Delta y} -(\tau_{yx}v_x + \sigma_{yy}v_y + \tau_{yz}v_z)\,|_y\right\} +$$

$$+ \Delta x\,\Delta y\,\left\{(\tau_{zx}v_x + \tau_{zy}v_y + \sigma_{zz}v_z)\,|_{z+\Delta z} -(\tau_{zx}v_x + \tau_{zy}v_y + \sigma_{zz}v_z)\,|_z\right\}.$$

The work is produced by the viscous and pressure forces and can be calculated as the product of the force component acting in a given direction (e.g. $\sigma_{xx}\,\Delta y\,\Delta z$) times the displacement in that direction at a time dt (e.g. $v_x dt$). The work per unit time is given by the product of the force by the velocity (e.g. $\sigma_{xx}\,v_x\,\Delta y\,\Delta z$).

Given that $\boldsymbol{\sigma} = -p\mathbf{I} + \boldsymbol{\tau}$, where $\mathbf{I}$ is the identity matrix (in which all the elements on the principal diagonal are unitary and all the elements above and below the diagonal are zeros), the energy balance can be written as:

$$\rho\frac{D}{Dt}\left(e + \frac{1}{2}v^2 + gh\right) = -\nabla \cdot \mathbf{q} - \nabla\cdot p\mathbf{v} + \nabla \cdot (\boldsymbol{\tau} \cdot \mathbf{v}) \quad . \tag{3.51}$$

In deriving this equation, the concept of potential energy ϕ was used, where ϕ is defined such that:

$$\mathbf{g} = -\nabla\phi \quad , \tag{3.52}$$

with $\phi = gh$. It can be noticed that:

$$\rho\frac{D\phi}{Dt} = \rho\frac{\partial\phi}{\partial t} + \rho\mathbf{v}\cdot\nabla\phi = \rho\mathbf{v}\cdot\mathbf{g} \quad , \tag{3.53}$$

as ϕ is independent of time. The term $\rho\mathbf{v}\cdot\mathbf{g}$ can be interpreted as the work performed by the body force $\mathbf{g}$.

3.5.1 Mechanical Energy Equation

In a thermodynamic system, only the sum of the three energy terms, $e + v^2/2 + gh$, is conserved. This sum will be referred to as *total energy* hereinafter, in order to distinguish it from the mechanical energy, $v^2/2 + gh$, which is not conserved. It is, however, possible to derive from the conservation of momentum a scalar equation commonly referred to as the *mechanical energy equation*.

To derive such equation, the scalar product of the equation of momentum by the velocity vector $\mathbf{v}$ must be performed:

$$\rho\mathbf{v} \cdot \frac{D\mathbf{v}}{Dt} = -\mathbf{v} \cdot \nabla p + \mathbf{v} \cdot (\nabla \cdot \boldsymbol{\tau}) + \rho\mathbf{v} \cdot \mathbf{g} \quad . \tag{3.54}$$

It can be shown that the term on the left-hand side of Eq. (3.54) is equal to:

$$\rho\mathbf{v} \cdot \frac{D\mathbf{v}}{Dt} = \rho\frac{D}{Dt}\left(\frac{1}{2}v^2\right) \quad , \tag{3.55}$$

namely to the substantial derivative of the kinetic energy associated to the control volume. It has already been noted that:

$$-\rho\mathbf{v} \cdot \mathbf{g} = \rho\frac{D}{Dt}(gh) \quad , \tag{3.56}$$

and therefore Eq. (3.54), once rewritten as:

$$\rho\frac{D}{Dt}\left(\frac{1}{2}v^2 + gh\right) = -\mathbf{v} \cdot \nabla p + \mathbf{v} \cdot \nabla \cdot \boldsymbol{\tau} \quad , \tag{3.57}$$

relates the change in mechanical energy (kinetic and potential) to the action of the forces applied on the surface of the control volume.

The equation expressing the conservation of the total energy can be rewritten in the form:

$$\rho\frac{D}{Dt}\left(e + \frac{1}{2}v^2 + gh\right) = -\nabla \cdot \mathbf{q} - p\nabla \cdot \mathbf{v} +$$
$$-\mathbf{v} \cdot \nabla p + \mathbf{v} \cdot \nabla \cdot \boldsymbol{\tau} + \boldsymbol{\tau} : \nabla\mathbf{v} \quad , \tag{3.58}$$

where the two terms $\nabla \cdot p\mathbf{v}$ and $\nabla \cdot (\boldsymbol{\tau} \cdot \mathbf{v})$ have been already expanded. Subtracting Eq. (3.58) from Eq. (3.57) yields:

$$\rho\frac{De}{Dt} = -\nabla \cdot \mathbf{q} - p\nabla \cdot \mathbf{v} + \boldsymbol{\tau} : \nabla\mathbf{v} \quad . \tag{3.59}$$

This equation may be referred to as the *thermal energy equation* as it allows to calculate the changes of temperature caused by the conductive heat transfer ($\nabla \cdot \mathbf{q}$), by reversible compression ($p\nabla \cdot \mathbf{v}$) and by irreversible viscous dissipation ($\boldsymbol{\tau} : \nabla\mathbf{v}$). The last two terms represent the conversion of mechanical energy into thermal energy.

For a Newtonian fluid, the viscous dissipation term $\boldsymbol{\tau} : \nabla\mathbf{v}$ is given by:

$$\boldsymbol{\tau} : \nabla\mathbf{v} = \mu\Phi_v = 2\mu \left\{ \left(\frac{\partial v_x}{\partial x}\right)^2 + \left(\frac{\partial v_y}{\partial y}\right)^2 + \left(\frac{\partial v_z}{\partial z}\right)^2 + \right.$$

$$+ \frac{1}{2}\left[\left(\frac{\partial v_y}{\partial x} + \frac{\partial v_x}{\partial y}\right)^2 + \left(\frac{\partial v_z}{\partial y} + \frac{\partial v_y}{\partial z}\right)^2 + \left(\frac{\partial v_x}{\partial z} + \frac{\partial v_z}{\partial x}\right)^2\right] +$$

$$\left. + \frac{1}{3}\left(\frac{\partial v_x}{\partial x} + \frac{\partial v_y}{\partial y} + \frac{\partial v_z}{\partial z}\right)^2\right\} \; . \tag{3.60}$$

Viscous dissipation causes an increase of the temperature of the fluid, which in practical applications is often negligible.

3.5.2 Bernoulli Equation

For the case of steady flow of an incompressible and Newtonian fluid, it is quite easy to derive from the mechanical energy equation the *Bernoulli equation*, named after the Swiss mathematician and physicist Daniel Bernoulli (1700–1782). To start, let us take the scalar product between the equation of momentum conservation and the velocity vector. Under the above-mentioned assumptions, taking the product yields:

$$\rho\mathbf{v}[(\mathbf{v} \cdot \nabla)\mathbf{v}] = -\mathbf{v} \cdot \nabla(p + \rho gh) + \mu\mathbf{v} \cdot \nabla^2\mathbf{v} \; . \tag{3.61}$$

Since $\nabla \cdot \mathbf{v} = 0$ for an incompressible fluid, the following identities can be written:

$$\mathbf{v}[(\mathbf{v} \cdot \nabla)\mathbf{v}] = \frac{1}{2}\nabla \cdot (\mathbf{v}v^2) \tag{3.62}$$

$$\mathbf{v} \cdot \nabla(p + \rho gh) = \nabla \cdot \mathbf{v}(p + \rho gh) \; . \tag{3.63}$$

Replacing Eqs. (3.62) and (3.63) into Eq. (3.61) and integrating over the control volume yields:

$$\int_V \nabla \cdot \mathbf{v}\left(\frac{1}{2}\rho v^2 + p + \rho gh\right) dV = \mu \int_V \mathbf{v} \cdot \nabla^2\mathbf{v}\, dV \; . \tag{3.64}$$

and, applying Gauss' theorem to the left-hand side of Eq. (3.64):

$$\int_S \left(\frac{1}{2}\rho v^2 + p + \rho g h \right) \mathbf{v} \cdot \mathbf{n} dS = \mu \int_V \mathbf{v} \cdot \nabla^2 \mathbf{v} dV \quad . \tag{3.65}$$

When the control volume is bounded by the surface that is impermeable to the fluid and the fluid can only enter the control volume through the surface A_1 and exit through the surface A_2, the left-hand side of Eq. (3.65) can be written as:

$$\int_S \left(\frac{1}{2}\rho v^2 + p + \rho g h \right) \mathbf{v} \cdot \mathbf{n} dS = \Delta \left[\left\langle v_n \left(\frac{1}{2}\rho v^2 + p + \rho g h \right) \right\rangle A \right] \quad , \tag{3.66}$$

where the angle brackets denote the mean value on the surface A, and $v_n = \mathbf{v} \cdot \mathbf{n}$. The viscous dissipation term on the right-hand side of Eq. (3.65) can be set equal to:

$$\mu \int_V \mathbf{v} \cdot \nabla^2 \mathbf{v} dV = -\mu \int_V \Phi_v dV = -l_v w \quad , \tag{3.67}$$

where Φ_v is the dissipation function and l_v must be properly evaluated depending on the geometry of the flow domain and on the other relevant flow parameters. Combining Eqs. (3.66) and (3.67) yields:

$$\Delta \left[\left\langle v_n \left(\frac{1}{2}\rho v^2 + p + \rho g h \right) \right\rangle A \right] = -l_v w \quad . \tag{3.68}$$

This is one of the forms in which the *Bernoulli equation* can be written. A simplified form can be written when v_n is equal to the magnitude of the velocity vector v and v, ρ and p are uniform on the surfaces A_1 and A_2. Recalling that, for steady flow, the mass flowrate through the control volume is $w = \rho_1 v_1 A_1 = \rho_2 v_2 A_2$, Eq. (3.68) becomes:

$$\Delta \left(\frac{1}{2}v^2 + \frac{p}{\rho} + g h \right) = -l_v \quad . \tag{3.69}$$

The Bernoulli equation will be discussed again in Chap. 9.

3.5.3 Examples

1 An incompressible Newtonian fluid of viscosity μ and density ρ flows radially outward from a porous sphere of radius R with volumetric flowrate Q. The flow is purely radial both inside and outside the sphere. The value of the fluid pressure at infinite radial distance is $p(\infty) = p_\infty$.

1. Determine the velocity distribution inside and outside the porous sphere.
2. Inside the sphere, the pressure gradient is related to the volumetric flowrate by the following relation:

$$Q = -\frac{kr^2}{\mu}\frac{dp}{dr} \quad , \tag{3.70}$$

known as the *Darcy equation*, derived experimentally by the French engineer Henry Darcy (1803–1858). Determine the change of pressure with r both inside and outside the sphere.

Solution   Since the flow is purely radial, both velocity and pressure depend solely on r.

1. The mass balance for the spherical shell of thickness dr, located at radial distance r from the center of the sphere, can be expressed as:

$$\underbrace{\frac{\partial}{\partial t}(4\pi r^2 dr)\rho}_{\substack{\text{Mass} \\ \text{accumulation}}} = \underbrace{4\pi r^2 v_r(r)\rho}_{\substack{\text{Inward} \\ \text{flow}}} - \underbrace{4\pi (r+dr)^2 v_r(r+dr)\rho}_{\substack{\text{Outward} \\ \text{flow}}} \quad . \tag{3.71}$$

In the limit $dr \to 0$, expressing $v_r(r+dr)$ by means of a Taylor series expansion yields:

$$4\pi r^2 dr\frac{\partial\rho}{\partial t} =$$

$$= 4\pi\rho\left\{r^2 v_r(r) - (r^2 + dr^2 + 2r dr)\left[v_r(r) + \frac{dv_r(r)}{dr}dr\right]\right\} \quad . \tag{3.72}$$

For incompressible fluid and steady flow, and neglecting the higher order terms in the expansion, Eq. (3.72) becomes:

$$0 = -2r v_r(r) - r^2\frac{dv_r(r)}{dr} = -\frac{d}{dr}[r^2 v_r(r)] = 0 \quad . \tag{3.73}$$

Upon integration:

$$r^2 v_r(r) = C \quad\longrightarrow\quad v_r(r) = \frac{C}{r^2} \quad . \tag{3.74}$$

As boundary condition to calculate the integration constant C, it is possible to impose that the volumetric flowrate through the surface of the sphere of radius R be equal to Q:

$$Q = 4\pi R^2 v_r(R) = 4\pi R^2\frac{C}{R^2} \quad\rightarrow\quad C = \frac{Q}{4\pi} \quad\rightarrow\quad v_r(r) = \frac{Q}{4\pi r^2} \quad . \tag{3.75}$$

2. Outside the sphere $(r > R)$, the pressure distribution can be determined using the Bernoulli equation. Along the radial direction, only the inertial term and the pressure term are non-zero. Taking the balance between a generic value $r > R$ of the radial coordinate and $r = r(\infty)$ yields:

$$\frac{p(r)}{\rho} + \frac{1}{2}v_r^2(r) = \frac{p(\infty)}{\rho} + \frac{1}{2}v_r^2(\infty) \quad . \tag{3.76}$$

However, for $r \to \infty$, $v_r(r) \to 0$:

$$p(r) = p_\infty - \frac{\rho}{2}v_r^2(r) = p_\infty - \frac{\rho Q^2}{32\pi^2 r^4} \quad . \tag{3.77}$$

On the surface of the sphere, the pressure is:

$$p(R) = p_\infty - \frac{\rho Q^2}{32\pi^2 R^4} \quad . \tag{3.78}$$

Inside the porous sphere $(r < R)$, the pressure drop is related to the volumetric flowrate by Eq. (3.70). Therefore:

$$dp = -\frac{\mu}{k}\frac{Q}{r^2}dr \quad \to \quad p(r) = \frac{\mu Q}{k}\frac{1}{r} + D \quad . \tag{3.79}$$

The integration constant D is determined imposing the condition for $r = R$ given by Eq. (3.78). Hence, the pressure distribution inside the sphere is obtained as:

$$p(r) = p_\infty - \frac{\mu Q}{kR}\left(1 - \frac{R}{r}\right) - \frac{\rho Q^2}{32\pi^2 R^4} \quad . \tag{3.80}$$

Problems

a Consider the purely radial flow of a liquid through the porous wall of an infinitely long cylinder with outer radius equal to $R = 0.15$ m. The liquid flowrate per unit length of the cylinder is $Q/L = 0.13$ m^2/s.

1. Compute the radial velocity v_r at a distance $r = 0.15$ m from the axis and at an infinite radial distance.
2. If the pressure inside the cylinder is equal to $1.2 \cdot 10^5$ Pa and the thickness of the porous wall is 0.1 m, calculate the pressure distribution from the center of the cylinder to the outside of the cylinder.

$\boxed{b}$ An incompressible Newtonian fluid is pumped isothermally through a long horizontal duct. Because of the very high value of the pressure inside the duct, the viscosity of the fluid depends on the pressure according to the relation $\mu = \mu_0 \exp(kp)$, with k known constant. Derive the pressure behavior as a function of the volumetric flowrate.

Chapter 4
Exact Solutions for Unidirectional Steady Flows

4.1 Unidirectional Flows

An exact solution of the Navier–Stokes equations can be obtained only in a limited number of cases. These include steady unidirectional flows in which the fluid moves in one direction only and, therefore, the velocity vector has just one non-zero component. The steady-state assumption limits the analysis to laminar flows. This chapter surveys a number of different unidirectional flows. In particular, flows in which the motion of the fluid is forced by a moving wall, known as *Couette flows*, after the French physicist Maurice Couette (1858–1943), and flows in which the motion of the fluid is driven by an imposed pressure gradient, known as *Poiseuille flows*, after the French physician, physiologist and physicist Jean Léonard Marie Poiseuille (1799–1869). We will also examine combined *Couette–Poiseuille flows*, like the flow in a closed rectangular cavity, which can be described by exploiting the superposition of effects thanks to the linearity of the equations that describe the fluid motion.

4.1.1 Plane Couette Flow

The problem is illustrated in Fig. 4.1. An incompressible Newtonian fluid flows between two flat, parallel walls placed at a distance H. The flow is determined by the relative motion of one wall with respect to the other, characterized by a constant velocity difference U. The length L and width W of the two walls are much larger than their distance H.

The velocity field does not depend on the coordinates x and z. Therefore, the following derivatives are zero:

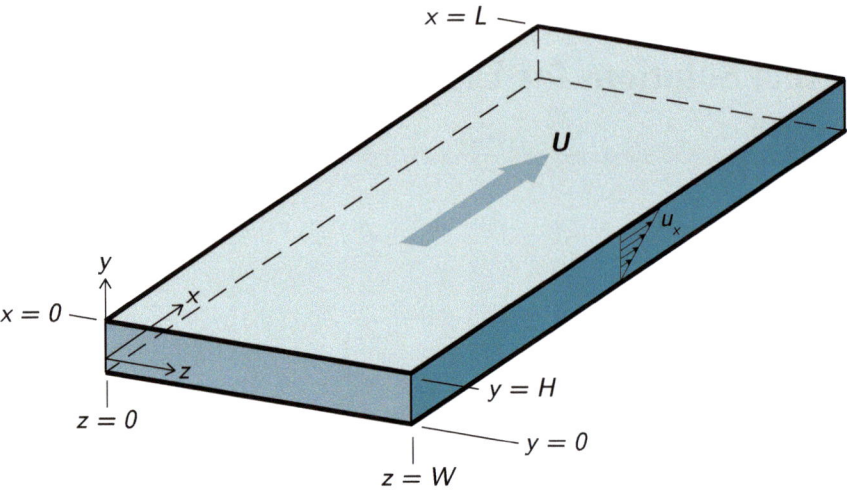

Fig. 4.1 Plane Couette flow

$$\frac{\partial v_x}{\partial x} = \frac{\partial v_y}{\partial x} = \frac{\partial v_z}{\partial x} = 0 \quad , \tag{4.1}$$

$$\frac{\partial v_x}{\partial z} = \frac{\partial v_y}{\partial z} = \frac{\partial v_z}{\partial z} = 0 \quad . \tag{4.2}$$

As a consequence, the components of $\mathbf{v}$ depend only on the y coordinate. It can also be intuitively assumed that: $v_z = 0$, $v_y = 0$, $v_x = v_x(y)$. These assumptions satisfy the continuity equation.

The *N-S* equations simplify to:

$$-\frac{\partial P}{\partial x} + \mu \frac{\partial^2 v_x}{\partial y^2} = 0 \quad , \tag{4.3}$$

$$-\frac{\partial P}{\partial y} = 0 \quad , \tag{4.4}$$

$$-\frac{\partial P}{\partial z} = 0 \quad , \tag{4.5}$$

In a Couette flow, the motion occurs in the absence of an imposed pressure gradient. If we assume $\partial P/\partial x = 0$, Eq. (4.3) reduces to:

$$\frac{d^2 v_x}{dy^2} = 0 \quad , \tag{4.6}$$

and thus, $v_x = c_1 y + c_2$. The boundary conditions:

$$y = 0 \quad \longrightarrow \quad v_x = 0, \tag{4.7}$$

$$y = H \quad \longrightarrow \quad v_x = U \quad , \tag{4.8}$$

lead to the solution of Eq. (4.6), which is:

$$v_x = \frac{U}{H} y \quad . \tag{4.9}$$

Equation (4.9) can be used to calculate the shear stress, obtaining:

$$\tau_{yx} = \mu \frac{U}{H} \quad . \tag{4.10}$$

Equation (4.10) shows that the shear stress is uniform between the walls and is also equal to the coefficient used to define viscosity in Sect. 1.1.2.

4.1.2 Plane Poiseuille Flow

The problem is illustrated in Fig. 4.2. An incompressible Newtonian fluid flows between two flat, parallel walls placed at a distance H. The length L and width W of the two walls are much larger than their distance H. The flow is determined by a pressure gradient and is directed along the x-axis. It is assumed that the flow is fully developed, namely that inlet effects can be neglected. As for the Couette flow, the independence of the flow field from the x and z coordinates leads to:

$$\frac{\partial v_x}{\partial x} = \frac{\partial v_y}{\partial x} = \frac{\partial v_z}{\partial x} = 0 \quad , \tag{4.11}$$

$$\frac{\partial v_x}{\partial z} = \frac{\partial v_y}{\partial z} = \frac{\partial v_z}{\partial z} = 0 \quad . \tag{4.12}$$

It follows that the components of $\mathbf{v}$ are only a function of the y coordinate. It can also be intuitively assumed that $v_z = 0$, $v_y = 0$, $v_x = v_x(y)$.

The N-S equations simplify to:

$$-\frac{\partial \mathcal{P}}{\partial x} + \mu \frac{\partial^2 v_x}{\partial y^2} = 0 \quad , \tag{4.13}$$

$$-\frac{\partial \mathcal{P}}{\partial y} = 0 \quad , \tag{4.14}$$

$$-\frac{\partial \mathcal{P}}{\partial z} = 0 \quad , \tag{4.15}$$

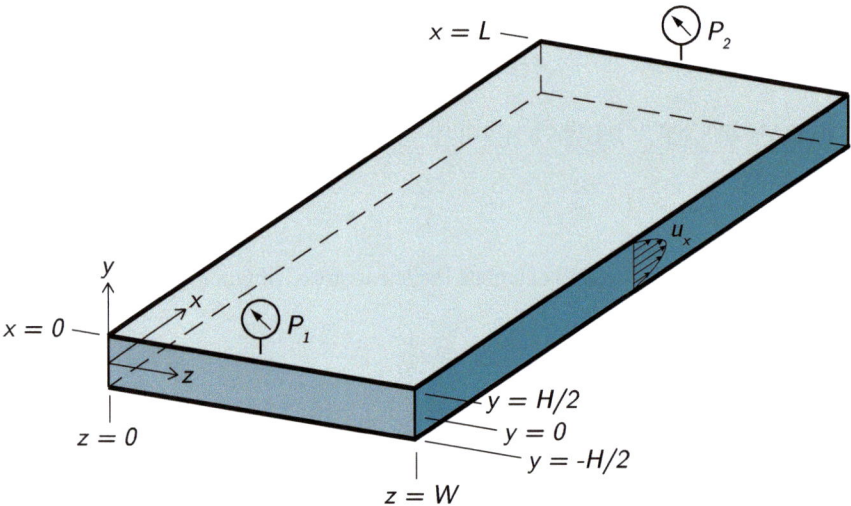

Fig. 4.2 Plane Poiseuille flow

where $\mathcal{P} = p + \rho g h$. From Eqs. (4.14) and (4.15) it is straightforward to conclude that $\mathcal{P} = \mathcal{P}(x)$. Equation (4.13) can be written as:

$$\mu \frac{d^2 v_x}{dy^2} = \frac{d\mathcal{P}}{dx} \quad . \tag{4.16}$$

In this equation, the term on the left-hand side is only a function of y and the one on the right-hand side is only a function of x. These terms can be equal if and only if they are constant. The condition $d\mathcal{P}/dx = $ constant along x implies that the equivalent pressure must vary linearly with x. This makes it possible to set $d\mathcal{P}/dx = (\mathcal{P}_2 - \mathcal{P}_1)/L = \Delta\mathcal{P}/L$. In the present case, the vertical elevation of the fluid does not change in the flow direction and, therefore, $\Delta\mathcal{P}$ coincides with the pressure drop imposed between the inlet and outlet sections of the channel. This implies that the imposed pressure gradient is known.

With the above considerations, we can easily integrate Eq. (4.16) and obtain:

$$v_x = \frac{1}{2\mu} \frac{d\mathcal{P}}{dx} y^2 + c_1 y + c_2 \quad . \tag{4.17}$$

By taking into account the boundary conditions:

$$y = \pm\frac{H}{2} \quad \longrightarrow \quad v_x = 0 \quad , \tag{4.18}$$

equation (4.17) becomes:

$$v_x = \frac{H^2}{8\mu}\left(-\frac{d\mathcal{P}}{dx}\right)\left[1 - \left(\frac{2y}{H}\right)^2\right] \quad . \tag{4.19}$$

The volumetric flowrate, Q, can be calculated as:

$$Q = W \int_{-H/2}^{H/2} v_x dy = \frac{WH^3}{12\mu}\left(-\frac{d\mathcal{P}}{dx}\right) \quad . \tag{4.20}$$

4.1.3 Poiseuille Flow in a Pipe

Let us now consider the flow of an incompressible Newtonian fluid in a pipe, as shown in Fig. 4.3. To study the problem, it is convenient to adopt a cylindrical coordinate system using the transformation:

$$r = (x^2 + y^2)^{1/2}, \quad \theta = \tan^{-1}(y/x), \quad z = z \quad . \tag{4.21}$$

In this coordinate system, considerations similar to those made in the previous examples allow us to write:

$$v_z = v_z(r), \quad v_r = v_\theta = 0 \quad . \tag{4.22}$$

These conditions satisfy the continuity equation. The N-S equations for this case are:

$$-\frac{\partial \mathcal{P}}{\partial r} = 0 \quad , \tag{4.23}$$

$$-\frac{1}{r}\frac{\partial \mathcal{P}}{\partial \theta} = 0 \quad , \tag{4.24}$$

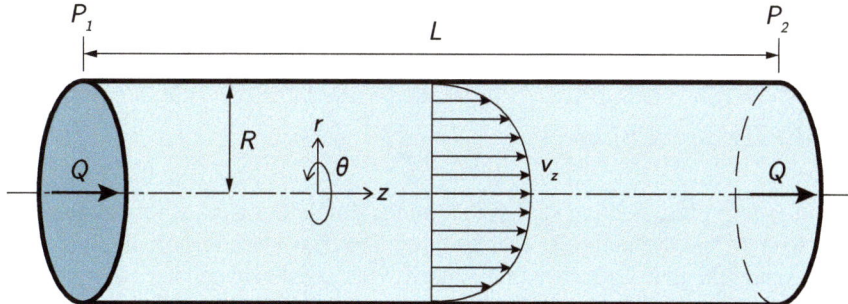

Fig. 4.3 Poiseuille flow in a pipe

$$-\frac{\partial \mathcal{P}}{\partial z} + \mu \frac{1}{r} \frac{\partial}{\partial r} \left(r \frac{\partial v_z}{\partial r} \right) = 0 \quad . \tag{4.25}$$

Equations (4.23) and (4.24) lead to conclude that $\mathcal{P}$ is only a function of z. Equation (4.25) also dictates $\partial \mathcal{P}/\partial z$ = constant. With this assumption, the solution of Eq. (4.25) reads as:

$$v_z = \frac{1}{4\mu} \frac{d\mathcal{P}}{dz} r^2 + c_1 \ln r + c_2 \quad . \tag{4.26}$$

The boundary conditions that Eq. (4.25) must satisfy are:

$$r = R \quad \longrightarrow \quad v_z = 0 \quad , \tag{4.27}$$

$$r = 0 \quad \longrightarrow \quad \frac{dv_z}{dr} = \frac{1}{2\mu} \frac{d\mathcal{P}}{dz} r + c_1 \frac{1}{r} = 0 \quad . \tag{4.28}$$

The second condition is obtained by considering that the fluid velocity is maximum precisely on the symmetry axis of the pipe, and allows us to set $c_1 = 0$ when the condition is applied for $r \to 0$ (note that $r = 0$ corresponds to singular points for dv_z/dr). The condition $c_1 = 0$ allows us to remove the logarithmic term in Eq. (4.26), which would otherwise determine a singularity even for the function v_z when $r = 0$.

The first boundary condition allows us to derive the following expression for v_z:

$$v_z(r) = \frac{R^2}{4\mu} \left(-\frac{d\mathcal{P}}{dz} \right) \left[1 - \left(\frac{r}{R} \right)^2 \right] \quad . \tag{4.29}$$

The average velocity is:

$$\langle v_z \rangle = \frac{R^2}{8\mu} \left(-\frac{d\mathcal{P}}{dz} \right) \quad . \tag{4.30}$$

and the corresponding flowrate is equivalent to the Hagen–Poiseuille equation (2.14).

4.1.4 Torsional Flow

In the system illustrated in Fig. 4.4, the inner cylinder of radius R rotates about its axis with angular velocity Ω. The resulting flow is a Couette flow instance. The rotation sets the fluid contained in the outer cylinder of diameter $\gg R$ in motion. It is further assumed that the height of the inner cylinder is also much larger than R. In this case the motion is purely tangential, i.e. only the component v_θ of the fluid velocity vector $\mathbf{v}$ is different from zero and is a function of r only. The N-S equations in cylindrical coordinates read as:

Fig. 4.4 Sketch of torsional flow

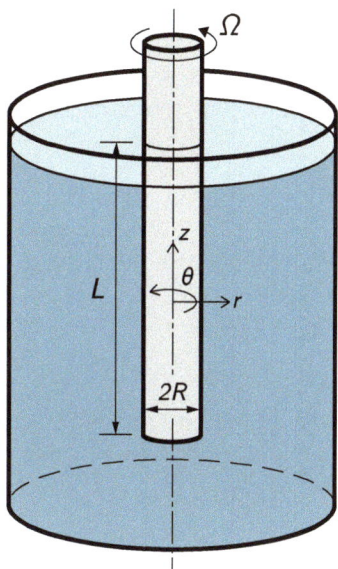

$$-\rho\frac{v_\theta^2}{r} = -\frac{\partial \mathcal{P}}{\partial r} \quad , \tag{4.31}$$

$$\mu\frac{\partial}{\partial r}\left[\frac{1}{r}\frac{\partial}{\partial r}(rv_\theta)\right] = 0 \quad , \tag{4.32}$$

$$-\frac{\partial \mathcal{P}}{\partial z} = 0 \quad , \tag{4.33}$$

where it is assumed that there is no pressure gradient in the direction of motion. Equation (4.31) shows that there must be a pressure gradient in the radial direction. Otherwise, this equation would reduce to $-\rho v_\theta^2/r = 0$ and would be verified only if $v_\theta = 0 \; \forall \; r$. This conditions, which implies absence of fluid motion everywhere in between the cylinders, is certainly false (at least, near the inner cylinder). In the present problem, the fluid is subject to a radial pressure gradient that is not imposed to produce the motion but, rather, is produced once the motion is triggered to balance the centrifugal force that acts on the fluid.

Equation (4.33) allows to establish that $\mathcal{P}$ is only a function of r. Equation (4.32) can be integrated and gives:

$$v_\theta = \frac{1}{2}c_1 r + \frac{c_2}{r} \quad . \tag{4.34}$$

The boundary conditions for this problem are:

$$r = R \;\longrightarrow\; v_\theta = R\Omega \quad , \tag{4.35}$$

$$r \rightarrow \infty \longrightarrow v_\theta = 0 \quad . \tag{4.36}$$

The second boundary condition indicates that the motion of the inner cylinder is not transmitted to large distances. With these conditions, Eq. (4.34) becomes:

$$v_\theta = \frac{R^2\Omega}{r} \quad . \tag{4.37}$$

It is now possible to calculate the torque exerted by the viscous force acting on the rotating cylinder:

$$G = 2\pi L R^2 \tau_{r\theta} \quad , \tag{4.38}$$

where $\tau_{r\theta}$ depends on the velocity gradients according to the constitutive equation for Newtonian fluids, which in cylindrical coordinates reads as:

$$\tau_{r\theta} = \mu \left[r \frac{\partial}{\partial r} \left(\frac{v_\theta}{r} \right) + \frac{1}{r} \frac{\partial v_r}{\partial \theta} \right] = -2\mu\Omega \frac{R^2}{r^2} \quad . \tag{4.39}$$

From Eq. (4.31), the pressure gradient $dP(r)/dr$ can be calculated as:

$$\frac{dP(r)}{dr} = \frac{\rho(R^2\Omega)^2}{r^3} \quad , \tag{4.40}$$

and, in turn, the equivalent pressure as:

$$P = P_0 - \rho \frac{(R^2\Omega)^2}{2r^2} \quad . \tag{4.41}$$

The pressure changes along r since it is necessary to balance the centrifugal force and reaches its minimum value for $r = R$. In this position, it may be necessary to verify that no cavitation conditions are established, i.e. the pressure does not fall below the vapor pressure, P_{vap}, of the fluid. To prevent this from happening, the following condition must apply: $P_{min} = P(r = R) > P_{vap}$. This condition yields:

$$R\Omega < \sqrt{\frac{2(P_0 - P_{vap})}{\rho}} \quad . \tag{4.42}$$

4.1.5 Unidirectional Free-Surface Flow

A special class of unidirectional flows is the flow of a thin layer of liquid over an inclined wall under the action of gravity. These flows are driven by a pressure gradient generated by a height difference. The film is also in contact with a gas and the liquid-gas interface is called *free surface*. The wall is very wide and sufficiently long with respect to the thickness of the film so that unidirectional, fully developed flow can be assumed. Also, the velocity is only a function of the wall-normal coordinate. As an example, consider the flow of a liquid of density ρ and viscosity μ down an inclined wall (β being the inclination angle with respect to the horizontal direction), illustrated in Fig. 4.5. In this case, since the free surface is flat and parallel to the wall, the thickness of the film does not vary along the flow direction. Neglecting end effects along the spanwise direction and considering that x is the wall-normal coordinate ($x = 0$ at the free film surface and $x = \delta$ at the wall) while z is the wall-parallel coordinate, the flow field can be obtained by considering the following simplifying assumptions:

1. Steady-state flow.
2. Fully-developed laminar flow.
3. Wall-parallel streamlines.

With these assumptions, the *N-S* equations in x, y and z read as:

$$0 = -\frac{\partial \mathcal{P}}{\partial x} \quad , \tag{4.43}$$

$$0 = -\frac{\partial \mathcal{P}}{\partial y} \quad , \tag{4.44}$$

$$\mu \frac{\partial^2 v_z}{\partial x^2} = \frac{\partial \mathcal{P}}{\partial z} \quad . \tag{4.45}$$

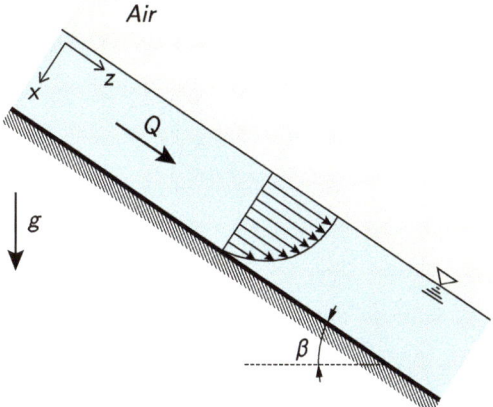

Fig. 4.5 Thin liquid film flowing down an inclined wall

Equations (4.43) and (4.44) allow to conclude that $\mathcal{P} = \mathcal{P}(z)$. The expression of $\mathcal{P}$ can be derived by considering that the pressure at the free surface, denoted as p_0, is constant. Knowing p_0 (e.g., equal to the atmospheric pressure), and considering that there is no flow in the x direction, along which the equation of fluid statics applies, it is possible to conclude that the pressure at any point with the same value of the x coordinate within the film is also constant. Therefore, the corresponding hydrostatic pressure distribution is $p = p_0 + \rho(g \cos \beta)x$. Recalling the definition of equivalent pressure, we can then set $\mathcal{P} = p_0 + \rho g h$, where $h = h_0 - z \sin \beta + x \cos \beta$ is the vertical elevation (measured with respect to a horizontal reference plane) of a generic point within the film and h_0 is the vertical coordinate associated with the origin of the Cartesian reference system. Assuming h_0 to be known, the equivalent pressure gradient can be written as:

$$\frac{\partial \mathcal{P}(z)}{\partial z} = \frac{\mathrm{d}\mathcal{P}(z)}{\mathrm{d}z} = \frac{\mathrm{d}p_0}{\mathrm{d}z} + \rho g \frac{\mathrm{d}h}{\mathrm{d}z} = -\rho g \sin \beta \quad, \tag{4.46}$$

and Eq. (4.45) becomes:

$$-\mu \frac{\mathrm{d}^2 v_z}{\mathrm{d}x^2} = \rho g \sin \beta \quad. \tag{4.47}$$

Integration of Eq. (4.47) yields:

$$v_z = -\frac{\rho g \sin \beta}{2\mu} x^2 + C_1 x + C_2 \quad. \tag{4.48}$$

A no-slip boundary condition can be imposed at the wall, while at the gas-liquid interface it can be assumed that the gas exerts no stress on the liquid:

$$v_z|_{x=\delta} = 0, \qquad \tau_{xz}|_{x=0} = \mu \frac{\mathrm{d}v_z}{\mathrm{d}x}\bigg|_{x=0} = 0 \quad. \tag{4.49}$$

These boundary conditions yield $C_1 = 0$ e $C_2 = (\rho g \sin \beta \delta^2)/(2\mu)$, and the velocity profile becomes:

$$v_z = \frac{\rho g \sin \beta \delta^2}{2\mu} \left[1 - \left(\frac{x}{\delta}\right)^2\right] \quad. \tag{4.50}$$

The mean velocity is:

$$\langle v_z \rangle = \frac{1}{\delta} \int_0^\delta v_z \mathrm{d}x = \frac{\rho g \sin \beta \delta^2}{3\mu} \quad. \tag{4.51}$$

The volumetric flow rate can be calculated as:

$$Q = \int_0^W \int_0^\delta v_z \, dx \, dy = \frac{\rho g \sin \beta W \delta^3}{3\mu} \quad . \tag{4.52}$$

The specific flowrate per unit width, Γ, can be calculated as:

$$\Gamma = \frac{1}{W} \int_0^W \int_0^\delta \rho v_z \, dx \, dy = \frac{\rho^2 g \sin \beta \delta^3}{3\mu} \quad . \tag{4.53}$$

The film thickness can be expressed as function of the average velocity, volumetric flowrate, and specific flowrate per unit width as follows:

$$\delta = \sqrt{\frac{3\mu \langle v_z \rangle}{\rho g \sin \beta}} = \sqrt[3]{\frac{3\mu Q}{\rho g W \sin \beta}} = \sqrt[3]{\frac{3\mu \Gamma}{\rho^2 g \sin \beta}} \quad . \tag{4.54}$$

4.1.6 Examples

1 The viscosity of a Newtonian liquid depends on the temperature according to the relation:

$$\mu(T) = \frac{\mu_0}{1 + \beta(T - T_0)} \quad , \tag{4.55}$$

where T_0 is a known reference temperature. The liquid flows between two flat plates, driven by an imposed pressure gradient as shown in Fig. 4.6. The walls are maintained at temperatures T_0 and T_1 with $T_0 < T_1$. It is assumed that the temperature changes linearly with position according to the relation:

$$T = T_0 + (T_1 - T_0)\frac{y}{H} \quad . \tag{4.56}$$

Determine the relation between the volumetric flowrate and the pressure drop.

Solution Since the viscosity is not constant, it is not possible to use the Navier–Stokes equations: The Cauchy equations must be used.

Denoting by • a generic flow variable, the following assumptions hold for the case under consideration:

1. Steady-state flow ($\rightarrow \partial \bullet / \partial t = 0$);
2. Fully-developed laminar flow ($\rightarrow \partial \bullet / \partial x = 0$);
3. Wall-parallel streamlines ($\rightarrow v_y = v_z = 0$, $\partial v_x / \partial z = 0$).

From these assumptions, it follows that the non-zero velocity component is $v_x = v_x(y)$. The continuity equation is satisfied. The Cauchy equations in x and y are:

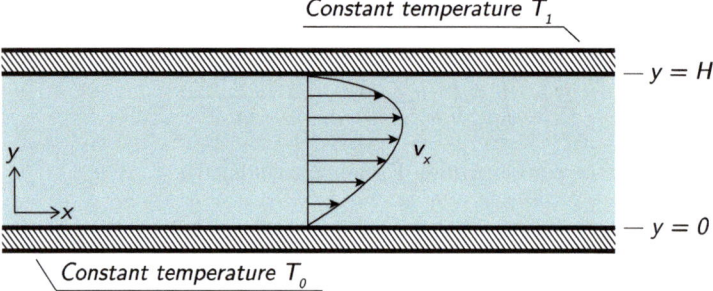

Fig. 4.6 Flow between two parallel plates with a constant temperature gradient

$$0 = -\frac{\partial P}{\partial x} + \frac{\partial}{\partial y}\left(\mu\frac{\partial v_x}{\partial y}\right) \quad , \tag{4.57}$$

$$0 = -\frac{\partial P}{\partial y} \quad , \tag{4.58}$$

where the pressure gradient is imposed and assumed to be constant. Integration of Eq. (4.57) yields:

$$\mu\frac{\partial v_x}{\partial y} = \frac{\partial P}{\partial x}y + C_1' \rightarrow v_x = \frac{\partial P}{\partial x}\int_0^y \frac{y}{\mu}dy + C_1'\int_0^y \frac{dy}{\mu} \quad . \tag{4.59}$$

Substituting Eqs. (4.55) and (4.56) for μ and integrating, yields:

$$v_x = \frac{1}{\mu_0}\frac{dP}{dx}y^2\left(\frac{1}{2} + \frac{\lambda y}{3H}\right) + C_1 y\left(1 + \frac{\lambda y}{2H}\right) \quad , \tag{4.60}$$

where $\lambda = \beta(T_1 - T_0)$. The integration constant C_1 is determined using the condition $v_x = 0$ per $y = H$.[1] The constant is equal to:

$$C_1 = -\frac{H}{\mu_0}\frac{dP}{dx}\frac{1/2 + \lambda/3}{1 + \lambda/2} \quad . \tag{4.61}$$

The volumetric flowrate is given by:

$$Q = \int_0^H v_x(y)W\,dy \quad , \tag{4.62}$$

where W is the width of the plates within which the liquid flows. Solving the integral provides the relationship between the volumetric flowrate and the pressure drop:

---

[1] The other boundary condition, $v_x = 0$ for $y = 0$, was implicitly used when the integral was calculated from 0.

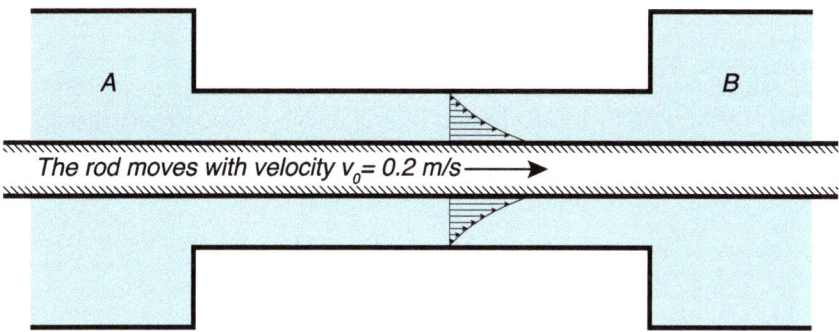

Fig. 4.7 Transport of a fluid between two reservoirs by means of a moving rod

$$Q = \frac{H^3 W}{\mu_0} \frac{\mathrm{d}\mathcal{P}}{\mathrm{d}x} \left[\frac{1}{6} + \frac{\lambda}{12} - \frac{(1/2 + \lambda/3)(1/2 + \lambda/6)}{1 + \lambda/2} \right] . \tag{4.63}$$

Note that Eq. (4.63) reduces to that already found for plane Poiseuille flow when $\lambda = 0$.

2 In the system shown in Fig. 4.7, a pipe of diameter equal to 40 mm connects two reservoirs, A and B, which are kept at the same pressure $\mathcal{P}_0$. Inside the pipe, a cylindrical rod of diameter equal to 20 mm moves with constant velocity 0.2 m/s. The fluid contained in the reservoirs has density $\rho = 800$ kg/m^3 and viscosity $\mu = 10^{-2}$ Pa · s. Determine the flowrate transported between the two reservoirs.

Solution The initial hypotheses are:

1. Incompressible Flow.
2. Steady-state flow.
3. Cylindrical symmetry $\longrightarrow v_r = v_\theta = 0$.
4. Fully-developed flow $\longrightarrow \partial \bullet / \partial z = 0$.

The continuity equation simplifies as follows:

$$0 = \frac{\partial}{\partial z} \rho v_z \longrightarrow v_z = v_z(r) . \tag{4.64}$$

Since the pressure in the two reservoirs is the same, there is no pressure gradient in the axial direction. Therefore, the N-S equations become:

$$0 = \frac{\partial \mathcal{P}}{\partial r}, \, 0 = \frac{1}{r} \frac{\partial \mathcal{P}}{\partial \theta}, \, 0 = \mu \frac{1}{r} \frac{\partial}{\partial r} \left(r \frac{\partial v_z}{\partial r} \right) . \tag{4.65}$$

The mass flowrate is $w = \langle v_z \rangle A \rho$, where $\langle v_z \rangle$ is the average velocity along z. By integrating twice the z-component of the N-S equations, the velocity profile can be obtained (note that the pressure does not depend on r):

$$v_z(r) = C_1 \ln r + C_2 \quad , \tag{4.66}$$

where the constants C_1 and C_2 can be calculated imposing the no-slip boundary conditions at the walls of the annular duct (R is the radius of the outer cylinder and kR the radius of the inner cylinder): For $r = R \rightarrow v_z = 0$, and for $r = kR \rightarrow v_z = v_0$, with v_0 velocity of the rod. These conditions lead to the following equations:

$$0 = C_1 \ln R + C_2, \quad v_0 = C_1 \ln(kR) + C_2 \quad , \tag{4.67}$$

which give:

$$C_1 = \frac{v_0}{\ln k}, \quad C_2 = -v_0 \frac{\ln R}{\ln k} \quad . \tag{4.68}$$

The resulting velocity profile is:

$$v_z(r) = v_0 \frac{\ln r}{\ln k} - v_0 \frac{\ln R}{\ln k} = \frac{v_0}{\ln k} \ln\left(\frac{r}{R}\right) \quad , \tag{4.69}$$

while the flowrate is:

$$w = \rho A \langle v_z \rangle = \rho A \frac{1}{A} 2\pi \int_{kR}^{R} v_z(r) r \, dr =$$

$$= 2\pi \rho \frac{v_0}{\ln k} \int_{kR}^{R} r \ln \frac{r}{R} dr = 2\pi \rho \frac{v_0}{\ln k} \left[\frac{1}{2} r^2 \ln \frac{r}{R} - \frac{1}{4} r^2 \right]_{kR}^{R} =$$

$$= \rho \frac{\pi v_0}{2 \ln k} R^2 \left[k^2 (1 - 2 \ln k) - 1 \right] = 5.85 \cdot 10^{-2} \text{kg/s} \quad . \tag{4.70}$$

3 A high viscosity fluid is transported between two reservoirs by means of a friction pump consisting of a conveyor belt with length L and width W, as shown in Fig. 4.8. The two reservoirs are at atmospheric pressure and have a difference in fluid height equal to h. Assume negligible inlet and outlet effects and $m \ll L$. This is an example of Couette–Poiseuille flow in which the Couette flow produced by the moving belt has to counteract the pressure gradient imposed by the different height of the fluid in the reservoirs.

1. Simplify and integrate the N-S equations in the flow direction imposed by the conveyor belt.
2. Determine the volumetric flowrate Q and the theoretical power P_M of the pump engine as a function of the belt speed U.
3. Assuming that the power transferred by the pump to the fluid can be written as $P_F = Q\rho g h$, determine the value of h for which P_F is maximum.

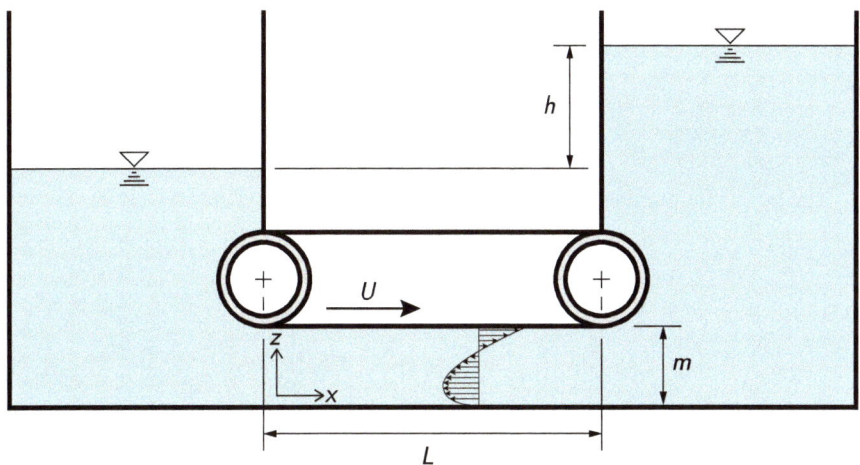

Fig. 4.8 Transport of fluid between two reservoirs kept at different fluid heights by means of a friction pump consisting of a conveyor belt

Solution

1. With the following assumptions:

   (a) Steady-state flow ($\rightarrow \partial \bullet /\partial t = 0$);
   (b) Fully-developed laminar flow ($\rightarrow \partial \bullet /\partial x = 0$);
   (c) Wall-parallel streamlines ($\rightarrow v_y = v_z = 0$, $\partial v_x/\partial y = 0$);
   (d) $h \simeq$ constant,

   it is easy to find that $v_x = v_x(z)$, and the x-component of the N-S equations takes the form:

   $$0 = -\frac{\partial \mathcal{P}}{\partial x} + \mu \frac{\partial^2 v_x}{\partial z^2} \quad \rightarrow \quad \frac{\Delta \mathcal{P}}{\mu L} = \frac{\rho g h}{\mu L} = \frac{d^2 v_x}{dz^2} \quad . \tag{4.71}$$

   Imposing the boundary conditions:

   $$v_x(0) = 0, \quad v_x(m) = U \quad , \tag{4.72}$$

   the velocity profile is:

   $$v_x(z) = \frac{\rho g h}{2\mu L} z(z - m) + \frac{U}{m} z = \frac{\rho g h}{2\mu L} z^2 + \left(\frac{U}{m} - \frac{\rho g h}{2\mu L} m \right) z \quad . \tag{4.73}$$

2. The volumetric flowrate of the pump is given by $Q = mW \langle v_x \rangle$. The average velocity can be calculated as:

$$\langle v_x \rangle = \frac{1}{m} \int_0^m \left[\frac{\rho g h}{2\mu L} z^2 + \left(\frac{U}{m} - \frac{\rho g h}{2\mu L} m \right) z \right] dz =$$

$$= \frac{U}{2} - \frac{1}{12} \frac{\rho g h}{\mu L} m^2 \quad . \tag{4.74}$$

The power required to keep the belt in motion at constant speed U is given by:

$$P_M = \tau_w LWU = \mu \left. \frac{dv_x(z)}{dz} \right|_{z=m} LWU = \mu \left(\frac{1}{2} \frac{\rho g h m}{\mu L} + \frac{U}{m} \right) LWU =$$

$$= \left(\frac{1}{2} \frac{\rho g h m}{L} + \mu \frac{U}{m} \right) LWU \quad . \tag{4.75}$$

3. From its definition, the power supplied to the fluid can be rewritten as:

$$P_F = \rho g h Q = \rho g h m W \langle v_x \rangle = \rho g h m W \left(\frac{U}{2} - \frac{1}{12} \frac{\rho g h}{\mu L} m^2 \right) =$$

$$= \rho g m W \frac{U}{2} h - \frac{1}{12} \frac{\rho^2 g^2 m^3 W}{\mu L} h^2 \quad . \tag{4.76}$$

The value of h for which $P_F(h)$ is maximum can be obtained by setting the derivative dP_F/dh equal to zero and solving with respect to h:

$$\frac{dP_F}{dh} = \rho g m W \frac{U}{2} - \frac{1}{6} \frac{\rho^2 g^2 m^3 W}{\mu L} h = 0 \quad , \tag{4.77}$$

and:

$$\bar{h} = 3U \frac{\mu L}{\rho g m^2} \quad . \tag{4.78}$$

4   Consider the flow inside a lid-driven rectangular cavity, shown in Fig. 4.9. This flow mimics the one that occurs between two consecutive teeth of a gear pump. Assume that the flow is at steady state, laminar and fully developed and that the streamlines are parallel to the wall. Consider $L \gg H$ and examine only the central region of the cavity, neglecting the flow recirculation effects near the vertical walls. In the central region of the cavity, the flow is unidirectional and produces a further example of Couette–Poiseuille flow: In this case, the Couette flow is producing a balancing Poiseuille flow necessary to comply with the mass conservation.

1. Simplify the N-S equations for the case under consideration.
2. Derive an expression for the pressure gradient and the velocity profile.
3. Derive an expression for the shear stress exerted on the moving lid.

Fig. 4.9 Flow in a
rectangular cavity

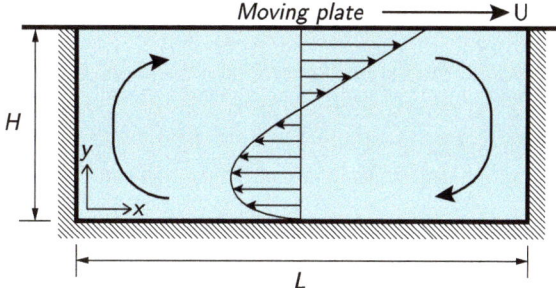

Solution The problem can be solved by assuming:

1. Incompressible Newtonian fluid.
2. Steady-state flow ($\rightarrow \partial \bullet / \partial t = 0$).
3. Fully-developed laminar flow ($\rightarrow \partial \bullet / \partial x = 0$).
4. Wall-parallel streamlines ($\rightarrow v_y = v_z = 0$, $\partial v_x / \partial z = 0$).

1. The *N-S* equations can be simplified as:

$$0 = -\frac{\partial \mathcal{P}}{\partial x} + \mu \frac{\partial^2 v_x}{\partial y^2}; \qquad 0 = -\frac{\partial \mathcal{P}}{\partial y}; \qquad 0 = -\frac{\partial \mathcal{P}}{\partial z} \quad . \tag{4.79}$$

2. It follows that $\mathcal{P} = \mathcal{P}(x)$ and $v_x = v_x(y)$. The velocity can be obtained from integration of the *x*-component of the *N-S* equations:

$$v_x = \frac{1}{2\mu}\frac{\mathrm{d}\mathcal{P}}{\mathrm{d}x} y^2 + C_1 y + C_2 \quad . \tag{4.80}$$

The constants C_1 and C_2 can be calculated from the following boundary conditions:

$$v_x|_{y=0} = 0; \qquad v_x|_{y=H} = U \quad , \tag{4.81}$$

as:

$$C_2 = 0; \qquad C_1 = -\frac{H}{2\mu}\frac{\mathrm{d}\mathcal{P}}{\mathrm{d}x} + \frac{U}{H} \quad . \tag{4.82}$$

The resulting velocity profile is:

$$v_x = \frac{1}{2\mu}\frac{\mathrm{d}\mathcal{P}}{\mathrm{d}x} y^2 + \left(\frac{U}{H} - \frac{H}{2\mu}\frac{\mathrm{d}\mathcal{P}}{\mathrm{d}x}\right) y \quad . \tag{4.83}$$

The pressure gradient is still unknown, but can be obtained noting that the total flowrate through each cross-section of the cavity must be zero by mass conservation. If the net flux is zero, then the average velocity must be zero:

$$\langle v_x \rangle = \frac{1}{H} \int_0^H v_x dy = \frac{1}{H} \int_0^H \left[\frac{1}{2\mu} \frac{dP}{dx} y^2 + \left(\frac{U}{H} - \frac{H}{2\mu} \frac{dP}{dx} \right) y \right] dy =$$

$$= \frac{U}{2} - \frac{dP}{dx} \frac{H^2}{12\mu} = 0 \quad \longrightarrow \quad \frac{dP}{dx} = \frac{6U\mu}{H^2} \ . \tag{4.84}$$

Substituting Eq. (4.84) into the velocity profile yields:

$$v_x = \frac{1}{2\mu} \frac{6U\mu}{H^2} y^2 + \left(\frac{U}{H} - \frac{H}{2\mu} \frac{6U\mu}{H^2} \right) y \quad \rightarrow \quad v_x = \frac{3U}{H^2} y^2 - \frac{2U}{H} y \ . \tag{4.85}$$

3. The shear stress acting on the moving lid is $\tau_{yx}(y = H)$, where:

$$\tau_{yx} = \mu \left(\frac{dv_y}{dx} + \frac{dv_x}{dy} \right) = \mu \frac{dv_x}{dy} \ . \tag{4.86}$$

It follows that:

$$\tau_{yx}(y = H) = \mu \left. \frac{dv_x}{dy} \right|_H = \frac{6\mu U}{H^2} H - \frac{2\mu U}{H} = \frac{4U\mu}{H} .$$

---

5   Consider the laminar flow of two immiscible Newtonian liquids A and B in a plane channel. Is it possible to find a condition for which the velocity profile has the shape shown in Fig. 4.10?

Solution   The velocity profile shown in Fig. 4.10 is unphysical since the derivative of the velocity profile cannot change sign at the interface between the liquids. At the interface, it must be:

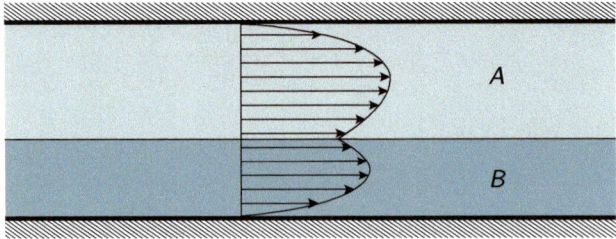

Fig. 4.10   Velocity profile for laminar flow of two immiscible fluids

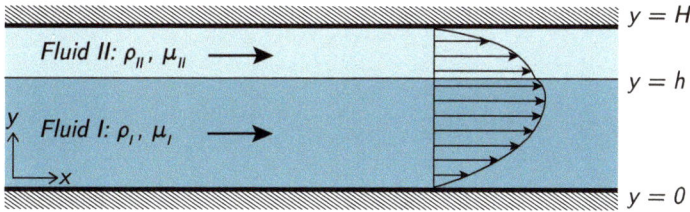

Fig. 4.11 Flow of two incompressible and immiscible Newtonian fluids in a horizontal channel

$$\tau_+ = \mu_A \left(\frac{dv_x}{dy}\right)_+ \quad , \tag{4.87}$$

$$\tau_- = \mu_B \left(\frac{dv_x}{dy}\right)_- \quad , \tag{4.88}$$

in which subscript $_+$ denotes the quantities referring to the upper surface of the interface, and subscript $_-$ denotes the quantities referring to the lower surface of the interface. Here, x is the longitudinal coordinate and y the wall-normal coordinate. For the equilibrium of the interface, it must be:

$$\tau_+ = \tau_- \quad . \tag{4.89}$$

Since μ_A and μ_B are both positive, it must be:

$$\left(\frac{dv_x}{dy}\right)_+ \cdot \left(\frac{dv_x}{dy}\right)_- > 0 \quad . \tag{4.90}$$

$\boxed{6}$ Consider the laminar flow of two incompressible and immiscible Newtonian fluids (denoted by I and II) in a plane channel of height H, as shown in Fig. 4.11. This flow is a typical simplified example of what can occur in a pipeline transporting a mixture of water and oil in the oil extraction and processing industry. Determine the velocity profile of the two fluids.

Solution Since the flow is laminar, fully developed and characterized by wall-parallel streamlines, v_x is the only non-zero velocity component for both phases and it only depends on y:

$$v_y = v_z = 0 \qquad v_x = v_x(y) \quad . \tag{4.91}$$

The x and y components of the N-S equations for each phase are (the y component reads the same for the two fluids):

$$\frac{d\mathcal{P}}{dy} = 0 \quad , \tag{4.92}$$

$$\mu_I \frac{d^2 v_{Ix}}{dy^2} = \frac{dP}{dx} = \text{constant}, \tag{4.93}$$

$$\mu_{II} \frac{d^2 v_{IIx}}{dy^2} = \frac{dP}{dx} = \text{constant} \quad, \tag{4.94}$$

where Eqs. (4.93) and (4.94) take into account the conditions (4.91) and (4.92), namely that the pressure is only a function of x and the velocities v_I and v_{II} are only functions of y. Integrating twice Eqs. (4.93) and (4.94), the following expression are obtained:

$$v_{Ix} = \frac{1}{2\mu_I} \frac{dP}{dx} y^2 + C_1 y + C_2 \quad, \tag{4.95}$$

$$v_{IIx} = \frac{1}{2\mu_{II}} \frac{dP}{dx} y^2 + C_3 y + C_4 \quad. \tag{4.96}$$

To evaluate the four constants C_i $(i = 1, 2, 3, 4)$, it is necessary to impose two boundary conditions for each phase:

no-slip at the walls:

$$y = 0 \quad \longrightarrow \quad v_{Ix} = 0 \quad, \tag{4.97}$$

$$y = H \quad \longrightarrow \quad v_{IIx} = 0 \quad. \tag{4.98}$$

no-slip at the interface:

$$y = h \quad \longrightarrow \quad v_{Ix} = v_{IIx} \quad. \tag{4.99}$$

equal shear stress at the interface:

$$y = h \quad \longrightarrow \quad \tau_{Ixy} = \tau_{IIxy} \rightarrow \mu_I \frac{dv_{Ix}}{dy} = \mu_{II} \frac{dv_{IIx}}{dy} \quad. \tag{4.100}$$

By imposing these boundary conditions, the following system of equations is obtained:

$$C_2 = 0 \quad, \tag{4.101}$$

$$\frac{1}{2\mu_{II}} \frac{dP}{dx} H^2 + C_3 H + C_4 = 0 \quad, \tag{4.102}$$

$$\frac{1}{2\mu_I} \frac{dP}{dx} h^2 + C_1 h = \frac{1}{2\mu_{II}} \frac{dP}{dx} h^2 + C_3 h + C_4 \quad, \tag{4.103}$$

$$\mu_I C_1 = \mu_{II} C_3 \quad. \tag{4.104}$$

Equation (4.102) yields:

$$C_4 = -\frac{1}{2\mu_{II}}\frac{d\mathcal{P}}{dx}H^2 - C_3 H \quad . \tag{4.105}$$

Substituting Eqs. (4.104) and (4.105) into Eq. (4.103) yields:

$$C_3 = -\frac{H}{2}\frac{d\mathcal{P}}{dx}\frac{1}{\mu_{II}}\frac{(1-k)^2\mu_I + k^2\mu_{II}}{(1-k)\mu_I + k\mu_{II}} \quad , \tag{4.106}$$

where $k = h/H$. The remaining constants read as:

$$C_1 = -\frac{H}{2}\frac{d\mathcal{P}}{dx}\frac{1}{\mu_I}\frac{(1-k)^2\mu_I + k^2\mu_{II}}{(1-k)\mu_I + k\mu_{II}} \quad , \tag{4.107}$$

$$C_4 = -\frac{H^2}{2\mu_{II}}\frac{d\mathcal{P}}{dx}\left[1 + \frac{(1-k)^2\mu_I + k^2\mu_{II}}{(1-k)\mu_I + k\mu_{II}}\right] \quad . \tag{4.108}$$

Therefore, the velocity profiles in each layer of fluid are:

$$v_{Ix} = \frac{H^2}{2\mu_I}\left(-\frac{d\mathcal{P}}{dx}\right)\left[\frac{y}{H}\frac{(1-k)^2\mu_I + k^2\mu_{II}}{(1-k)\mu_I + k\mu_{II}} - \left(\frac{y}{H}\right)^2\right] \quad , \tag{4.109}$$

$$v_{IIx} = \frac{H^2}{2\mu_{II}}\left(-\frac{d\mathcal{P}}{dx}\right)\left[\left(\frac{y}{H} + 1\right)\frac{(1-k)^2\mu_I + k^2\mu_{II}}{(1-k)\mu_I + k\mu_{II}} - \left(\frac{y}{H}\right)^2 + 1\right] \quad . \tag{4.110}$$

---

7 Two viscous fluids (viscosities μ' and μ'', respectively) are injected in a vertical pipe of diameter D. At a certain distance from the inlet, a laminar flow is established with one fluid moving near the wall and the other fluid moving in the central region of the pipe.

1. Write the equations of motion of the two fluids and the boundary conditions, stating the simplifying assumptions used.
2. Given the flowrates of the two fluids, Q' and Q'', evaluate the pressure gradient $\Delta P/L$ and the fraction of pipe cross-sectional area occupied by each fluid.

Solution The problem can be solved assuming:

1. Steady-state flow ($\rightarrow \partial\bullet/\partial t = 0$).
2. Fully-developed laminar flow ($\rightarrow \partial\bullet/\partial z = 0$).
3. Wall-parallel streamlines ($\rightarrow v_r = v_\theta = 0$, $\partial v_z/\partial z = \partial v_z/\partial\theta = 0$).

Therefore, the only non-zero velocity component is $v_z = v_z(r)$.

1. The continuity equation is identically satisfied ($\partial v_z/\partial z = 0$), and the *N-S* equations take the same form as in the Poiseuille flow problem:

$$0 = -\frac{\partial P}{\partial r} \quad , \tag{4.111}$$

$$0 = -\frac{1}{r}\frac{\partial P}{\partial \theta} \quad , \tag{4.112}$$

$$0 = -\frac{\partial P}{\partial z} + \mu \left[\frac{1}{r}\frac{d}{dr}\left(r\frac{dv_z}{dr}\right)\right] \quad . \tag{4.113}$$

This system of equations gives the velocity distribution for both fluids. The boundary conditions are:

$$r = R \to v_z' = 0 \quad ; \quad r = \lambda R \to \tau_{rz}' = \mu'\frac{dv_z'}{dr} = \mu''\frac{dv_z''}{dr} = \tau_{rz}'' \quad ;$$

$$r = \lambda R \to v_z' = v_z'' \quad ; \quad r = 0 \to= \frac{dv_z''}{dr} = 0 \quad . \tag{4.114}$$

2. Equation (4.26) can be applied to both fluids:

$$v_z'(r) = \frac{1}{4\mu'}\frac{\Delta P}{L}r^2 + C_1 \ln r + C_2 \quad , \tag{4.115}$$

$$v_z''(r) = \frac{1}{4\mu''}\frac{\Delta P}{L}r^2 + C_3 \ln r + C_4 \quad . \tag{4.116}$$

The boundary conditions allow to determine the four integration constants as:

$$C_1 = 0 \quad ; \quad C_2 = -\frac{1}{4\mu'}\frac{\Delta P}{L}R^2 \quad ,$$

$$C_3 = 0; \quad C_4 = -\frac{1}{4\mu''}\frac{\Delta P}{L}R^2\left[\lambda^2 - \frac{\mu''}{\mu'}(\lambda^2 - 1)\right].$$

Therefore, the velocity profiles are:

$$v_z'(r) = \frac{1}{4\mu'}\frac{\Delta P}{L}R^2\left[\left(\frac{r}{R}\right)^2 - 1\right] \quad , \tag{4.117}$$

$$v_z''(r) = \frac{1}{4\mu''}\frac{\Delta P}{L}R^2\left\{\left(\frac{r}{R}\right)^2 - \left[\lambda^2 - \frac{\mu''}{\mu'}(\lambda^2 - 1)\right]\right\} \quad . \tag{4.118}$$

The volumetric flowrate is $Q = \langle v_z \rangle A$. Leaving out the details of integration, it is found that:

$$Q' = 2\pi \int_{\lambda R}^{R} v_z'(r)r dr = \cdots = \frac{\pi R^4}{4\mu'}\frac{\Delta P}{L}\left[\lambda^2\left(1 - \frac{\lambda^2}{2}\right) - \frac{1}{2}\right] \quad , \tag{4.119}$$

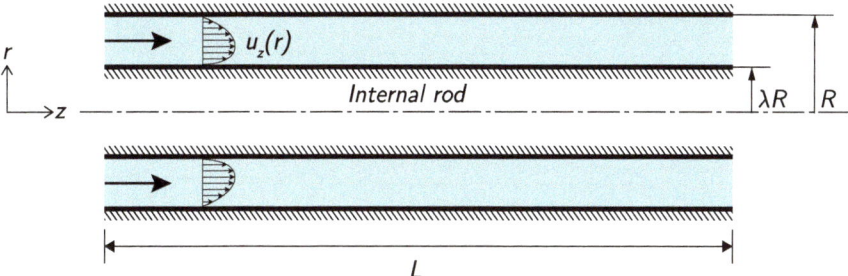

Fig. 4.12 Flow of a Newtonian fluid driven by a pressure gradient in an annular duct

and for the fluid in the central region of the pipe:

$$Q'' = 2\pi \int_0^{\lambda R} v_z''(r) r \, dr = \cdots =$$

$$= \frac{\pi R^4}{4\mu''} \frac{\Delta \mathcal{P}}{L} \left[-\frac{\lambda^4}{2} + \lambda^2 \frac{\mu''}{\mu'} (\lambda^2 - 1) \right] . \tag{4.120}$$

Since the flow rates Q' and Q'' are known, solving the system of Eqs. (4.119) and (4.120) gives the values of $\Delta \mathcal{P}/L$ and λ. From the value of λ, the fraction of pipe area occupied by each fluid can be determined.

$\boxed{8}$ An incompressible Newtonian fluid flows through an annulus consisting of two cylinders of radii λR and R, respectively, with $\lambda < 1$, as illustrated in Fig. 4.12. The pressure drop over a length L is $\Delta \mathcal{P}$.

1. Calculate the velocity distribution and volumetric flowrate.
2. Prove that, in the limit $\lambda \to 1$, the solutions for the velocity and the volumetric flowrate are those calculated for the flow between parallel plates.
3. Compare the behavior of the solution for $\lambda \to 0$ with that obtained for flows in circular pipes.

Solution

1. Using a procedure similar to the one adopted in the previous problem, it can be verified that the continuity equation is identically satisfied, and by integrating the N-S equations, the following equation for the velocity profile can be obtained:

$$v_z(r) = \frac{1}{4\mu} \frac{\Delta \mathcal{P}}{L} r^2 + C_1 \ln r + C_2 . \tag{4.121}$$

Applying the following boundary conditions:

$$r = \lambda R \to v_z = 0; \quad r = R \to v_z = 0 , \tag{4.122}$$

to Eq. (4.121) yields:

$$v_z(\lambda R) = 0 \rightarrow 0 = \frac{1}{4\mu}\frac{\Delta P}{L}\lambda^2 R^2 + C_1 \ln \lambda R + C_2 \quad,$$

$$v_z(R) = 0 \rightarrow 0 = \frac{1}{4\mu}\frac{\Delta P}{L}R^2 + C_1 \ln R + C_2 \quad. \tag{4.123}$$

Subtracting the first equation from the second equation gives:

$$0 = \frac{1}{4\mu}\frac{\Delta P}{L}R^2(1 - \lambda^2) + C_1 \ln \frac{R}{\lambda R} \quad, \tag{4.124}$$

and thus:

$$C_1 = \frac{1}{4\mu}\frac{\Delta P}{L}R^2\frac{1 - \lambda^2}{\ln \lambda} \quad. \tag{4.125}$$

The second constant reads as:

$$C_2 = -\frac{1}{4\mu}\frac{\Delta P}{L}\left[R^2 + R^2(1 - \lambda^2)\frac{\ln R}{\ln \lambda}\right] \quad, \tag{4.126}$$

and the velocity profile is:

$$v_z(r) = \frac{1}{4\mu}\frac{\Delta P}{L}\left[r^2 - R^2 + R^2\frac{1 - \lambda^2}{\ln \lambda} \ln \frac{r}{R}\right] \quad. \tag{4.127}$$

The volumetric flowrate is defined as $Q = \langle v_z \rangle A = \langle v_z \rangle \pi R^2(1 - \lambda^2)$. Leaving out the details of the integration, it is found that:

$$\langle v_z \rangle = \frac{1}{\pi R^2(1 - \lambda^2)}\int_{\lambda R}^{R}\int_0^{2\pi} v_z(r)r\mathrm{d}\theta\mathrm{d}r =$$

$$= \left(-\frac{\Delta P}{L}\right)\frac{R^2}{8\mu}\left(1 + \lambda^2 + \frac{1 - \lambda^2}{\ln \lambda}\right) \quad, \tag{4.128}$$

and:

$$Q = \frac{\pi R^4}{8\mu}\left(-\frac{\Delta P}{L}\right)\left[1 - \lambda^4 + \frac{(1 - \lambda^2)^2}{\ln \lambda}\right] \quad. \tag{4.129}$$

2. *Solution for $\lambda \rightarrow 1$.* Adopting a change of variables, denoting by $H = R(1 - \lambda)$ the space between the two cylinders and setting $y = (r - \lambda R) - H/2$, the dimensionless variable r/R can be expressed as:

$$\frac{r}{R} = \left(\frac{y}{H} + \frac{1}{2}\right) - \lambda\left(\frac{y}{H} - \frac{1}{2}\right) \ . \tag{4.130}$$

Consider first the expression for the average velocity. Substituting the variable H into the expression of the average velocity, yields:

$$\langle v_z \rangle = -\frac{\Delta P}{L} \frac{R^2}{8\mu} \frac{(1-\lambda)^2}{(1-\lambda)^2}\left(1 + \lambda^2 + \frac{1-\lambda^2}{\ln\lambda}\right) =$$

$$= \left(-\frac{\Delta P}{L}\right)\frac{H^2}{8\mu}\frac{1+\lambda}{1-\lambda}\left(\frac{1+\lambda^2}{1-\lambda^2} + \frac{1-\lambda^2}{\ln\lambda}\right) \ . \tag{4.131}$$

To find the limit of this function for $\lambda \to 1$, it is appropriate to set $\lambda = 1 - \epsilon$ and perform a Taylor series expansion. Omitting some algebraic steps, it is found that:

$$\lim_{\lambda\to 1}\langle v_z\rangle = \cdots = \lim_{\epsilon\to 0}\left(-\frac{\Delta P}{L}\right)\frac{H^2}{8\mu}\left(\frac{2-\epsilon}{\epsilon}\right)\left[\frac{\epsilon^2 - 2\epsilon + 2}{2\epsilon - \epsilon^2} + \frac{1}{\ln(1-\epsilon)}\right] = \cdots$$

$$\cdots = \lim_{\epsilon\to 0}\left(-\frac{\Delta P}{L}\right)\frac{H^2}{8\mu}\left(\frac{\frac{2}{3} - \frac{1}{6}\epsilon + \frac{1}{3}\epsilon^2 + \cdots}{1 + \frac{1}{2}\epsilon + \frac{1}{3}\epsilon^2 + \frac{1}{4}\epsilon^3 + \cdots}\right) =$$

$$= \left(-\frac{\Delta P}{L}\right)\frac{H^2}{12\mu} \ , \tag{4.132}$$

which is precisely the mean velocity for laminar flow between two parallel horizontal plates.

Consider now the velocity distribution and substitute the expression for H in the equation of the velocity profile. The following expression can be written for $v_z(r)$:

$$v_z(r) = -\frac{\Delta P}{L}\frac{R^2}{4\mu}\left[1 - \left(\frac{r}{R}\right)^2\frac{1-\lambda^2}{\ln\lambda}\ln\frac{r}{R}\right] =$$

$$= -\frac{\Delta P}{L}\frac{R^2}{4\mu}\frac{(1-\lambda)^2}{(1-\lambda)^2}\left\{1 - \left[\frac{y}{H}(1-\lambda) + \frac{1}{2}(1+\lambda)\right]^2 + \right.$$

$$\left. -\frac{1-\lambda^2}{\ln\lambda}\ln\left[\frac{y}{H}(1-\lambda) + \frac{1}{2}(1+\lambda)\right]\right\} \ . \tag{4.133}$$

Proceeding as above and defining $\lambda = 1 - \epsilon$, one obtains:

$$\lim_{\lambda\to 1}v_z(r) = \lim_{\epsilon\to 0}v_z(y) = \cdots = -\frac{\Delta P}{L}\frac{H^2}{8\mu}\left[1 - \left(2\frac{y}{H}\right)^2\right] \ , \tag{4.134}$$

which is precisely the velocity distribution for laminar flow between two parallel horizontal plates.

3. *Solution for* $\lambda \to 0$. The solution for this case is straightforward:

$$\lim_{\lambda \to 0} \langle v_z \rangle = \lim_{\lambda \to 0} \left(-\frac{\Delta P}{L}\right) \frac{R^2}{8\mu} \left(1 + \lambda^2 + \frac{1-\lambda^2}{\ln \lambda}\right) =$$

$$= \left(-\frac{\Delta P}{L}\right) \frac{R^2}{8\mu} \quad , \tag{4.135}$$

which is precisely the equation for the mean fluid velocity in laminar pipe flow. Similarly, for the velocity profile we have:

$$\lim_{\lambda \to 0} v_z(r) = \lim_{\lambda \to 0} \left(-\frac{\Delta P}{L}\right) \frac{R^2}{4\mu} \left[1 - \left(\frac{r}{R}\right)^2 \frac{1-\lambda^2}{\ln \lambda} \ln \frac{r}{R}\right] =$$

$$= \left(-\frac{\Delta P}{L}\right) \frac{R^2}{4\mu} \left[1 - \left(\frac{r}{R}\right)^2\right] \quad , \tag{4.136}$$

which is the fluid velocity distribution in laminar pipe flow (Poiseuille flow).

9 Consider the laminar flow of a liquid with viscosity μ in the space between two vertical cylinders of inner radius r_1 and outer radius r_2. The outer cylinder moves with angular velocity Ω. Determine the velocity profile.

Solution The following assumptions can be made:

1. Steady-state flow ($\to \partial \bullet / \partial t = 0$).
2. Tangential flow (no radial or axial flow) ($\to v_z = v_r = 0$).
3. Wall-parallel streamlines ($\to v_\theta = v_\theta(r), \partial v_\theta / \partial \theta = \partial v_\theta / \partial z = 0$).

The continuity equation imposes that v_θ does not depend on θ and the *N-S* equations are:

$$-\rho \frac{v_\theta^2}{r} = -\frac{\partial P}{\partial r} \quad , \tag{4.137}$$

$$0 = -\frac{1}{r}\frac{\partial P}{\partial \theta} + \mu \frac{\partial}{\partial r}\left[\frac{1}{r}\frac{\partial}{\partial r}(r v_\theta)\right] \quad , \tag{4.138}$$

$$0 = -\frac{\partial P}{\partial z} \quad . \tag{4.139}$$

Due to the symmetry of the problem with respect to the θ coordinate, there is no pressure gradient in the flow direction ($\partial P / \partial \theta = 0$) and hence $P = P(r)$. Integration of the *N-S* equation along the θ component gives:

$$v_\theta = \frac{C_1}{2}r + \frac{C_2}{r} \quad . \tag{4.140}$$

The boundary conditions are:

$$v_\theta|_{r=r_1} = 0; \qquad v_\theta|_{r=r_2} = \Omega r_2 \quad ; \tag{4.141}$$

which give the following expressions of the integration constants:

$$C_1 = 2\Omega\frac{r_2^2}{r_2^2 - r_1^2}; \qquad C_2 = -\Omega\frac{r_1^2 r_2^2}{r_2^2 - r_1^2} \quad . \tag{4.142}$$

Finally, the velocity profile is:

$$v_\theta = \Omega\frac{r_2^2}{r_2^2 - r_1^2}r - \Omega\frac{r_1^2 r_2^2}{r_2^2 - r_1^2}\frac{1}{r} \quad . \tag{4.143}$$

[10] As shown in Fig. 4.13, a friction pump consists of a cylinder of diameter D and length W rotating at an angular velocity Ω in the clockwise direction inside a hollow coaxial cylinder of inner diameter $D + 2h$. A liquid flows in the thin space between the two cylinders and a septum separates the incoming flow at pressure $\mathcal{P}_{in}$ from the outgoing flow at pressure $\mathcal{P}_{out}$. This is an example of Couette–Poiseuille flow.

1. Write the governing equations by adopting the appropriate simplifications and boundary conditions. Assume **non** negligible thickness h with respect to the diameter D.
2. Assuming negligible thickness h with respect to the diameter D, derive an expression for the volumetric flowrate Q through the pump.
3. Derive an expression for the torque T that must be applied to the rotor at steady-state as a function of the pressure jump $\Delta\mathcal{P}$.

Solution

1. If the thickness h cannot be neglected with respect to the diameter D, the problem must be solved in cylindrical coordinates. The following assumptions can be made:

   (a) Steady-state flow ($\rightarrow \partial\bullet/\partial t = 0$);
   (b) Tangential flow (no radial or axial flow) ($\rightarrow v_z = v_r = 0$);
   (c) Wall-parallel streamlines ($\rightarrow v_\theta = v_\theta(r)$, $\partial v_\theta/\partial\theta = \partial v_\theta/\partial z = 0$).

   The continuity equation imposes that v_θ does not depend on θ and the N-S equations are:

$$-\rho\frac{v_\theta^2}{r} = -\frac{\partial \mathcal{P}}{\partial r} \quad , \tag{4.144}$$

$$0 = -\frac{1}{r}\frac{\partial \mathcal{P}}{\partial \theta} + \mu\frac{\partial}{\partial r}\left[\frac{1}{r}\frac{\partial}{\partial r}(rv_\theta)\right] \quad , \tag{4.145}$$

$$0 = -\frac{\partial \mathcal{P}}{\partial z} \quad . \tag{4.146}$$

The boundary conditions are

$$v_\theta|_{r=D/2} = v_p = \frac{\Omega D}{2}; \qquad v_\theta|_{r=h+D/2} = 0 \quad . \tag{4.147}$$

Note that, in this problem, $\mathcal{P} = \mathcal{P}(r, \theta)$.

2. If the thickness h is much smaller than the diameter D, the problem can be solved in Cartesian coordinates, and the governing equations can be easily integrated. The assumptions made in the previous point are still valid, and setting the x coordinate parallel to the circumference, the N-S equations in the x direction take the form:

$$0 = -\frac{\partial \mathcal{P}}{\partial x} + \mu\frac{\partial^2 v_x}{\partial z^2} \rightarrow \frac{\Delta \mathcal{P}}{\mu\pi D} = \frac{\mathrm{d}^2 v_x}{\mathrm{d}z^2} \quad , \tag{4.148}$$

where $\Delta \mathcal{P}/\pi D$ is the pressure gradient. The velocity profile is obtained by integrating twice:

$$v_x(z) = \frac{1}{\mu}\frac{\Delta \mathcal{P}}{\pi D}\frac{z^2}{2} + C_1 z + C_2 \quad . \tag{4.149}$$

Fig. 4.13 Schematic of a friction pump. The outer cylinder is fixed while the inner cylinder rotates

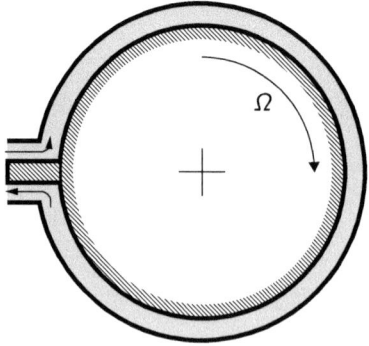

With the boundary conditions:

$$v_x(0) = v_p = \frac{\Omega D}{2}; \quad v_x(h) = 0 \quad , \tag{4.150}$$

the integration constants C_1 and C_2 read as:

$$C_1 = -\frac{\Omega D}{2h} - \frac{\Delta \mathcal{P}}{\mu \pi D}\frac{h}{2}; \quad C_2 = \Omega\frac{D}{2} \quad , \tag{4.151}$$

and the velocity profile is:

$$v_x(z) = \frac{\Delta \mathcal{P}}{2\mu \pi D}h^2\left[\left(\frac{z}{h}\right)^2 - \frac{z}{h}\right] - \frac{\Omega D}{2}\left(\frac{z}{h} - 1\right) \quad . \tag{4.152}$$

3. The volumetric flowrate is:

$$Q = W\int_0^h v_z(x)\mathrm{d}z = \cdots = \frac{Wh}{2}\left(\frac{\Omega D}{2} - \frac{\Delta \mathcal{P}}{\mu \pi D}\frac{h^2}{6}\right) \quad . \tag{4.153}$$

The torque T that must be applied to the inner cylinder is given by:

$$T = -\tau_w W\pi D\frac{D}{2} \quad , \tag{4.154}$$

where the shear stress at the wall is:

$$\tau_w = \mu\frac{\mathrm{d}v_x}{\mathrm{d}z}\bigg|_0 = -\mu\frac{\Omega D}{2h} - \frac{\Delta \mathcal{P}h}{2\pi D} \quad . \tag{4.155}$$

Hence, the torque is:

$$T = \pi\frac{D^3 W\mu\Omega}{4h}\left(1 + \frac{\Delta \mathcal{P}}{\pi \mu \Omega D^2}h^2\right) \quad . \tag{4.156}$$

---

11 The shaft of a friction pump rotates at 3600 rpm. The radius of the shaft is $a = 0.05\ m$, and its length is $W = 0.1$ m. The thin space between the shaft and the hub is $h = 5\cdot 10^{-3}$ m. The pumped fluid is oil with viscosity $\mu = 2\cdot 10^{-2}$ Pa $\cdot$ s.

1. Given the power P_a required to keep the shaft rotating ($P_a = 60\ W$), determine the pressure gradient in the fluid.
2. Determine the volumetric flowrate per unit length of the shaft.

Solution In the previous problem, it was found for the velocity profile:

$$v_x(z) = \frac{1}{2\mu}\frac{d\mathcal{P}}{dx}h^2\left[\left(\frac{z}{h}\right)^2 - \frac{z}{h}\right] - \Omega a\left(\frac{z}{h} - 1\right) , \qquad (4.157)$$

and, for the shear stress at the wall:

$$\tau_w = \mu\left.\frac{dv_x(z)}{dz}\right|_{z=0} = -\left(\frac{d\mathcal{P}}{dx}\frac{h}{2} + \mu\frac{\Omega a}{h}\right) . \qquad (4.158)$$

The power required to rotate the shaft is given by the product of the applied force (given, in turn, by the wall-shear stress times the area over which the stress acts) and the velocity of the wall:

$$P_a = \tau_w\,2\pi a W\,\Omega a = \left(\frac{d\mathcal{P}}{dx}\frac{h}{2} + \mu\frac{\Omega a}{h}\right)2\pi a W\,\Omega a = 60\ W \quad . \qquad (4.159)$$

From the data of the problem, the pressure gradient is equal to:

$$\frac{d\mathcal{P}}{dx} = 1.04\cdot 10^4\ \mathrm{Pa/m} , \qquad (4.160)$$

and, therefore, the volumetric flowrate per unit width is:

$$\frac{Q}{W} = \frac{\Omega a h}{2} - \frac{1}{12\mu}\frac{d\mathcal{P}}{dx}h^3 = 0.042\ \mathrm{m^2/s} \quad . \qquad (4.161)$$

$\boxed{12}$ A liquid film flows downward in contact with the outer wall of a vertical tube of radius R. Determine the velocity profile for an assigned and constant value, Q, of the volumetric flowrate of the film.

Solution Using the same assumptions as in Sect. 4.1.5, and adopting a cylindrical coordinate system (r, θ, z), it can be concluded that:

$$v_r = v_\theta = 0 \qquad v_z = v_z(r) \quad . \qquad (4.162)$$

The *N-S* equations read as:

$$\frac{\partial p}{\partial\theta} = \frac{\partial p}{\partial r} = 0 , \qquad (4.163)$$

$$\frac{1}{r}\frac{\partial}{\partial r}\left(r\mu\frac{\partial v_z}{\partial r}\right) = -\rho g \quad . \qquad (4.164)$$

The velocity profile can be obtained by integrating Eq. (4.164) twice:

$$\frac{\partial v_z}{\partial r} = -\frac{1}{2\mu}\rho g r + \frac{C_1}{\mu r} \quad , \tag{4.165}$$

$$v_z = -\frac{1}{4\mu}\rho g r^2 + \frac{C_1}{\mu}\ln r + C_2 \quad . \tag{4.166}$$

The constant C_1 can be determined by imposing that the shear stress is zero at the free surface of the liquid:

$$r = R + \delta \quad \rightarrow \quad \tau_{zr} = \mu\frac{\partial v_z}{\partial r} = 0 \quad , \tag{4.167}$$

with δ the thickness of the liquid film. The constant C_2 can be determined by imposing the no-slip condition at the wall:

$$r = R \quad \rightarrow \quad v_z = 0 \quad . \tag{4.168}$$

The resulting velocity profile is:

$$v_z = \frac{\rho g R^2}{4\mu}\left[1 - \left(\frac{r}{R}\right)^2 + 2a^2\ln\left(\frac{r}{R}\right)\right] \quad , \tag{4.169}$$

in which $(R + \delta) = aR$. The unknown coefficient a must be derived as a function of the assigned value Q of the volumetric flowrate:

$$Q = \pi(a^2 - 1)R^2\langle v_z \rangle \quad , \tag{4.170}$$

$$\langle v_z \rangle = \frac{1}{\pi R^2(a^2 - 1)}\int_0^{2\pi}\int_R^{aR} v_z(r)r\,d\theta dr \quad . \tag{4.171}$$

The flowrate reads as:

$$Q = \pi\frac{\rho g R^4}{2\mu}\left(a^4\ln a - \frac{3}{4}a^4 + a^2 - \frac{1}{4}\right) \quad .$$

From this equation, it is possible to derive a.

13 A liquid film flows downward on the vertical wall shown in Fig. 4.14. The specific flowrate of the liquid ($\rho = 800\,\mathrm{kg/m^3}$, $\mu = 1.2 \cdot 10^{-1}\,\mathrm{Pa \cdot s}$) is $\Gamma = 0.3\,\mathrm{kg/m\,s}$. This is an example of Couette–Poiseuille flow.

Fig. 4.14 Falling liquid film
in counter-current gas flow

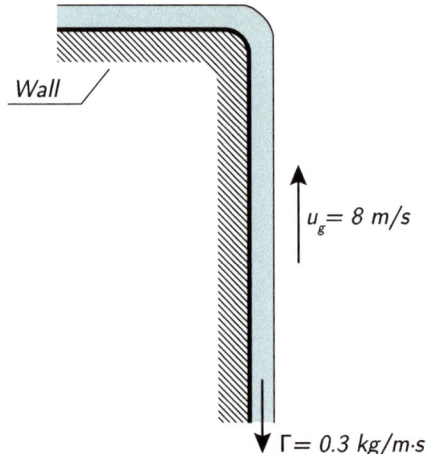

Wall

$u_g = 8\ m/s$

$\Gamma = 0.3\ kg/m\cdot s$

1. Calculate the thickness of the liquid film.
2. Consider the situation in which a gas ($\rho_g = 1.2\ \text{kg/m}^3$) with undisturbed velocity $v_g = 8$ m/s generates a counter-current flow near the interface of the film, determine the film thickness assuming that the shear stress exerted by the gas on the liquid can be expressed as $\tau_i = 0.5 f \rho_g v_g^2$ with $f = 5 \cdot 10^{-3}$.

Solution   The solution of this problem is similar to that of a liquid film moving down an inclined wall. Equation (4.48), written here for vertical wall ($\sin \beta = 1$), becomes:

$$v_z = -\frac{\rho g}{2\mu} x^2 + C_1 x + C_2 \quad . \tag{4.172}$$

1. If the gas has zero velocity, the boundary conditions are:

$$v_z|_{x=\delta} = 0, \qquad \tau_{xz}|_{x=0} = \mu \frac{dv_z}{dx}\bigg|_{x=0} = \tau_i = 0 \quad , \tag{4.173}$$

for which the integration constants are $C_1 = 0$ and $C_2 = (\rho g \delta^2)/(2\mu)$. The velocity profile is:

$$v_z = \frac{\rho g \delta^2}{2\mu} \left[1 - \left(\frac{x}{\delta}\right)^2 \right] \quad . \tag{4.174}$$

The specific flowrate (per unit width of the wall) is defined as $\Gamma = \rho \delta \langle v_z \rangle$. The mean velocity is:

$$\langle v_z \rangle = \frac{1}{\delta} \int_0^\delta v_z dx = \frac{\rho g \delta^2}{2\mu} \int_0^1 \left[1 - \left(\frac{x}{\delta}\right)^2 \right] d\left(\frac{x}{\delta}\right) = \frac{\rho g \delta^2}{3\mu} \quad . \tag{4.175}$$

Therefore, the thickness of the film is:

$$\delta = \sqrt[3]{\frac{3\mu\Gamma}{\rho^2 g}} = 2.58 \text{ mm} \quad . \tag{4.176}$$

2. When the gas exerts a non-negligible tangential stress on the liquid film, the velocity profile is modified: The location at which the maximum velocity is reached moves toward the wall, and is not at the free surface anymore, Since the flowrate is imposed, an increase in the film thickness is expected.

   To calculate the new velocity profile, the following boundary conditions must be used:

$$v_z|_{x=\delta} = 0; \quad \tau_{xz}|_{x=0} = \mu \frac{dv_z}{dx}\bigg|_{x=0} = \tau_i = \frac{5 \cdot 10^{-3}}{2} \rho_g v_g^2 \quad , \tag{4.177}$$

from which the integration constants can be obtained as $C_1 = \tau_i/\mu$, $C_2 = \delta[(\rho g \delta)/2 - \tau_i]/\mu$. The velocity profile is:

$$v_z = \frac{\rho g \delta^2}{2\mu}\left[1 - \left(\frac{x}{\delta}\right)^2\right] - \frac{\tau_i \delta}{\mu}\left(1 - \frac{x}{\delta}\right) \quad . \tag{4.178}$$

The mean velocity is:

$$\langle v_z \rangle = \frac{\rho g \delta^2}{3\mu} - \frac{\tau_i \delta}{2\mu} \quad , \tag{4.179}$$

and the specific mass flowrate Γ is:

$$\Gamma = \frac{\rho^2 g}{3\mu}\delta^3 - \frac{\tau_i \rho}{2\mu}\delta^2 \quad . \tag{4.180}$$

Since Γ is known, the thickness can be obtained from the following equation:

$$\delta^3 - \frac{3}{2}\frac{\tau_i}{\rho g}\delta^2 = \frac{3\mu}{\rho^2 g}\Gamma \quad . \tag{4.181}$$

Examining the magnitude of the coefficients in the equation, it can be seen that the quadratic term leads to very small variations on the value of the thickness, which is only slightly larger than that calculated for $\tau_i = 0$. Neglecting the quadratic term, it is found that $\delta = 2.6 \cdot 10^{-3}$ m.

Problems

[a] Two immiscible liquids A and B, having densities $\rho_A > \rho_B$ and viscosities $\mu_A < \mu_B$, flow between two parallel plates of width W, which are infinitely long. The lower plate is stationary while the upper plate moves with velocity U. The thicknesses of the two liquid layers are H_A and H_B, respectively. Assume constant pressure everywhere.

1. Determine the velocity profile within each layer.
2. Calculate the shear stress per unit length on the two plates.

[b] The volume between two cylinders of radii $r_1 = 500$ mm and $r_2 = 520$ mm and height 10 mm is filled with oil ($\rho_o = 800$ kg/m^3, $\mu_o = 1.5 \cdot 10^{-2}$ Pa $\cdot$ s) and water ($\rho = 10^3$ kg/m^3, $\mu = 10^{-3}$ Pa $\cdot$ s) in equal proportions. The inner cylinder rotates at peripheral velocity $v_{\theta,0} = 0.2$ m/s. Neglecting the effects due to gravity and assuming laminar flow regime, calculate:

1. the average velocity in the two films with respect to the stationary cylinder.
2. The power required to maintain the rotation of the inner cylinder.

[c] A conveyor belt transports a viscous fluid (viscosity $\mu = 0.1$ Pa $\cdot$ s, density $\rho = 800$ kg/m^3) between two reservoirs, as shown in Fig. 4.15.

1. Assuming that the belt is horizontal, its velocity is equal to 0.2 m/s, and the shear stress on the upper surface of the interface is zero, calculate the transferred flowrate for a value of the film thickness equal to 2 mm and a belt width of 0.4 m;
2. Assuming that the belt is tilted downward by 10°, calculate the thickness of the film if the transferred flowrate is equal to $2 \cdot 10^{-3}$ m^3/s.

[d] A viscous fluid ($\rho = 800$ kg/m^3, $\mu = 0.1$ Pa $\cdot$ s) moves along a conveyor belt of width W and length L, inclined by an angle $\alpha = 30°$ with respect to the horizontal. The belt moves upward with velocity $U = 1.5$ m/s.

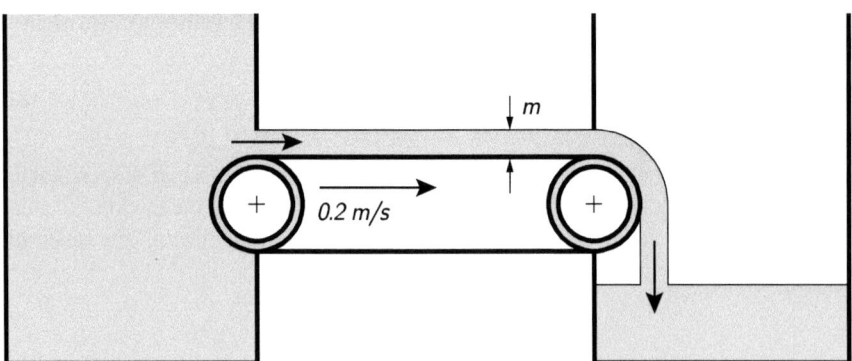

Fig. 4.15 Transport of fluid between two reservoirs by means of a friction pump consisting of a conveyor belt

1. Simplify the continuity and Navier–Stokes equations.
2. Determine the velocity profile in the film and the flowrate as a function of the thickness δ.
3. Calculate the thickness δ that, for a given belt speed, maximizes the flowrate transferred upward.
4. Calculate the thickness δ that, for a given belt speed, generates a zero net flow along the inclined belt.
5. Calculate the power required to keep the belt in motion.

$\boxed{e}$ Consider the flow solved in Sect. 4.1.6, Problem 6.

1. Determine the ratio between the volumetric flowrates of the two fluids I and II.
2. For $H = 50$ mm, $w_I = 0.24$ kg/s, $w_{II} = 0.017$ kg/s (both flowrates refer to a channel of unit width, $\rho_I = 10^3$ kg/m³, $\rho_{II} = 1.2$ kg/m³, $\mu_I = 10^{-3}$ Pa · s and $\mu_{II} = 1.7 \cdot 10^{-5}$ Pa · s), determine the ratio h/H and compare the ratio of the volumetric flowrates with the ratio between the volumes occupied by the two fluids.

$\boxed{f}$ Two immiscible and incompressible Newtonian fluids flow down a plane inclined by an angle α with respect to the horizontal direction, as in Fig. 4.16. Assuming laminar flow and streamlines parallel to the wall, determine the velocity distribution within the two fluids and determine the layer thickness if the specific flowrates per unit width, Q_a and Q_b, are given.

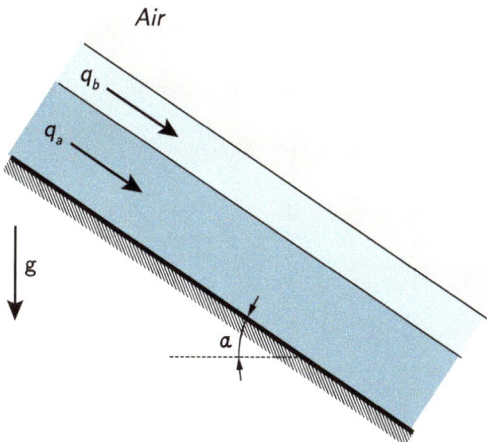

Fig. 4.16 Newtonian fluids flowing superimposed on an inclined plane

Chapter 5
Approximate Solutions for Low Reynolds Number Flows

5.1 Dimensionless Form of the Conservation Equations

Approximate solutions of the *N-S* equations can be obtained in many cases of practical interest, provided that suitable simplifications can be made. Simplifications stem from the evaluation of the order of magnitude of the different terms appearing in the *N-S* equations. To assess the order of magnitude of each term, it is useful to write the equations in dimensionless form so that the terms of similar magnitude can be identified and the dimensionless groups that determine the structure of the flow field can be determined.

The non-dimensionalization of the equations is carried out by normalizing each variable with respect to the characteristic velocity, V, characteristic length, L, characteristic time, T, and characteristic pressure, Π, of the problem under consideration. In dimensionless form, the continuity equation for an incompressible fluid:

$$\nabla \cdot \mathbf{v} = 0 \quad , \tag{5.1}$$

becomes:

$$\frac{V}{L} \tilde{\nabla} \cdot \tilde{\mathbf{v}} = 0 \quad , \tag{5.2}$$

or:

$$\tilde{\nabla} \cdot \tilde{\mathbf{v}} = 0 \quad , \tag{5.3}$$

where $\tilde{\mathbf{v}} = \mathbf{v}/V$, $\tilde{x}, \tilde{y}, \tilde{z} = x/L, y/L, z/L$ e $\tilde{\nabla} = L\nabla$. By performing these substitutions in the *N-S* equations:

$$\rho \frac{\partial \mathbf{v}}{\partial t} + \rho \mathbf{v} \cdot \nabla \mathbf{v} = -\nabla \mathcal{P} + \mu \nabla^2 \mathbf{v} \quad , \tag{5.4}$$

© The Author(s), under exclusive license to Springer Nature Switzerland AG 2024
A. Soldati and C. Marchioli, *Fluid Mechanics for Mechanical Engineers*,
https://doi.org/10.1007/978-3-031-53950-3_5

it is possible to write:

$$\rho \frac{V}{T} \frac{\partial \tilde{\mathbf{v}}}{\partial \tilde{t}} + \left(\frac{\rho V^2}{L}\right) \tilde{\mathbf{v}} \cdot \tilde{\nabla} \tilde{\mathbf{v}} = -\frac{\Pi}{L} \tilde{\nabla} \tilde{P} + \frac{\mu V}{L^2} \tilde{\nabla}^2 \tilde{\mathbf{v}} \quad . \tag{5.5}$$

When a characteristic time is not imposed on the system, as it would happen when the system is excited at a given frequency, it is possible to assume $T = L/V$. Furthermore, recalling the definition of Reynolds number:

$$Re = \frac{\rho V L}{\mu} \quad , \tag{5.6}$$

the N-S equations become:

$$Re \left(\frac{\partial \tilde{\mathbf{v}}}{\partial \tilde{t}} + \tilde{\mathbf{v}} \cdot \tilde{\nabla} \tilde{\mathbf{v}}\right) = -\frac{\Pi L}{\mu V} \tilde{\nabla} \tilde{P} + \tilde{\nabla}^2 \tilde{\mathbf{v}} \quad . \tag{5.7}$$

The characteristic pressure can be conveniently defined in two different ways, depending on whether the system is dominated by inertia:

$$\Pi = \rho V^2 \quad , \tag{5.8}$$

or by viscous forces:

$$\Pi = \frac{\mu V}{L} \quad . \tag{5.9}$$

In these two limiting cases, Eq. (5.7) becomes:

$$\text{Inertia-dominated}: \quad Re \left(\frac{\partial \tilde{\mathbf{v}}}{\partial \tilde{t}} + \tilde{\mathbf{v}} \cdot \tilde{\nabla} \tilde{\mathbf{v}} + \tilde{\nabla} \tilde{P}\right) = \tilde{\nabla}^2 \tilde{\mathbf{v}} \quad , \tag{5.10}$$

$$\text{Viscous forces-dominated}: \quad Re \left(\frac{\partial \tilde{\mathbf{v}}}{\partial \tilde{t}} + \tilde{\mathbf{v}} \cdot \tilde{\nabla} \tilde{\mathbf{v}}\right) = -\tilde{\nabla} \tilde{P} + \tilde{\nabla}^2 \tilde{\mathbf{v}} \quad . \tag{5.11}$$

These two equations are the starting point for the derivation of an approximate solution of the N-S equations, and nicely highlight the role played by the Reynolds number in the two limiting cases $Re \to 0$ and $Re \to \infty$.

5.2 Creeping Flow

In the limiting case $Re \to 0$, referred to as *creeping flow*, viscous forces are dominant and Eq. (5.11) can be simplified as:

$$- \tilde{\nabla}\tilde{\mathcal{P}} + \tilde{\nabla}^2\tilde{\mathbf{v}} = 0 \quad . \tag{5.12}$$

It may be observed that this partial differential equation is linear, unlike the full *N-S* equation. Furthermore, since the acceleration terms turn out to be negligible, the creeping flow solutions are not time-dependent.

5.2.1 Flow Between Coaxial Disks in Relative Rotation

A first problem that can be solved in the limit $Re \to 0$ is the flow of a Newtonian incompressible fluid between two disks in relative rotational motion. The problem is illustrated in Fig. 5.1: The lower disk (or plate) rotates and the upper disk is stationary. In cylindrical coordinates, the symmetry of the problem imposes that all derivatives $\partial/\partial\theta$ are zero. Furthermore, given that v_r and v_z are zero for $z = 0$ and $z = H$ and that there is no net flow along these two directions, it must be $v_r = 0$ and $v_z = 0$ in the entire velocity field, which also leads to conclude that $\mathcal{P}$ must be uniform. The latter result is a direct consequence of the hypothesis $Re \to 0$, which allows to neglect the inertial terms in the *N-S* equations. In conclusion, the problem reduces to:

$$\mu \left[\frac{\partial}{\partial r} \left(\frac{1}{r} \frac{\partial}{\partial r} (r v_\theta) \right) + \frac{\partial^2 v_\theta}{\partial z^2} \right] = 0 \quad , \tag{5.13}$$

with boundary conditions:

$$v_\theta = r\Omega, \ \text{for} \ z = 0 \ \text{and} \ v_\theta = 0, \ \text{for} \ z = H \quad . \tag{5.14}$$

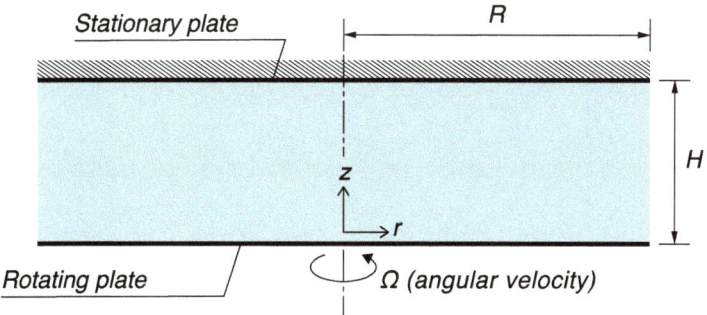

Fig. 5.1 Creeping flow between two disks in relative rotation

The boundary condition for $z = 0$ suggests that v_θ is everywhere proportional to r:

$$v_\theta = r f(z) \quad , \tag{5.15}$$

with $f(z = 0) = \Omega$ and $f(z = H) = 0$. Substituting Eq. (5.15) into Eq. (5.13) yields:

$$\frac{d^2 f}{dz^2} = 0 \quad , \tag{5.16}$$

which, with the boundary conditions (5.14), leads to:

$$f(z) = \Omega \left(1 - \frac{z}{H} \right) \quad , \tag{5.17}$$

$$v_\theta = r\Omega \left(1 - \frac{z}{H} \right) \quad . \tag{5.18}$$

The maximum velocity is $v_{\theta,max} = v_\theta(r = R, z = 0) = R\Omega$. The Reynolds number based on this velocity is equal to:

$$Re = \frac{\rho R \Omega H}{\mu} = \frac{\rho H^2 \Omega}{\mu} \frac{R}{H} \quad . \tag{5.19}$$

Equation (5.19) shows how the condition of creeping flow, namely $Re \ll 1$, is not trivially verified for the present problem, in which $R \gg H$. To verify the $Re \ll 1$ condition, it must be $\rho H^2 \Omega / \mu \ll 1$. From a practical point of view, e.g. in the case of motion between rotating disks in a viscometer with $\rho \simeq 10^3$ kg/m^3 and $H \simeq 10^{-3}$ m, this condition sets a limit on the angular velocity Ω.

The term $\rho H^2 \Omega / \mu$ has a clear physical meaning. Since $v_r = 0$ and $v_z = 0$, the dominant inertial term in the N-S equations is:

$$\frac{\rho v_\theta^2}{r} \simeq \frac{\rho (r\Omega)^2}{r} = \rho r \Omega^2 \quad , \tag{5.20}$$

while the dominant viscous term is:

$$\mu \frac{\partial^2 v_\theta}{\partial z^2} \simeq \mu \frac{r\Omega}{H^2} \quad . \tag{5.21}$$

The ratio between these two terms, i.e. their relative order of magnitude, is:

$$\frac{\frac{\rho v_\theta^2}{r}}{\mu \frac{\partial^2 v_\theta}{\partial z^2}} \propto \frac{\rho r \Omega^2}{\mu \frac{r\Omega}{H^2}} = \frac{\rho H^2 \Omega}{\mu} \quad . \tag{5.22}$$

To find the torque that must be applied to the disk to keep its rotation at velocity Ω, it is first necessary to calculate the shear stress $\tau_{z\theta}$ produced by the fluid on the disk:

$$\tau_{z\theta}\,|_0 = \mu \left(\frac{\partial v_\theta}{\partial z} + \frac{1}{r}\frac{\partial v_z}{\partial \theta} \right)\bigg|_0 = -\frac{\mu r \Omega}{H} \quad . \tag{5.23}$$

The local torque associated to this shear stress, evaluated on an infinitesimal area $dA = r\,d\theta dr$ is:

$$dG = dF\,r = \left(\tau_{z\theta}\,|_0\; r d\theta dr \right) r = -\frac{\mu r^2 \Omega}{H}(r d\theta dr) \quad . \tag{5.24}$$

The total torque exerted by the fluid on the rotating disk is, therefore:

$$G = \int_A dG = -\frac{\pi \mu \Omega R^4}{2H} \quad . \tag{5.25}$$

The torque T that must be applied to keep the rotation of the bottom disk has the same magnitude but opposite sign. Note that Eq. (5.25) can be used to measure the viscosity of a fluid, once G is known.

5.2.2 Flow Past a Sphere (Stokes Problem)

A second problem that can be solved analytically in the limit $Re \to 0$ is the creeping flow around a sphere. The analytical solution was found by Sir George Gabriel Stokes, and is therefore known as the *Stokes solution*. In this case, the Reynolds number is defined as $Re = \rho U D/\mu$, where D is the diameter of the sphere and U is the (uniform) velocity of the fluid, as shown in Fig. 5.2.

In spherical coordinates, the problem is symmetrical with respect to the azimuthal angle φ, and thus $\partial \bullet /\partial \varphi = 0$, $v_\varphi = 0$. The continuity equation becomes:

$$\frac{1}{r^2}\frac{\partial}{\partial r}(r^2 v_r) + \frac{1}{r \sin\theta}\frac{\partial}{\partial \theta}(\sin\theta v_\theta) = 0 \quad . \tag{5.26}$$

The r and θ components of the *N-S* equations are:

$$0 = -\frac{\partial P}{\partial r} + \mu \left[\frac{1}{r^2}\frac{\partial}{\partial r}\left(r^2\frac{\partial v_r}{\partial r}\right) + \frac{1}{r^2 \sin\theta}\frac{\partial}{\partial \theta}\left(\sin\theta \frac{\partial v_r}{\partial \theta}\right) + \right.$$

$$\left. -\frac{2v_r}{r^2} - \frac{2}{r^2}\frac{\partial v_\theta}{\partial \theta} - \frac{2}{r^2}v_\theta \cot\theta \right] \quad , \tag{5.27}$$

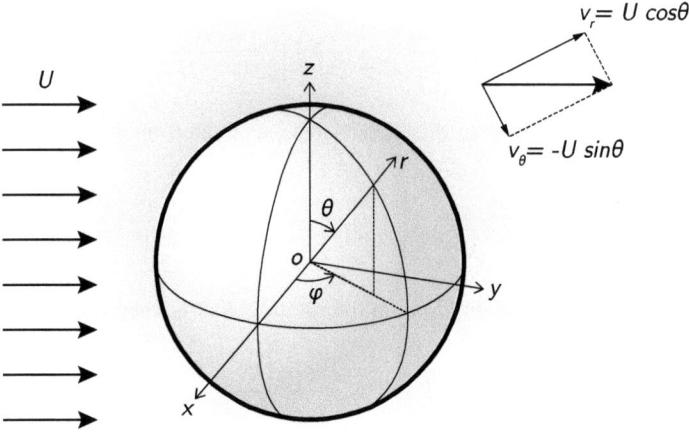

Fig. 5.2 Stokes problem: Flow at uniform velocity past an immersed sphere

$$0 = -\frac{1}{r}\frac{\partial \mathcal{P}}{\partial \theta} + \mu\left[\frac{1}{r^2}\frac{\partial}{\partial r}\left(r^2\frac{\partial v_\theta}{\partial r}\right) + \frac{1}{r^2 \sin\theta}\frac{\partial}{\partial \theta}\left(\sin\theta\frac{\partial v_\theta}{\partial \theta}\right) +\right.$$

$$\left.+\frac{2}{r^2}\frac{\partial v_r}{\partial \theta} - \frac{v_\theta}{r^2 sin^2\theta}\right] \quad, \tag{5.28}$$

where R is the radius of the sphere. The no-slip condition on the surface of the sphere requires $v_r = v_\theta = 0$ for $r = R$. In addition, far away from the sphere, the velocity field is unaffected by the presence of the sphere and the fluid velocity components can be written as $v_r = U \cos\theta$ and $v_\theta = -U \sin\theta$ for $r \to \infty$.

To solve the problem, it is assumed that v_r and v_θ have the same dependence on θ in proximity to the sphere as they do for $r \to \infty$:

$$v_r = A(r)\cos\theta \quad, \tag{5.29}$$

$$v_\theta = B(r)\sin\theta \quad, \tag{5.30}$$

where $A(r)$ and $B(r)$ are unknown functions of r. Substituting these expressions for v_r and v_θ in the continuity equation yields:

$$\frac{\cos\theta}{r^2}\left[2r A(r) + r^2\frac{\mathrm{d}A(r)}{\mathrm{d}r}\right] + 2\frac{\cos\theta}{r}B(r) = 0 \quad. \tag{5.31}$$

Dividing all terms by $(2 \cos \theta)/r$, an expression that does not depend on θ is obtained. One can proceed in a similar way for the two components of the N-S equations (5.27) and (5.28). From Eq. (5.27), it follows that $\mathcal{P}(r)$ must have the form:

$$\mathcal{P} = \mathcal{P}_0 + \mu C(r) \cos \theta \quad . \tag{5.32}$$

Substituting this equation into Eqs. (5.27) and (5.28) yields:

$$-\frac{dC(r)}{dr} + \frac{1}{r^2}\frac{d}{dr}\left(r^2\frac{dA(r)}{dr}\right) - 4\frac{A(r)}{r^2} - 4\frac{B(r)}{r^2} = 0 \quad , \tag{5.33}$$

$$C(r) + \frac{1}{r}\frac{d}{dr}\left(r^2\frac{dB(r)}{dr}\right) - 2\frac{B(r)}{r} - 2\frac{A(r)}{r} = 0 \quad . \tag{5.34}$$

The solution of this system of linear ordinary differential equations (5.31), (5.33) and (5.34) for the functions $A(r)$, $B(r)$ and $C(r)$ allows to determine the velocity field and the pressure distribution around the sphere:

$$v_r = U\left[1 - \frac{3}{2}\frac{R}{r} + \frac{1}{2}\left(\frac{R}{r}\right)^3\right]\cos \theta \quad , \tag{5.35}$$

$$v_\theta = -U\left[1 - \frac{3}{4}\frac{R}{r} - \frac{1}{4}\left(\frac{R}{r}\right)^3\right]\sin \theta \quad , \tag{5.36}$$

$$\mathcal{P} = \mathcal{P}_0 - \frac{3\mu U}{2R}\left(\frac{R}{r}\right)^2 \cos \theta \quad , \tag{5.37}$$

where $\mathcal{P} = p + \rho g h$ (with h vertical elevation of the fluid with respect to an arbitrary plane: for instance, $h = r \cos \theta$ with respect to the origin in Fig. 5.2) and $\mathcal{P}_0 = $ pressure in the $z = 0$ plane, away from the sphere.

Equations (5.35), (5.36) and (5.37) were derived under the hypothesis $Re \to 0$, namely when the viscous terms in the N-S equations are negligible compared to the inertial terms. The order of magnitude of the viscous terms can be estimated from the following term:

$$\mu\frac{1}{r^2}\frac{\partial}{\partial r}\left(r^2\frac{\partial v_r}{\partial r}\right) = \mu\frac{1}{r^2}\frac{\partial}{\partial r}\left[r^2 U\left(\frac{3}{2}\frac{R}{r^2} - \frac{3}{2}\frac{R^3}{r^4}\right)\cos \theta\right] =$$

$$= \mu\frac{U}{r^2}\frac{\partial}{\partial r}\left[\left(\frac{3}{2}R - \frac{3}{2}\frac{R^3}{r^2}\right)\cos \theta\right] \quad . \tag{5.38}$$

Assuming $\cos \theta \simeq 1$ and expanding the derivative:

$$\mu \frac{1}{r^2} \frac{\partial}{\partial r} \left(r^2 \frac{\partial v_r}{\partial r} \right) = \mu \frac{U}{r^2} 3 \frac{R^3}{r^3} = 3\mu U \frac{R^3}{r^5} \quad . \tag{5.39}$$

The order of magnitude of the inertial terms can be estimated from the term:

$$\rho v_r \frac{\partial v_r}{\partial r} = \rho \left[U \left(1 - \frac{3}{2}\frac{R}{r} + \frac{1}{2}\frac{R^3}{r^3} \right) \cos \theta \right] \left[U \left(\frac{3}{2}\frac{R}{r^2} - \frac{3}{2}\frac{R^3}{r^4} \right) \cos \theta \right] . \tag{5.40}$$

Since $r \geq R$, it is possible to assume $R^3/r^3 << R/r$ and $R^3/r^4 << R/r^2$. Therefore, taking again $\cos \theta \simeq 1$ and considering $R^2/r^3 << R/r^2$, Eq. (5.40) becomes:

$$\rho v_r \frac{\partial v_r}{\partial r} \simeq \rho U^2 \left(1 - \frac{3}{2}\frac{R}{r} \right) \frac{3}{2}\frac{R}{r^2} = \frac{3}{2}\rho U^2 \left(\frac{R}{r^2} - \frac{3}{2}\frac{R^2}{r^3} \right) \simeq \frac{3}{2}\rho U^2 \frac{R}{r^2} \quad , \tag{5.41}$$

and the ratio between the inertial and viscous terms is:

$$\frac{\rho v_r \dfrac{\partial v_r}{\partial r}}{\mu \dfrac{1}{r^2} \dfrac{\partial}{\partial r} \left(r^2 \dfrac{\partial v_r}{\partial r} \right)} = \frac{\dfrac{3}{2}\rho U^2 \dfrac{R}{r^2}}{3\mu U \dfrac{R^3}{r^5}} = \frac{1}{4} Re \left(\frac{r}{R} \right)^3 \quad . \tag{5.42}$$

Equation (5.42) shows how the creeping flow hypothesis is also acceptable for $Re <<$ 1 but only near the sphere, where $r/R \simeq 1$. Away from the sphere, and particularly when $r >> R$, this assumption is inaccurate and the Eqs. (5.35), (5.36) and (5.37) no longer apply: in this case, the velocity field and the pressure field tend asymptotically to those obtained in the absence of the sphere.

Once the velocity components and the pressure distribution around the sphere are known, it is possible to calculate the total force F exerted by the fluid on the sphere. For a system in which U is directed along the z axis and g has the same direction but opposite orientation, this force is also directed along z and results from the combined force components normal to the surface, F_n and tangential to the surface, F_t:

$$F_n = \int_0^{2\pi} \int_0^{\pi} (-p \mid_{r=R} \cos \theta) R^2 \sin \theta d\theta d\varphi \quad , \tag{5.43}$$

$$F_t = \int_0^{2\pi} \int_0^{\pi} (-\tau_{r\theta} \mid_{r=R} \sin \theta) R^2 \sin \theta d\theta d\varphi \quad , \tag{5.44}$$

where:

$$p \mid_{r=R} = \mathcal{P}_0 - \rho g R \cos \theta - \frac{3}{2}\frac{\mu U}{R} \cos \theta \quad , \tag{5.45}$$

and:

$$\tau_{r\theta}\mid_{r=R}= \mu \left[r\frac{\partial}{\partial r}\left(\frac{v_\theta}{r}\right) + \frac{1}{r}\frac{\partial v_r}{\partial \theta}\right]_{r=R} = -\frac{3\mu U}{2R}\sin\theta \quad . \tag{5.46}$$

Solving for the integrals yields:

$$F_n = \underbrace{\frac{4}{3}\pi R^3\rho g}_{\text{Archimedes' Thrust}} + \underbrace{2\pi\mu RU}_{\text{Form Drag}} \quad , \tag{5.47}$$

$$F_t = \underbrace{4\pi\mu RU}_{\text{Skin Friction}} \quad , \tag{5.48}$$

and:

$$F = F_n + F_t = \frac{4}{3}\pi R^3\rho g + \underbrace{6\pi\mu RU}_{\text{Stokes Drag}} \quad . \tag{5.49}$$

This result, widely confirmed by experimental measurements at low Reynolds number, shows that friction (or flow resistance) is the sum of a contribution, called *Form Drag*, that is due to the change in pressure around the immersed body, and a contribution, called *Skin Friction*, that is due to viscous stresses generated by the interaction between the fluid and the immersed body. The *Stokes Law* $F_D = 6\pi\mu RU$ coincides with equation (2.43), derived in Chap. 3. Indeed, once rewritten as a function of the diameter D and the Reynolds number Re, it reads as:

$$F_D = 6\pi\mu\frac{D}{2}U\frac{4\rho U D}{4\rho U D} = \frac{1}{2}\frac{24}{\frac{\rho U D}{\mu}}\frac{\pi D^2}{4}\rho U^2 = \frac{1}{2}C_D A_p\rho U^2 \quad , \tag{5.50}$$

with the drag coefficient C_D given by Eq. (2.39) since the flow past the sphere is purely viscous.

5.2.3 Examples

1 The torque applied to the disk shown in Fig. 5.3, which rotates at angular velocity $\Omega = 0.1$ rev/min, is $G = 1.4\cdot 10^{-3}$ N · m. Calculate the viscosity of the fluid knowing that its density is $\rho = 800$ kg/m^3.

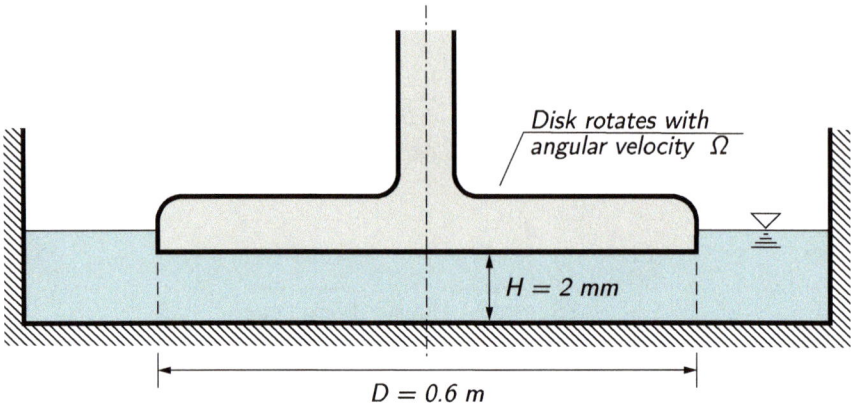

Fig. 5.3 Schematic of rotating viscometer for very viscous fluids

Solution From the results obtained in Sect. 5.2.1, the shear stress that must be applied to keep the disk rotating at angular velocity Ω is given by:

$$\tau_{z\theta} = \mu \left[\frac{\partial v_\theta}{\partial z} + \frac{1}{r} \frac{\partial v_z}{\partial \theta} \right] = \mu r \frac{\Omega}{H} \quad , \tag{5.51}$$

and the viscous torque can be calculated as:

$$T = \int_0^{\frac{D}{2}} 2\pi \mu \frac{\Omega}{H} r^3 \, dr = \frac{\pi \mu \Omega D^4}{32H} \quad . \tag{5.52}$$

Solving for the viscosity yields:

$$\mu = \frac{32TH}{\pi \Omega D^4} = 2.01 \cdot 10^{-2} \text{ Pa} \cdot \text{s} \quad . \tag{5.53}$$

To confirm the validity of this result, it is necessary to verify that the Reynolds number of the flow inside the viscometer is sufficiently small. In the present problem, the reference Reynolds number is based on the maximum tangential velocity and the distance between the disks:

$$Re = \rho \frac{HR\Omega}{\mu} = 2.4 \cdot 10^{-2} \quad . \tag{5.54}$$

This value is low enough to consider as reliable the viscosity measurement provided by Eq. (5.53).

2 Consider an incompressible Newtonian fluid flowing radially between two disks in the outward direction, as illustrated in Fig. 5.4. Neglecting the flow redistribution near the inlet, determine:

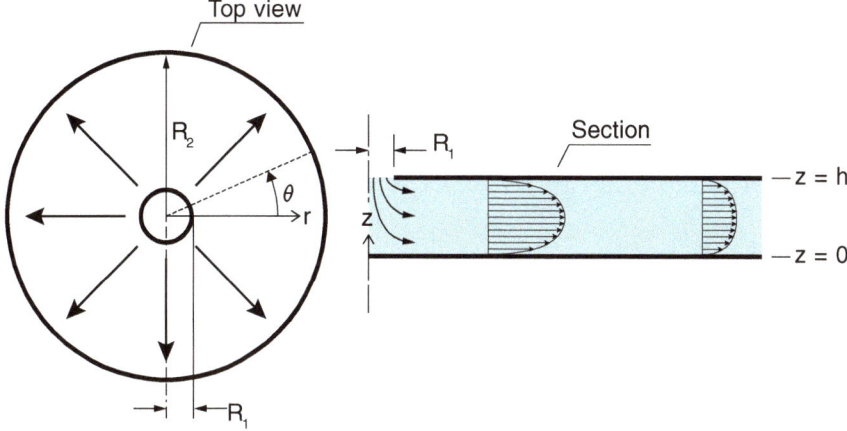

Fig. 5.4 Example of radial creeping flow

1. the equations of conservation of mass and momentum, under the assumption that the inertial terms can be neglected;
2. the velocity profile and the flowrate as a function of the pressure drop between R_1 and R_2.

Solution The following assumptions can be made:

1. Newtonian incompressible fluid.
2. Creeping flow ($\rightarrow \partial \bullet / \partial t = 0$).
3. Negligible inlet effects.
4. Flow only in radial direction with wall-parallel streamlines ($\rightarrow v_z = v_\theta = 0$, uniform v_r profile along the θ direction).

1. The continuity equation simplifies as:

$$\frac{1}{r}\frac{\partial}{\partial r}(rv_r) = 0 \quad \rightarrow \quad rv_r = C_1(z) \quad , \tag{5.55}$$

and:

$$v_r = \frac{C_1(z)}{r} \quad \rightarrow \quad \frac{\partial^2 v_r}{\partial z^2} = \frac{1}{r}\frac{d^2 C_1(z)}{dz^2} \quad . \tag{5.56}$$

The *N-S* equations become:

$$0 = -\frac{\partial \mathcal{P}}{\partial r} + \mu \left(\frac{\partial^2 v_r}{\partial z^2} \right) \quad , \tag{5.57}$$

$$0 = -\frac{\partial \mathcal{P}}{\partial \theta} \quad , \tag{5.58}$$

$$0 = -\frac{\partial \mathcal{P}}{\partial z} \quad . \tag{5.59}$$

From the last two equations, it follows that $\mathcal{P} = \mathcal{P}(r)$.

2. Combining the continuity equation with the N-S equation along r yields:

$$0 = -\frac{\partial \mathcal{P}}{\partial r} + \mu \frac{1}{r} \frac{d^2 C_1(z)}{dz^2} \quad . \tag{5.60}$$

Since $\mathcal{P}$ is only a function of r, from Eq. (5.60) it follows necessarily that $d^2 C_1(z)/dz^2 = $ constant. Furthermore, observing that $dr = r\, d(\ln r)$, it is easy to write:

$$dP = \mu \frac{1}{r} \frac{d^2 C_1(z)}{dz^2} dr = \mu \frac{d^2 C_1(z)}{dz^2} d(\ln r) \quad \rightarrow$$

$$\rightarrow \quad \underbrace{\frac{dP}{d(\ln r)}}_{f(r)} = \underbrace{\mu \frac{d^2 C_1(z)}{dz^2}}_{g(z)} = \text{constant} \quad . \tag{5.61}$$

Equations (5.56) and (5.62) lead to the following expression of the fluid velocity:

$$v_r = \frac{z^2}{2r} \frac{d^2 C_1(z)}{dz^2} + \frac{C_2}{r} z + \frac{C_3}{r} \quad . \tag{5.62}$$

Given the boundary conditions:

$$z = 0 \rightarrow v(r, 0) = 0 \rightarrow C_3 = 0 \quad , \tag{5.63}$$

$$z = h \rightarrow v(r, h) = 0 \rightarrow C_2 = -\frac{d^2 C_1(z)}{dz^2} \frac{h^2}{2} \frac{1}{h} =$$

$$= -\frac{d^2 C_1(z)}{dz^2} \frac{h}{2} \quad , \tag{5.64}$$

the resulting velocity profile is:

$$v_r(r, z) = \frac{d^2 C_1(z)}{dz^2} \left(\frac{z^2}{2r} - \frac{hz}{2r} \right) \quad . \tag{5.65}$$

Using Eq. (5.62), it is possible to recast equation (5.65) as:

$$v_r(r, z) = \frac{1}{2r\mu} \frac{-dP}{d(\ln r)} (hz - z^2) \quad , \tag{5.66}$$

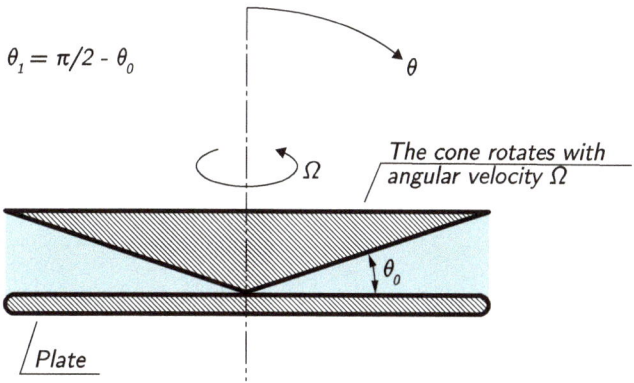

$\theta_1 = \pi/2 - \theta_0$

Ω

The cone rotates with angular velocity Ω

θ_0

Plate

Fig. 5.5 Schematic of the cone-and-plate viscometer

or, equivalently:

$$v_r(r, z) = \frac{1}{2r\mu} \frac{-\Delta\mathcal{P}}{\ln(R_2/R_1)}(hz - z^2) \ . \tag{5.67}$$

The volumetric flowrate is given by

$$Q = 2\pi r \int_0^h v_r(r, z)\mathrm{d}z = 2\pi r \int_0^h \frac{1}{2r\mu} \frac{-\Delta\mathcal{P}}{\ln(R_2/R_1)}(hz - z^2)\mathrm{d}z =$$

$$= \frac{\pi h^3}{6\mu} \frac{-\Delta\mathcal{P}}{\ln(R_2/R_1)} \ . \tag{5.68}$$

$\boxed{3}$ The cone-and-plate viscometer, shown in Fig. 5.5, consists of a stationary flat plate on which the fluid whose viscosity is to be measured is placed, and by an upside-down cone. The cone is lowered toward the plate until its vertex comes into contact with the plate. The cone rotates at angular velocity Ω and the viscosity of the fluid is determined by measuring the torque applied to keep the rotation.

Assume creeping flow conditions and purely tangential flow. Also, restrict the solution to small values of θ_0.

1. Simplify the *N-S* equations for the present case.
2. Determine the velocity distribution.
3. Derive the torque T exerted by the fluid on the cone.

Solution

1. For an incompressible Newtonian fluid in creeping flow conditions, if the motion is purely tangential ($v_r = v_\theta = 0$) in a spherical coordinate system, then the only non-zero velocity component is $v_\varphi = v_\varphi(r, \theta)$ and the *N-S* equations become:

$$0 = \frac{\partial P}{\partial r} \quad , \tag{5.69}$$

$$0 = \frac{\partial P}{\partial \theta} \quad , \tag{5.70}$$

$$0 = \mu \left[\frac{1}{r^2} \frac{\partial}{\partial r} \left(r^2 \frac{\partial v_\varphi}{\partial r} \right) + \frac{1}{r^2 \sin \theta} \frac{\partial}{\partial \theta} \left(\sin \theta \frac{\partial v_\varphi}{\partial \theta} \right) - \frac{v_\varphi}{r^2 \sin^2 \theta} \right] \quad , \tag{5.71}$$

with boundary conditions:

$$v_\varphi = 0 \ \text{for} \ \theta = \frac{\pi}{2} \ \text{and} \ v_\varphi = r\Omega \ \text{for} \ \theta = \theta_1 \quad . \tag{5.72}$$

2. The boundary condition for $\theta = \theta_1$ suggests that the velocity profile has the following form:

$$v_\varphi = r\Omega f(\theta) \quad . \tag{5.73}$$

The boundary conditions for f are:

$$f = 0 \ \text{for} \ \theta = \frac{\pi}{2} \ \text{e} \ f = 1 \ \text{for} \ \theta = \theta_1 \quad . \tag{5.74}$$

By substituting Eq. (5.73) into Eq. (5.71), the following differential equation for f can be obtained:

$$f'' + f' \cot \theta + \left(2 - \frac{1}{\sin^2 \theta} \right) f = 0 \quad . \tag{5.75}$$

The solution of this equation exists[1] but it is rather complex. However, as the angle θ_0 is always very small, it is convenient to consider the limit $\theta_0 \to 0$. Within this limit, Eq. (5.75) becomes:

$$f'' + f = 0 \quad \longrightarrow \quad f'' = -f \quad , \tag{5.76}$$

and is fairly simple to solve. It suffices to multiply both terms by $2f'$ to obtain:

$$2f'f'' = -2ff' \quad . \tag{5.77}$$

---

[1] The solution of equation (5.75) has the following form:

$$f = \text{constant} \cdot \left[\cot \theta + \frac{\sin \theta}{r} \ln \left(\frac{1 + \cos \theta}{1 - \cos \theta} \right) \right] \quad .$$

Integration yields:

$$f'^2 = -f^2 + C_1 \quad \longrightarrow \quad f' = \sqrt{C_1 - f^2} \ . \tag{5.78}$$

Further integration yields:

$$d\theta = \frac{df}{\sqrt{C_1 - f^2}} \quad \rightarrow \quad \theta = \sin^{-1}\left(\frac{f}{\sqrt{C_1}}\right) + C_2 \ , \tag{5.79}$$

or, equivalently:

$$\frac{f}{\sqrt{C_1}} = \sin(\theta - C_2) \ . \tag{5.80}$$

From the boundary conditions, it follows that $\sqrt{C_1} = 1/\sin(\theta_1 - \pi/2)$ and $C_2 = \pi/2$. Therefore, the complete solution for f is:

$$f(\theta) = \frac{\sin(\theta - \pi/2)}{\sin(\theta_1 - \pi/2)} \ . \tag{5.81}$$

Since $\theta_1 = \pi/2 - \theta_0$ and $\sin\theta \to \theta$ for $\theta \to 0$, the solution can be recast for small values of the angle θ_0 as:

$$f(\theta) = \frac{\theta - \pi/2}{\theta_1 - \pi/2} = \frac{\pi/2 - \theta}{\theta_0} \ , \tag{5.82}$$

and thus the velocity v_φ can be expressed as:

$$v_\varphi(r, \theta) = r\Omega\left(\frac{\pi/2 - \theta}{\theta_0}\right) \ . \tag{5.83}$$

This expression satisfies the boundary conditions.

3. The torque exerted by the fluid on the cone is equal to:

$$T = \int_A r \, \tau_{\varphi\theta} dA \ , \tag{5.84}$$

where r is the radial distance between the rotation axis (about which the torque is being measured) and the point of application of the force per unit area, $\tau_{\varphi\theta}$. From this definition, it follows that:

$$\tau_{\varphi\theta} = \mu\left[\frac{\sin\theta}{r}\frac{\partial}{\partial\theta}\left(\frac{v_\varphi}{\sin\theta}\right)\right] \ . \tag{5.85}$$

Substituting into Eq. (5.83), for small values of θ_0, it is possible to write:

$$T_{\varphi\theta} = -\mu \frac{\Omega}{\theta_0} \quad . \tag{5.86}$$

The resulting torque is:

$$T = \int_0^{2\pi} \int_0^R r \left(-\mu \frac{\Omega}{\theta_0}\right) r \mathrm{d}\theta \mathrm{d}r = -\frac{2}{3}\pi R^3 \mu \frac{\Omega}{\theta_0} \quad . \tag{5.87}$$

4 Determine the torque that must be applied to the cone of a cone-and-plate vis-cometer to rotate at an angular velocity $\Omega = 10$ rad/min, if the fluid viscosity is $\mu = 0.1$ Pa · s, its density $\rho = 750$ kg/m^3, the cone radius $R = 0.1$ m and the open-ing angle $\theta_0 = 0.5°$.

Solution The torque can be calculated using Eq. (5.87):

$$T = \frac{2}{3}\pi R^3 \mu \frac{\Omega}{\theta_0} = 4.0 \cdot 10^{-3} \, \mathrm{Nm} \quad . \tag{5.88}$$

However, it is necessary to verify that creeping flow conditions indeed apply. This can be done by calculating the Reynolds number and verifying that it is sufficiently close to zero. The flow is not characterized by a characteristic velocity or length; however, the velocity will not be greater than the maximum velocity, which is equal to ΩR, and the characteristic length can be assumed equal to the maximum distance between the plate and the cone, which is $R \sin \theta_0 \simeq R\theta_0$. Therefore, the resulting value of Reynolds number is:

$$Re = \frac{\rho R^2 \Omega \theta_0}{\mu} = 0.11 \quad , \tag{5.89}$$

which is sufficiently low to ensure that creeping flow conditions apply.

Also note that, by substituting $\theta_0 = h/R$ in Eq. (5.88), one obtains:

$$T = -\frac{2}{3}\pi R^4 \mu \frac{\Omega}{H} \quad . \tag{5.90}$$

This expression is formally identical to that given in Eq. (5.52), which expresses the torque exerted on the upper plate of a parallel-plate viscometer.

5.3 Lubrication Theory

5.3.1 Analysis of the Navier-Stokes Equations

The *Lubrication Theory* is an approximate theory, first derived by Osborne Reynolds in 1886, which applies to the flow between two nearly parallel walls and is based on the assumption that in each section between the walls the velocity profile can be described by the same equations that apply when the walls are parallel.

Consider, for example, the situation shown in Fig. 5.6 in which a viscous fluid flows in a gap of varying thickness formed by two flat walls in relative motion. Without loss of generality, it can be assumed that one wall is moving (with constant velocity equal to U) and the other wall is stationary. The relative motion allows the pressure within the gap to increase, thus becoming higher than the outer ambient pressure. The pressure increase allows to counteract the normal forces that are transmitted through the fluid by the walls, preventing them from coming into contact and thus realizing the lubrication effect.

The configuration shown in Fig. 5.6 is representative of the flow produced by a slider bearing. More generally, the solution of the lubrication problem is of particular interest when the flow of a thin film of lubricant in machinery components is considered.

Let H_1 and H_2 be the maximum and minimum gap thickness, respectively, and L its length. The lubrication theory can be applied if and only if:

$$\frac{H_1}{L} \ll 1 \quad , \quad \frac{H_1 - H_2}{L} \ll 1 \quad . \tag{5.91}$$

It is also assumed, subject to a-posteriori justification, that the inertial terms are negligible as for the case of creeping flow. As a result, the equations to be solved read as:

$$\frac{\partial v_x}{\partial x} + \frac{\partial v_y}{\partial y} = 0 \quad , \tag{5.92}$$

$$0 = -\frac{\partial \mathcal{P}}{\partial x} + \mu \left(\frac{\partial^2 v_x}{\partial x^2} + \frac{\partial^2 v_x}{\partial y^2} \right) \quad , \tag{5.93}$$

$$0 = -\frac{\partial \mathcal{P}}{\partial y} + \mu \left(\frac{\partial^2 v_y}{\partial x^2} + \frac{\partial^2 v_y}{\partial y^2} \right) \quad . \tag{5.94}$$

To check whether further simplifications are possible, it is convenient to write equations (5.92), (5.92) and (5.94) in dimensionless form, taking into account that two characteristic lengths can be defined, $H_1(H_2 \cong H_1)$ and L. This allows to define $\tilde{x}$ and $\tilde{y}$ as:

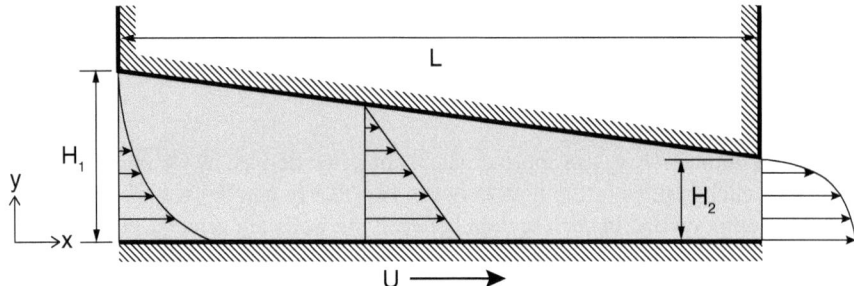

Fig. 5.6 Schematic of the flow of viscous fluid within a wall-bounded gap of varying thickness. The flow is induced by the relative motion between the lower wall (which moves with constant velocity U) and the upper wall, which is stationary. The schematic reproduces in a simplified way the behavior of a flat thrust pad immersed in an oil bath. The lubrication theory applies if $H_1/L \ll 1$ and $(H_1 - H_2)/L \ll 1$

$$\tilde{x} = \frac{x}{L} \quad , \quad \tilde{y} = \frac{y}{H_1} \quad . \tag{5.95}$$

Similarly, the relative velocity U can be taken as the characteristic velocity along x (see Fig. 5.6), while the characteristic velocity along y is denoted by V. In dimensionless form, the continuity equation becomes:

$$\frac{H_1 U}{V L} \frac{\partial \tilde{v}_x}{\partial \tilde{x}} + \frac{\partial \tilde{v}_y}{\partial \tilde{y}} = 0 \quad . \tag{5.96}$$

In this problem, $\partial \tilde{v}_x/\partial \tilde{x}$ is different from zero since the axial velocity changes along x. It follows that the order of magnitude of $H_1 U/V L$ must be unitary and, in turn, that V can be expressed as:

$$V = \frac{U H_1}{L} \quad , \tag{5.97}$$

which allows to write the continuity equation simply as:

$$\frac{\partial \tilde{v}_x}{\partial \tilde{x}} + \frac{\partial \tilde{v}_y}{\partial \tilde{y}} = 0 \quad . \tag{5.98}$$

In dimensionless form, Eq. (5.93) becomes:

$$\frac{\Pi H_1^2}{\mu L U} \frac{\partial \tilde{P}}{\partial \tilde{x}} = \left[\left(\frac{H_1}{L} \right)^2 \frac{\partial^2 \tilde{v}_x}{\partial \tilde{x}^2} + \frac{\partial^2 \tilde{v}_x}{\partial \tilde{y}^2} \right] \quad , \tag{5.99}$$

where Π denotes the characteristic pressure. In this equation, since $H_1/L \ll 1$, the first term on the right-hand side is negligible compared to the second term. Furthermore, expressing the characteristic pressure Π as:

$$\Pi = \frac{\mu U L}{H_1^2} \quad , \tag{5.100}$$

makes the coefficient in front of the pressure gradient, $\partial \tilde{\mathcal{P}}/\partial \tilde{x}$, equal to unity. Using the expressions of V and Π just given, Eq. (5.94) becomes:

$$\frac{\partial \tilde{\mathcal{P}}}{\partial \tilde{y}} = \left(\frac{H_1}{L}\right)^2 \frac{\partial^2 \tilde{v}_y}{\partial \tilde{y}^2} \approx 0 \quad . \tag{5.101}$$

This equation shows how, as a first approximation, the pressure is only a function of the axial coordinate, $\mathcal{P} = \mathcal{P}(x)$. Consider now also the inertial terms of the N-S equation. Using the dimensionless variables previously introduced, the x-component of the N-S equation reads as:

$$\frac{\rho U^2}{L}\left(\tilde{v}_x \frac{\partial \tilde{v}_x}{\partial \tilde{x}} + \tilde{v}_y \frac{\partial \tilde{v}_x}{\partial y}\right) = \frac{\mu U}{H_1^2}\left[-\frac{\partial \tilde{\mathcal{P}}}{\partial \tilde{x}} + \left(\frac{H_1}{L}\right)^2 \frac{\partial^2 \tilde{v}_x}{\partial \tilde{x}^2} + \frac{\partial^2 \tilde{v}_x}{\partial \tilde{y}^2}\right] \quad . \tag{5.102}$$

The inertial terms on the left-hand side are negligible if:

$$\frac{\rho U^2}{L} \ll \frac{\mu U}{H_1^2} \quad , \tag{5.103}$$

namely if $Re \cdot (H_1/L) \ll 1$. Taking into account that $H_1/L \ll 1$, this result allows to extend the solution just obtained even to cases in which $Re \geq 1$.

5.3.2 Velocity Distribution

The analysis of the conservation equations allows to reduce the problem to the following equations:

$$\mathcal{P} = \mathcal{P}(x) \quad , \tag{5.104}$$

$$\frac{\partial \mathcal{P}}{\partial x} = \mu \frac{\partial^2 v_x}{\partial y^2} \quad , \tag{5.105}$$

where $v_x = v_x(x, y)$. Integration of equation (5.105) gives:

$$v_x = \frac{1}{2\mu}\frac{d\mathcal{P}}{dx}y^2 + c_1 y + c_2 \quad,$$ (5.106)

since $\mathcal{P}$ is only a function of x. The boundary conditions:

$$y = 0 \;\rightarrow\; v_x = U \quad,$$ (5.107)

$$y = H(x) \;\rightarrow\; v_x = 0 \quad,$$ (5.108)

lead to the following expression for v_x:

$$v_x = U\left[1 - \frac{y}{H(x)}\right] - \frac{1}{2\mu}\left(\frac{d\mathcal{P}}{dx}\right)yH(x)\left[1 - \frac{y}{H(x)}\right] \quad.$$ (5.109)

This equation is also valid in the particular case $H(x) = H_1 = $ constant, namely in the case of Couette-Poiseuille flow driven by an imposed pressure gradient, $d P/dx$. The behavior of v_x within the gap is shown, for three different positions along the thrust pad (i.e. along the x coordinate) in Fig. 5.6. The velocity profile has a concave shape for $x = 0$, a convex shape for $x = L$ and a triangular shape only at the point where the pressure gradient vanishes.

5.3.3   Pressure Distribution

In Eq. (5.109), the pressure gradient is unknown, and so is the flowrate between the plates. However, since the fluid is incompressible, at any x location within the gap it must be:

$$\Gamma = \int_0^{H(x)} v_x dy = \text{constant} \quad.$$ (5.110)

This equation represents the conservation of mass, Eq. (5.92), written in integral form. Solving for the integral yields:

$$\Gamma = \frac{UH(x)}{2} - \frac{H^3(x)}{12\mu}\frac{d\mathcal{P}}{dx} \quad.$$ (5.111)

Equation (5.111) can be used to derive the following expression for $\mathcal{P}(x)$:

$$\mathcal{P}(x) = \mathcal{P}_0 + 6\mu U\int_0^x \frac{dx}{H^2(x)} - 12\mu\Gamma\int_0^x \frac{dx}{H^3(x)} \quad,$$ (5.112)

where $H(x)$ can be any function of x that satisfies the assumptions previously discussed.

With reference to Fig. 5.6, given that the increase of pressure occurs only within the gap of length L, it is possible to set $\mathcal{P}(L) = \mathcal{P}(0)$ and use Eq. (5.112) to calculate the flowrate per unit gap width, Γ:

$$\Gamma = \frac{1}{2}U\frac{\int_0^L H^{-2}(x)\mathrm{d}x}{\int_0^L H^{-3}(x)\mathrm{d}x} \quad , \tag{5.113}$$

with:

$$H(x) = H_1 + \frac{(H_2 - H_1)}{L}x \quad . \tag{5.114}$$

The two integrals in Eq. (5.112) have the following solutions:

$$\int_0^x \frac{\mathrm{d}x}{H^2(x)} = \frac{x}{H_1 H(x)} \xrightarrow{x=L} \int_0^L \frac{\mathrm{d}x}{H^2(x)} = \frac{L}{H_1 H_2} \quad , \tag{5.115}$$

$$\int_0^x \frac{\mathrm{d}x}{H^3(x)} = \frac{x\,(H_1 + H(x))}{2H_1^2 H^2(x)} \xrightarrow{x=L} \int_0^L \frac{\mathrm{d}x}{H^3(x)} = \frac{L}{2}\frac{H_1 + H_2}{H_1^2 H_2^2} \quad . \tag{5.116}$$

Substituting Eq. (5.113) into Eq. (5.112) yields:

$$\mathcal{P}(x) = 6\mu U \int_0^x \frac{H(x) - \mathcal{A}}{H^3(x)}\mathrm{d}x \quad , \tag{5.117}$$

where it is assumed $\mathcal{P}_0 = 0$ and the geometric aspect ratio $\mathcal{A}$ is defined as:

$$\mathcal{A} = \frac{\int_0^L H^{-2}(x)\mathrm{d}x}{\int_0^L H^{-3}(x)\mathrm{d}x} \quad . \tag{5.118}$$

If Eq. (5.114) holds, then it is also possible to write:

$$\mathcal{A} = \frac{2H_1 H_2}{H_1 H_2} \quad \rightarrow \quad \Gamma = \frac{1}{2}U\mathcal{A} \quad . \tag{5.119}$$

If the surface of the thrust pad is flat, then the following expression for $\mathcal{P}(x)$ can be derived:

$$\mathcal{P}(x) = \frac{6\mu U L}{H_1^2 - H_2^2}\frac{[H_1 - H(x)][H(x) - H_2]}{H^2(x)} \quad . \tag{5.120}$$

5.3.4 Calculation of Pressure Forces and Shear Forces

The pressure distribution given by Eq. (5.120) has a pronounced maximum and gives rise to a vertical thrust that can be calculated as:

$$\frac{F_N}{W} = \int_0^L \mathcal{P}(x)\mathrm{d}x = \frac{6\mu U L^2}{(H_1 - H_2)^2}\left[\ln\frac{H_1}{H_2} - \frac{2(H_1 - H_2)}{H_1 + H_2}\right] \quad . \tag{5.121}$$

The maximum value of F_N is obtained for $H_1 \simeq 2.2H_2$. The tangential force exerted on the thrust pad is due to the action of the fluid shear stress and can be calculated as:

$$F_S = -\mu \int_0^L \left.\left|\frac{\partial v_x}{\partial y}\right|\right|_{y=0} \mathrm{d}x \quad . \tag{5.122}$$

For a flat thrust pad, the force is equal to:

$$F_S = \frac{\mu U L}{H_1 - H_2}\left[4\ln\frac{H_1}{H_2} - \frac{6(H_1 - H_2)}{H_1 + H_2}\right] \quad . \tag{5.123}$$

For $H_1/H_2 = 2.2$, the ratio between F_S and F_N is:

$$\frac{F_S}{F_N} \simeq 5\frac{H_2}{L} \quad . \tag{5.124}$$

This implies that a large thrust along the wall-normal direction can be achieved at the expense of generating a limited amount of tangential friction.

5.3.5 Examples

1 A fluid of density $\rho = 800$ kg/m^3 and viscosity $\mu = 10^{-2}$Pa $\cdot$ s flows between two square flat walls of size $W = L = 1$ m. One of the walls is inclined with respect to the other, as shown in Fig. 5.7. The flowrate is equal to $w = 4$ kg/s. Determine the force acting on each wall.

Solution Using the coordinate system shown in Fig. 5.7, the forces acting on the two walls are:

1. Horizontal wall:

$$F_{1x} = W \int_0^L \tau(y = 0)\mathrm{d}x \quad , \tag{5.125}$$

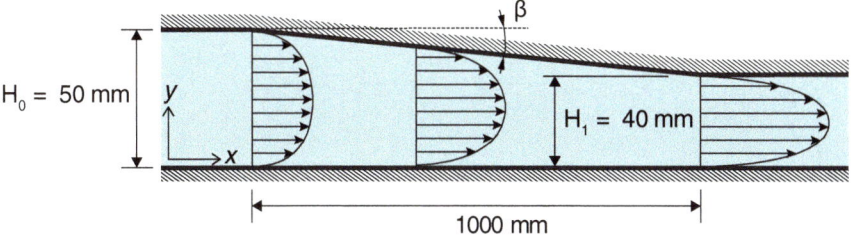

Fig. 5.7 Application of the lubrication theory: Flow between slightly-inclined flat walls. The velocity profile, while maintaining a parabolic shape, changes continuously along the flow direction

$$F_{1y} = -W \int_0^L [\mathcal{P}(x) - \mathcal{P}_0]\, dx \quad , \tag{5.126}$$

2. Inclined wall:

$$F_{2x} = W \int_0^L \tau(H)dx + W \int_0^L [\mathcal{P}(x) - \mathcal{P}_0]\tan\beta dx \quad , \tag{5.127}$$

$$F_{2y} = -W \int_0^L \tau(H)\tan\beta dx + W \int_0^L (\mathcal{P}(x) - \mathcal{P}_0)\, dx \quad , \tag{5.128}$$

where $\mathcal{P}_0$ is the pressure acting at the inlet section of the channel.

For the pressure gradient, Eq. (5.111) yields:

$$-\frac{d\mathcal{P}}{dx} = 12\frac{\Gamma\mu}{H^3(x)} \quad , \tag{5.129}$$

where $\Gamma = w/(\rho W)$, and the height $H(x)$ can be expressed as $H(x) = H_0(1 - \kappa x)$, with $\kappa = 0.2 = (H_0 - H_1)/(H_0 L)$. Pressure can be obtained by separating the variables and by integrating along x:

$$-d\mathcal{P} = \Gamma\frac{12\mu}{H^3(x)}dx \;\rightarrow\; \mathcal{P}_0 - \mathcal{P}(x) = 12\frac{\Gamma\mu}{H_0^3}\int_0^x \frac{1}{(1 - \kappa x)^3}dx \quad . \tag{5.130}$$

The resulting pressure distribution is:

$$\mathcal{P}(x) = \mathcal{P}_0 - 12\frac{\Gamma\mu}{H_0^3}\frac{1}{2\kappa}\left[\frac{1}{(1 - \kappa x)^2} - 1\right] \quad . \tag{5.131}$$

The shear stress can be derived from Eq. (5.109), setting $U = 0$. This equation can be rewritten in terms of the volumetric flowrate and derived with respect to y to obtain:

$$\tau(y) = \mu \frac{dv_x(y)}{dy} = 6\frac{\mu\Gamma}{H^3(x)}[H(x) - 2y] \quad , \tag{5.132}$$

and:

$$\tau(y = 0) = 6\frac{\mu\Gamma}{H^2(x)} \quad . \tag{5.133}$$

From Eqs. (5.125) and (5.132), the force F_{1x} can be calculated as:

$$F_{1x} = W \int_0^L 6\frac{\mu\Gamma}{H^2(x)}dx = 6\frac{\mu W\Gamma}{H_0^2} \int_0^L \frac{1}{(1 - \kappa x)^2}dx$$

$$= 6\frac{\mu W\Gamma}{H_0^2}\left[\frac{L}{(1 - \kappa L)}\right] = 0.15 \ N \quad . \tag{5.134}$$

The pressure force can be calculated as:

$$F_{1y} = -W \int_0^L 12\frac{\Gamma\mu}{H_0^3}\frac{1}{2\kappa}\left[\frac{1}{(\kappa x - 1)^2} - 1\right]dx =$$

$$= 6\frac{\Gamma\mu}{H_0^3\kappa}\left(\frac{\kappa L^2}{\kappa L - 1}\right) = -3.0 \ N \quad . \tag{5.135}$$

The relative inclination of the two walls corresponds to an angle $\beta = 0.01$, for which $\tan\beta \simeq 0.01$. Therefore, on the inclined wall, the forces are:

$$F_{2x} = F_{1x} - \tan\beta \ F_{1y} = 0.18 \ N \quad , \tag{5.136}$$

and:

$$F_{2y} = -\tan\beta F_{1x} - F_{1y} = 3.0 \ N \quad . \tag{5.137}$$

2   The vertical hopper shown in Fig. 5.8 is used to feed a horizontal pipe of length $L = 6$ m with a flowrate of 6 kg/s of viscous fluid ($\rho = 800$ kg/m^3, $\mu = 2$ Pa $\cdot$ s).

1. Determine the pressure drop in the pipe (horizontal section).
2. Simplify the Navier-Stokes equations for the hopper (vertical section).
3. Determine the liquid height in the hopper (vertical section).
4. Determine the time required for the height of the liquid in the hopper to decrease to 10% of its initial value if the supply of liquid is halted.

Discard inlet/outlet effects and assume negligible concentrated pressure losses.

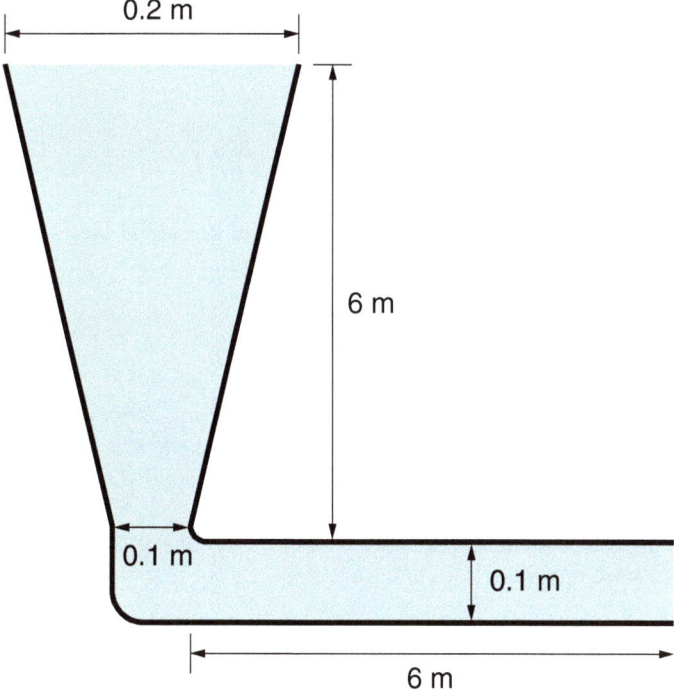

Fig. 5.8 Schematic of a hopper-fed horizontal pipe

Solution

1. Let section 2 be the outlet section of the pipe, section 1 be the section at the pipe bend, and section 0 be the section at the ridge of the hopper. In the horizontal section, the Reynolds number of the flow is equal to 38. This low value makes it is possible to use the Hagen-Poiseuille relation, given by Eq. (2.14), to calculate the pressure drop:

$$\Delta p = p_1 - p_0 = 8\frac{\mu Q}{\pi R^4}L = 3.67 \cdot 10^4 \, Pa \quad . \tag{5.138}$$

2. In the vertical section, the flow is laminar and there is no dependence of the flow variables on θ, given the cylindrical symmetry of the problem. Based on the assumption of the lubrication theory, the inertial terms of the N-S equations can be neglected. In addition, given the low tapering of the hopper, it can be assumed that the pressure is only a function of the axial coordinate, $\partial P/\partial r = 0$. In conclusion, the N-S equations for steady flow are:

$$0 = -\frac{\partial P}{\partial r} \quad , \tag{5.139}$$

$$0 = -\frac{1}{r}\frac{\partial \mathcal{P}}{\partial \theta} \quad , \tag{5.140}$$

$$0 = -\frac{\partial \mathcal{P}}{\partial z} + \mu \left[\frac{1}{r}\frac{\partial}{\partial r}\left(r\frac{\partial v_z}{\partial r}\right)\right] \quad , \tag{5.141}$$

which are the same equations that describe the Poiseuille flow in a pipe. The continuity equation can be used in its integral form:

$$\frac{\mathrm{d}Q}{\mathrm{d}z} = 0 \quad . \tag{5.142}$$

3. The radius of the vertical section can be expressed as $r = r(z) = R_1(1 + \kappa z)$, with R_0 the maximum radius, R_1 the minimum radius and $\kappa = (R_0 - R_1)/(R_1 L)$, where $L = 6$ m. To determine the height of the liquid in the vertical section, the equation of motion (5.141) along z must be solved. As already noted, this equation can be interpreted as the equation describing the Poiseuille flow in a tube of variable radius $r(z)$. Therefore:

$$\left(-\frac{\mathrm{d}\mathcal{P}}{\mathrm{d}z}\right) = 8\frac{\mu Q}{\pi r^4(z)} \quad \longrightarrow \quad \Delta\mathcal{P} = \frac{8\mu Q}{\pi R_1^4}\int_z^0 \frac{\mathrm{d}z}{(1 + \kappa z)^4} =$$

$$= \frac{8\mu Q}{\pi R_1^4}\frac{1}{3\kappa}\left[\frac{1}{(\kappa z + 1)^3} - 1\right] \quad . \tag{5.143}$$

The total pressure difference, $\Delta\mathcal{P}$, is equal to:

$$\Delta\mathcal{P} = p_1 - p_0 - \rho g z \quad , \tag{5.144}$$

where $p_0 = p_{atm}$ and $p_1 = p_{atm} + 3.67 \cdot 10^4\,Pa$. From Eqs. (5.143) and (5.144), the equilibrium height in the hopper can be derived:

$$z_{eq} = \frac{1}{\rho g}\left\{p_1 - p_0 - 8\frac{\mu Q}{\pi R_1^4}\frac{1}{3\kappa}\left[\frac{1}{(\kappa z + 1)^3} - 1\right]\right\} = 6.05\,m \quad . \tag{5.145}$$

4. A volumetric balance on the hopper yields:

$$\frac{\mathrm{d}V}{\mathrm{d}t} = -Q \quad , \tag{5.146}$$

where the differential volume $\mathrm{d}V$ is equal to $\pi r^2(z)\mathrm{d}z$. The flowrate depends on the liquid height according to Eq. (5.145). This equation, for a generic z and variable Q, can be written as:

$$Q = \frac{\pi R_1^4}{8\mu} \rho g z \left[L + \frac{1}{3\kappa} - \frac{1}{3\kappa(\kappa z + 1)^3} \right]^{-1} \quad . \tag{5.147}$$

Therefore, from Eqs. (5.146) and (5.147), it is possible to obtain:

$$\int_0^t dt = -\frac{8\mu}{\rho g R_1^2} \int_{z_{eq}}^{z_f} \frac{(1 + \kappa z)^2 dz}{z} \left[L + \frac{1}{3\kappa} - \frac{1}{3\kappa(\kappa z + 1)^3} \right] dz \quad . \tag{5.148}$$

Solving for this integral provides the time required:

$$t = -\frac{8\mu}{\rho g R_1^2} \left\{ L \ln \frac{z_f}{z_{eq}} + \frac{1}{3\kappa} \ln \frac{1 + \kappa z_f}{1 + \kappa z_{eq}} + \right.$$

$$\left. + \left(L + \frac{1}{3\kappa} \right) \left[\frac{\kappa^2}{2}(z_f^2 - z_{eq}^2) + 2\kappa(z_f - z_{eq}) \right] \right\} = 27.3 \ s \quad . \tag{5.149}$$

---

3 In the system shown in Fig. 5.9, a truncated-cone piston advances with constant velocity $U = 10$ mm/s inside a cylindrical chamber occupied by a fluid of density $\rho = 800$ kg/m^3 and viscosity $\mu = 50 \cdot 10^{-3}$Pa $\cdot$ s. Considering the position of the piston shown in Fig. 5.9, determine:

1. the fluid flowrate at the outlet,
2. the force required to advance the piston.

Solution

1. The change over time in the fraction of piston volume that is inside the cylinder corresponds to the flowrate that the piston pushes outside of the cylinder. For the given position, this flowrate is equal to:

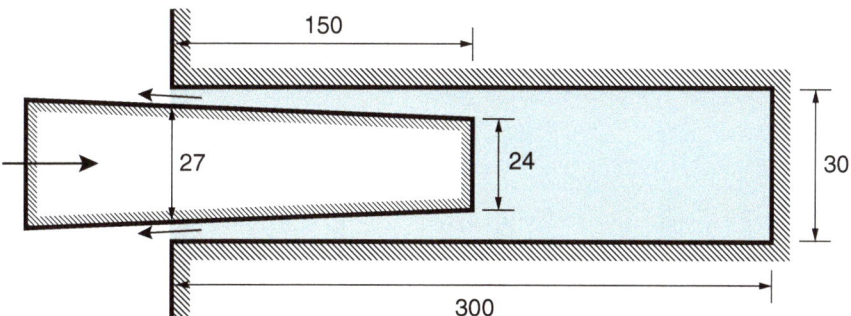

Fig. 5.9 Schematic of a truncated-cone piston that advances at a constant speed and removes fluid from a cylindrical cavity. All dimensions are expressed in mm

$$Q = \frac{dV}{dt} = \pi r^2 \frac{dx}{dt} = \pi r^2 U = 5.72 \cdot 10^{-6} \mathrm{m}^3/\mathrm{s} \quad , \tag{5.150}$$

where $r = 13.5$ mm.

2. For simplicity, it is assumed that the space between the cylinder and the piston can be considered as a plane channel of length $L = 0.15$ m, width equal to the circumference of the cylinder, $W = \pi D_c = 0.094$ m and heights $H_0 = 3 \cdot 10^{-3}$ m and $H_1 = 1.5 \cdot 10^{-3}$ m. The channel height, $H(x)$, changes linearly with the x coordinate, as shown in Fig. 5.9:

$$H(x) = H_0(1 - kx) \quad , \tag{5.151}$$

where $k = (H_0 - H_1)/LH_0 = 3.33$ m^{-1}.

The unidirectional flow in the channel is described by Eq. (5.106):

$$v_x = \frac{1}{2\mu} \frac{d\mathcal{P}}{dx} y^2 + C_1 y + C_2 \quad . \tag{5.152}$$

By applying the boundary conditions:

$$y = 0 \rightarrow v_x = 0 \quad , \tag{5.153}$$

$$y = \delta \rightarrow v_x = -U \quad , \tag{5.154}$$

it is possible to derive the fluid velocity v_x and the specific flowrate Q/W:

$$v_x = -U \frac{y}{H(x)} - \frac{1}{2\mu} \frac{d\mathcal{P}}{dx} y H(x) \left[1 - \frac{y}{H(x)} \right] \quad , \tag{5.155}$$

$$\frac{Q}{W} = \int_0^{H(x)} v_x dy = -\frac{U H(x)}{2} - \frac{H^3(x)}{12\mu} \frac{d\mathcal{P}}{dx} \quad . \tag{5.156}$$

From Eq. (5.156), the pressure gradient can be expressed as:

$$\frac{d\mathcal{P}}{dx} = -12\mu \left[\frac{U}{2H^2(x)} + \frac{Q}{W} \frac{1}{H^3(x)} \right] \quad , \tag{5.157}$$

while, for the pressure inside the cylinder:

$$\mathcal{P}_0 - \mathcal{P}(L) = \frac{12\mu}{H_0^3} \frac{Q}{W} \frac{1}{2k} \left[\frac{1}{(1 - kL)^2} - 1 \right] +$$

$$+ \frac{6U\mu}{H_0^2} \frac{1}{k} \left(\frac{1}{1 - kL} - 1 \right) = 708.5 \text{ Pa} \quad . \tag{5.158}$$

More simply, $\mathcal{P}_0 = 708.5$ Pa if pressure is evaluated with respect to the external pressure, $\mathcal{P}(L) = 0$. The x component of the force F_p acting on the piston can be obtained from the following force balance:

$$F_p = \mathcal{P}_0 A_c - \overline{\tau}_c S_c \quad , \tag{5.159}$$

where the term accounting for the rate of change of the momentum of the liquid has been neglected, $\overline{\tau}_c$ is the average shear stress acting on the cylinder wall, A_c is the cross-sectional area and S_c is the area of the lateral surface of the cylinder in the section wetted by the liquid. To derive $\overline{\tau}_c$, it is convenient to substitute Eq. (5.157) into Eq. (5.155) and obtain for v_x:

$$v_x = -U \frac{y}{H(x)} + 6y \left[1 - \frac{y}{H(x)} \right] \left[\frac{U}{2H(x)} + \frac{Q}{W} \frac{1}{H^2(x)} \right] \cdot \tag{5.160}$$

The shear stress on the cylinder wall is then given by:

$$\tau_c = \mu \frac{\partial v_x}{\partial y} \big|_{y=0} = -\mu \frac{U}{H(x)} + \frac{6\mu}{H(x)} \left[\frac{U}{2} + \frac{Q}{W} \frac{1}{H(x)} \right] \cdot \tag{5.161}$$

Integrating on the wall from 0 to L, the mean value of the shear stress, $\overline{\tau}_c$, can be obtained as:

$$\overline{\tau}_c = \frac{1}{L} \left[\frac{2\mu U}{k H_0} \ln \frac{1}{1 - kL} + \frac{6\mu}{k} \frac{Q}{W} \frac{1}{H_0^2} \left(\frac{1}{1 - kL} - 1 \right) \right] \quad , \tag{5.162}$$

from which the value $\overline{\tau}_c = 4.52$ N/m is calculated. The force acting on the piston is:

$$F_p = 708.5 \frac{\pi (0.03)^2}{4} - 4.52 \cdot 0.15 \cdot \pi \cdot 0.03 = 0.437 \ N \quad . \tag{5.163}$$

This force can be calculated also in a more complicated way by considering all the forces acting on the piston, which are the pressure force acting on the inner face of the piston, with section $A_p = \pi D_p^2/4$ where $D_p = 0.024$ m, the x component of the pressure force acting on the lateral (conical) surface of the piston, and finally the force due to the shear stress, also acting on the lateral surface of the piston:

$$F_p = \mathcal{P}_c A_p + \int_0^L \mathcal{P}(x) \pi \overline{D}_c \tan \alpha dx + \int_0^L \tau(x) \pi \overline{D}_c dx \quad , \tag{5.164}$$

where, for simplicity, $\overline{D}_c$ has been set equal to the average value of the diameter of the piston inside the cylinder, $\overline{D}_c = 0.0255$ m, and $\alpha = 1/100$ is the inclination angle measured with respect to the axis aligned with the lateral surface of the

piston. The expression of $\mathcal{P}(x)$ to be used in Eq. (5.164) can be obtained from
the integration of equation (5.157):

$$\mathcal{P}(x) = \mathcal{P}_0 - \frac{6\mu}{H_0^2 k} \left\{ \frac{Q}{H_0 W} \left[\frac{1}{(1-kx)^2} - 1 \right] + \right.$$

$$\left. + U \left[\frac{1}{1-kx} - 1 \right] \right\} \quad , \tag{5.165}$$

from which, the average value of the pressure, $\overline{\mathcal{P}}$, acting on the wall of the piston
is given by:

$$\overline{\mathcal{P}} = \frac{1}{L} \int_0^L \mathcal{P}(x)\mathrm{d}x \; =$$

$$= \mathcal{P}_0 - \frac{6\mu}{L H_0^2 k} \left[\frac{Q}{H_0 W} \left(\frac{1}{(1-kL)} \frac{1}{k} - L \right) + U \left(\frac{1}{k} \ln \frac{1}{1-kL} - L \right) \right] \; =$$

$$= 467 \, \text{Pa} \quad . \tag{5.166}$$

As shown in Eq. (5.164), the component of the pressure force acting along the x
direction is equal to $\overline{\mathcal{P}} \tan \alpha \simeq \overline{\mathcal{P}} \alpha$. The shear stress acting on the piston wall is
obtained by deriving the velocity profile, Eq. (5.160):

$$\tau_p = \mu \frac{\partial v_x}{\partial y} \big|_{y=H(x)} = \mu \left[\frac{4U}{H(x)} + 6 \frac{Q}{W} \frac{1}{H^2(x)} \right] \; . \tag{5.167}$$

Integrating this expression from 0 to L yields:

$$\overline{\tau}_p = \frac{1}{L} \left\{ \frac{4\mu U}{k H_0} \ln \frac{1}{1-kL} + \frac{6\mu}{k} \frac{Q}{W} \frac{1}{H_0^2} \left[\frac{1}{1-kL} - 1 \right] \right\} =$$

$$= 4.98 \, N/m \quad . \tag{5.168}$$

Finally, the force acting on the piston is:

$$F_p = 708.5 \frac{\pi(0.024)^2}{4} + 4.98 \cdot 0.15 \cdot \pi \cdot 0.0255 +$$

$$+ \; 467 \cdot 0.15 \cdot \pi \cdot 0.0255 \cdot \frac{1}{100} = 0.437 \, N \quad . \tag{5.169}$$

Fig. 5.10 The shaft of a
friction bearing rotates
with angular velocity Ω
and moves laterally to
sustain the applied load,
P. For visualization
purposes, the thickness of
the oil gap is magnified:
the percentage difference
between the radius a of
the shaft and the radius R
of the hub is usually of the
order of $0.1\% \div 0.2\%$

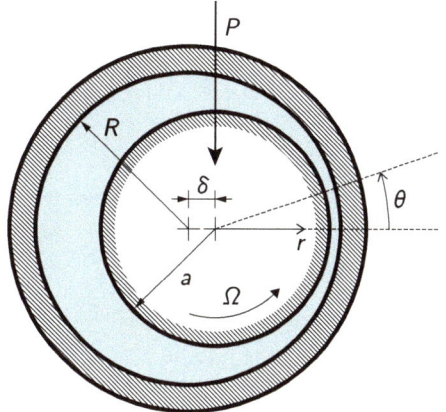

$\boxed{4}$ Consider the shaft-hub system with lubricant shown in Fig. 5.10. Calculate the
eccentricity δ between the shaft and the hub required to sustain the load P applied
on the shaft.

Solution In the system considered, the vertical (downward) load on the shaft is
balanced by a shift of the shaft to the right by an amount equal to δ. When this
happens, the shaft and the hub are not concentric, the gap they form does not have
uniform thickness and a local increase in pressure occurs in the region where the
gap gets thinner. If the system is designed correctly, there will be a specific non-
concentric position of the shaft, for a particular value of the eccentricity δ, for which
the resultant of the pressure forces becomes equal in magnitude and opposite in
direction to the load P without any contact occurring between the shaft and the hub:
in such position, the system is in equilibrium.

To determine an expression for δ at equilibrium, the hub can be considered as
stationary and the shaft as rotating counterclockwise. A Couette-Poiseuille flow is
thus established in the bearing. Since the average thickness of the gap, $\bar{h}$, is very
small compared to the radius a of the shaft,[2] it is possible to use the lubrication
theory solutions already found for the plane geometry, provided that the rotational
speed is sufficiently low. Denoting by Γ the volumetric flowrate of lubricant in the gap
per unit shaft length, the pressure gradient in the gap can be derived from Eq. (5.111):

$$\Gamma = \frac{U H(x)}{2} - \frac{H^3(x)}{12\mu} \frac{d\mathcal{P}}{dx} \quad , \tag{5.170}$$

in which $U = a\Omega$, where Ω is the angular velocity of the shaft, and $x = a\theta$, where
θ is the circumferential coordinate. It can also be assumed that the thickness of the

---

[2] Tipically, $\bar{h}/a \simeq 1/1000$.

gap, $H(x)$, changes according to the relation $H(x) = h(\theta) = \bar{h}(1 - \delta \cos \theta)$. For the pressure gradient, the following expression is obtained:

$$\frac{d\mathcal{P}}{d\theta} = \frac{12\mu a}{h^3} \left(\frac{\Omega a h}{2} - \Gamma \right) \quad . \tag{5.171}$$

Integrating between 0 and 2π yields:

$$\int_{\mathcal{P}(0)}^{\mathcal{P}(2\pi)} d\mathcal{P} = \int_0^{2\pi} \frac{12\mu a}{h^3} \left(\frac{\Omega a h}{2} - \Gamma \right) d\theta = 0 \quad , \tag{5.172}$$

with $\mathcal{P}(0) = \mathcal{P}(2\pi)$. From Eq. (5.172), the specific flowrate of lubricant in the gap can be calculated as:

$$\Gamma = \left[\int_0^{2\pi} \frac{a\Omega}{\bar{h}^2 (1 - \delta \cos \theta)^2} d\theta \right] \left[\int_0^{2\pi} \frac{2}{\bar{h}^3 (1 - \delta \cos \theta)^3} d\theta \right]^{-1} =$$

$$= \frac{\bar{h} a\Omega}{2} \left[\int_0^{2\pi} \frac{d\theta}{(1 - \delta \cos \theta)^2} \right] \left[\int_0^{2\pi} \frac{d\theta}{(1 - \delta \cos \theta)^3} \right]^{-1} =$$

$$= \ldots = \bar{h} a\Omega \frac{1 - \delta^2}{2 + \delta^2} \quad . \tag{5.173}$$

Using this expression for Γ, the pressure distribution in the gap, $\mathcal{P}(\theta)$, can be obtained from Eq. (5.171) as:

$$\int_{\mathcal{P}(0)}^{\mathcal{P}(\theta)} d\mathcal{P} = \frac{12\mu a}{h^3} \int_0^\theta \left(\frac{\Omega a h}{2} - \bar{h} a\Omega \frac{1 - \delta^2}{2 + \delta^2} \right) d\theta \quad , \tag{5.174}$$

from which:

$$\mathcal{P}(\theta) = \mathcal{P}(0) - 6\mu\Omega \left(\frac{a}{\bar{h}} \right)^2 \left[\frac{\delta \sin \theta (2 - \delta \cos \theta)}{(2 + \delta^2)(1 - \delta \cos \theta)^2} \right] \quad . \tag{5.175}$$

The pressure $\mathcal{P}(\theta)$ acts on both the surface of the shaft and the inner wall of the hub, generating a force (referred to as load-carrying capacity) that balances the applied external load.

The pressure distribution given by Eq. (5.175) is shown qualitatively in Fig. 5.11, while Fig. 5.12 shows the distribution as a function of θ for two different values of δ. The pressure in the bottom portion of the bearing, corresponding to values of the angle θ between 0 and π, always exceeds $\mathcal{P}(0)$ while always being lower in the top portion of the bearing, corresponding to values of θ between π and 2π. The pressure distribution in these two portions of the bearing is not symmetric, especially for high

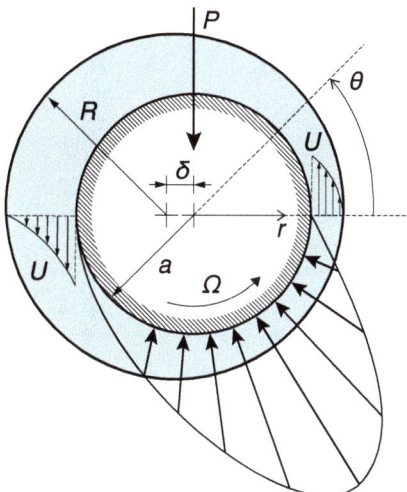

Fig. 5.11 Qualitative behavior of the pressure within the shaft-hub system of a friction bearing, Eq. (5.175). For simplicity, only the behavior on the inner wall of the hub, between $\theta = 0$ and $\theta = \pi$, is shown

values of δ, as also highlighted by the fact that the maximum and minimum values attained by pressure exhibit different magnitude.

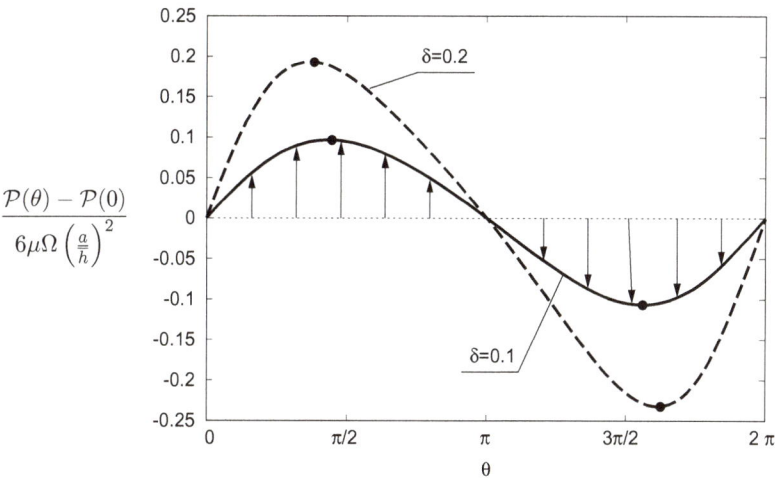

Fig. 5.12 Pressure distribution within the shaft-hub system of a friction bearing (Eq. 5.175). The lines refer to two different values of the displacement δ. Pressure maxima and minima are indicated by the filled circles

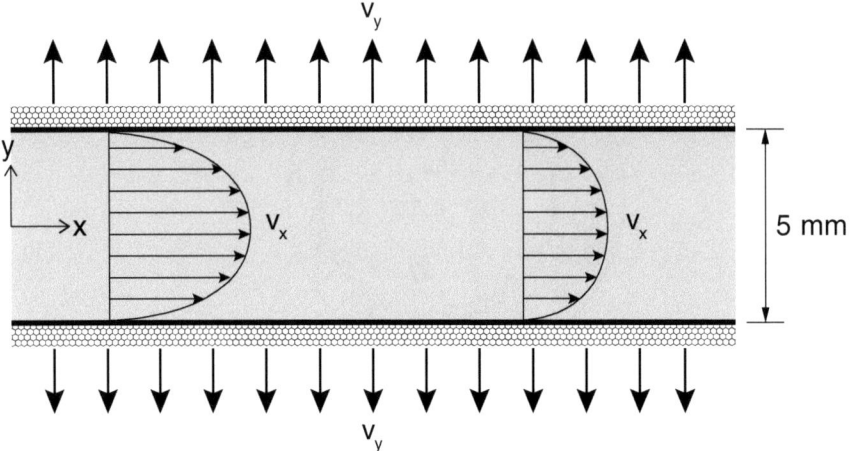

Fig. 5.13 Flow of a viscous fluid driven by an imposed pressure gradient between porous walls

Integrating the pressure distribution allows to calculate the upward force, F, that balances the applied load, P:

$$\frac{F}{W} = \int_0^{2\pi} \mathcal{P}(\theta) a \cos\theta \, d\theta = 12\pi\mu\Omega \frac{a^3}{h^2} \frac{\delta}{\sqrt{1 - \delta^2(2 + \delta^2)}} \quad . \tag{5.176}$$

This equation allows to derive the displacement δ needed to balance P.

⑤ A viscous liquid ($\rho = 800$ kg/m^3, $\mu = 1.2 \cdot 10^{-1}$ Pa $\cdot$ s) flows between two flat walls of width 1 m, placed at a distance of 5 mm, as shown in Fig. 5.13. The inlet flowrate is $Q_0 = 2.5 \cdot 10^{-3}$ m^3/s. The two walls are porous and allow for a lateral outflow that can be characterized by the relation: $v_y = k[\mathcal{P}(x) - \mathcal{P}_0]$ where $k = 2.7 \cdot 10^{-8}$ m^{2}s/kg, $\mathcal{P}(x)$ is the pressure at the position x with $\mathcal{P}(0) = 2 \cdot 10^5$ Pa and $\mathcal{P}_0$ is the atmospheric pressure $\mathcal{P}_0 \simeq 1 \cdot 10^5$ Pa). Determine the flowrate at $x = 1$ m.

Solution Considering the inlet flow conditions, the Reynolds number[3] is:

$$Re = \frac{v D_{eq} \rho}{\mu} = 2\frac{Q_0 \rho}{W \mu} = 35.4 \quad . \tag{5.177}$$

The mean velocity in the flow directions is:

$$\langle v_x \rangle = \frac{Q_0}{H \, W} = 0.5 \text{ m/s} \quad , \tag{5.178}$$

---

[3] In a plane channel with infinite width, the equivalent diameter is equal to twice the distance between the walls

and the velocity v_y at the porous wall is:

$$v_y(0) = k[\mathcal{P}(x) - \mathcal{P}_0] = 2.7 \cdot 10^{-3} \ m/s \ . \tag{5.179}$$

The low value of the Reynolds number and of the ratios H/x and $v_y/\langle v_x \rangle$ make it possible to use the approximations of the lubrication theory. Hence, the inertial terms can be neglected and it can be assumed that $\partial P/\partial y \simeq 0$. Considering also that the velocity and the derivatives with respect to z are zero, the mass conservation equation becomes:

$$\frac{\partial v_x}{\partial x} + \frac{\partial v_y}{\partial y} = 0 \ , \tag{5.180}$$

while the x component of the N-S equations is:

$$0 = -\frac{d\mathcal{P}}{dx} + \frac{\partial^2 v_x}{\partial y^2} \ . \tag{5.181}$$

Since $\mathcal{P} = \mathcal{P}(x)$, Eq. (5.181) can be integrated twice with the boundary conditions:

$$v_x = 0 \ \text{for} \ y = \pm\frac{H}{2} \ , \tag{5.182}$$

to find:

$$v_x = \frac{H^2}{8\mu}\left(-\frac{d\mathcal{P}}{dx}\right)\left[1 - \left(\frac{2y}{H}\right)^2\right] \ . \tag{5.183}$$

From this equation, the average velocity can be calculated as:

$$\langle v_x \rangle = \int_0^H v_x dy = \frac{Q}{HW} = \frac{H^2}{12\mu}\left(-\frac{d\mathcal{P}}{dx}\right) \ . \tag{5.184}$$

Formally, this is the solution of the Poiseuille flow problem, except that the gradient $d\mathcal{P}/dx$ depends on x. Considering the lateral outflow of liquid through the porous walls, the following mass balance equation can be written:

$$\frac{dQ}{dx} = -2Wk\left[\mathcal{P}(x) - \mathcal{P}_0\right] \ . \tag{5.185}$$

Substituting the expression of Q obtained from Eq. (5.184) into Eq. (5.185) yields:

$$\frac{H^3W}{12\mu}\left(-\frac{d^2\mathcal{P}}{dx^2}\right) = -2Wk\left[\mathcal{P}(x) - \mathcal{P}_0\right] \ , \tag{5.186}$$

and:

$$\frac{d^2\mathcal{P}}{dx^2} - \lambda^2\mathcal{P} = -\lambda^2\mathcal{P}_0 \quad , \tag{5.187}$$

with $\lambda^2 = (24\mu k)/H^3 = 0.62\ m^{-2}$.
   The solution of equation (5.187) is:

$$\mathcal{P}(x) = Ae^{-\lambda x} + Be^{\lambda x} + \mathcal{P}_0 \quad , \tag{5.188}$$

where $\mathcal{P}_0$ is a particular solution. The constant B must be zero to keep the solution bounded, and the constant A is determined by considering that $\mathcal{P}(0) = 2\mathcal{P}_0$. Therefore, the equation for the pressure has the following form:

$$\mathcal{P}(x) = \mathcal{P}_0(1 + e^{-\lambda x}) \quad . \tag{5.189}$$

The pressure tends asymptotically to the value $\mathcal{P}_0$. For the pressure gradient, one finds:

$$\frac{d\mathcal{P}(x)}{dx} = -\lambda\mathcal{P}_0 e^{-\lambda x} \quad . \tag{5.190}$$

The pressure gradient tends asymptotically to 0. From Eq. (5.184), the flowrate at $x = 1$ m can be calculated as:

$$Q(x = 1) = \frac{H^3 W}{12\mu}\left(-\frac{d\mathcal{P}(x)}{dx}\right) =$$

$$= \frac{H^3 W}{12\mu}\left(\lambda\mathcal{P}_0 e^{-\lambda x}\right) = 3.11 \cdot 10^{-8}\ m^3/s \quad . \tag{5.191}$$

6 A uniform flowrate of rain, $q = 2 \cdot 10^{-5}\ m^3/m^2 s$, falls on an inclined plane as shown in Fig. 5.14. The falling speed of the drops is $v_p = 10$ m/s and the inclination angle of the plane is $45°$. Assuming that a laminar film of thickness δ is formed from the top edge of the plane ($\delta = 0$ for $x = 0$), determine:

1. the specific flowrate of the film, Γ (in kg/ms), as a function of the x coordinate;
2. the film thickness at position $x = 0.5$ m, assuming zero tangential stresses at the interface between the liquid and the atmosphere;
3. the film thickness at position $x = 0.5$ m, taking into account the momentum transferred from the drops to the liquid film.

Solution

1. To determine the specific flowrate, Γ, along the inclined plane, it is convenient to write the following mass balance:

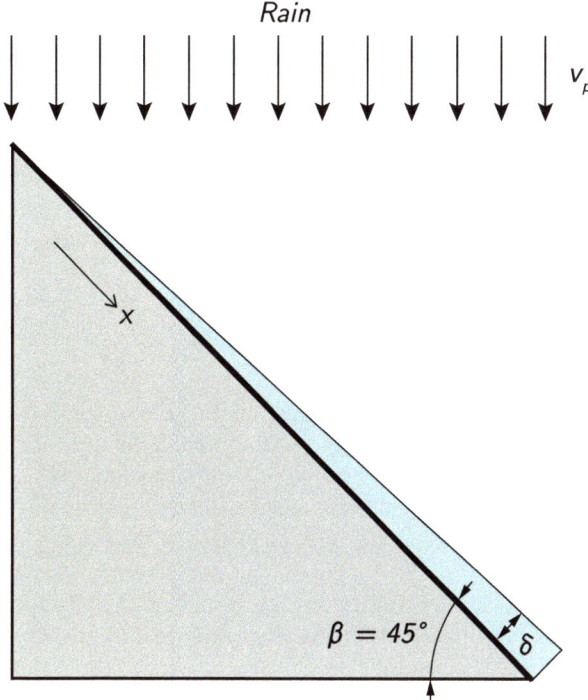

Fig. 5.14 Formation of a liquid film due to rain falling on a rooftop

$$\frac{d\Gamma(x)}{dx} = q \cos \beta \quad , \tag{5.192}$$

where the specific rainfall rate q is given with reference to a horizontal plane, β is the inclination angle of the plane with respect to the vertical direction, and $q \cos \beta$ is the fraction of q that actually contributes to the thickening of the film. The boundary condition that Eq. (5.192) must obey is:

$$x = 0 \quad \longrightarrow \quad \Gamma(0) = 0 \quad , \tag{5.193}$$

and the resulting flowrate of the film is found to increase linearly along the plane as:

$$\Gamma(x) = qx \cos \beta \quad . \tag{5.194}$$

2. In this problem, v_y is different from zero and v_x depends not only on y but also on x. However, it is also observed that the thickness of the film, $\delta(x)$, is such that $\delta(x)/L \ll 1$ and that $v_y/v_x \ll 1$, except in the limit $x \to 0$. it is therefore possible to assume that the relation between the flowrate and the film thickness

is, at each position along the plane, equal to that obtained for constant flowrate and fully-developed flow (Eq. 4.53):

$$\Gamma(x) = \frac{\rho g \delta^3(x) \sin \beta}{3\mu} \quad . \tag{5.195}$$

Recasting this equation with respect to $\delta(x)$ and substituting the expression for $\Gamma(x)$ provided by Eq. (5.194), the film thickness can be obtained as:

$$\delta(x) = \sqrt[3]{\frac{3\mu q x}{\rho g \tan \beta}} \quad . \tag{5.196}$$

For $x = 0.5$ m, the film thickness is equal to 0.145 mm.

3. As the drops fall onto the film, they exert a force in the direction parallel to the film that is equal to the rate of change of their momentum. This force tends to accelerate the film and can be taken into account by applying the resulting shear stress as a boundary condition for the calculation of the velocity profile.

   The momentum of the drops falling on the film in the x direction is equal to $m_p v_{p,x}$ with m_p mass of rain drops transferring momentum to the film and $v_{p,x} = v_p \sin \beta$. The resulting force is equal to:

$$F = \frac{d(m_p v_{p,x})}{dt} = v_{p,x} \frac{dm_p}{dt} = v_p \sin \beta \rho q \cos \beta L W \quad , \tag{5.197}$$

where W is the width of the film. Therefore, the shear stress applied by the drops to the film is:

$$\tau_i = \frac{F}{LW} = v_p \rho q \sin \beta \cos \beta \quad . \tag{5.198}$$

In Chap. 4, the flowrate per unit width for a film with non-zero shear stress at the liquid-air interface was calculated, Eq. (4.180). In the present case, Eq. (4.180) becomes:

$$\Gamma = \frac{\rho g \sin \beta \delta^3}{3\mu} + \frac{\tau_i \delta^2}{2\mu} \quad . \tag{5.199}$$

From this equation, it is straightforward to obtain $\delta = 0.139$ mm for $x = 0.5$ m.

⑦ In an experiment, a jet of oil (having density $\rho = 800$ kg/m^3 and viscosity $\mu = 1.2 \cdot 10^{-1}$ Pa · s) flows out of a vertical conduit, with circular cross section and diameter $D = 2$ cm, and subsequently drips onto the side wall of a cylinder, forming a horizontal film, as shown in Fig. 5.15. The mass flowrate of the jet is equal to $\dot{m} = 0.01$ kg/s and that the radius of the cylinder is $R = 5$ cm.

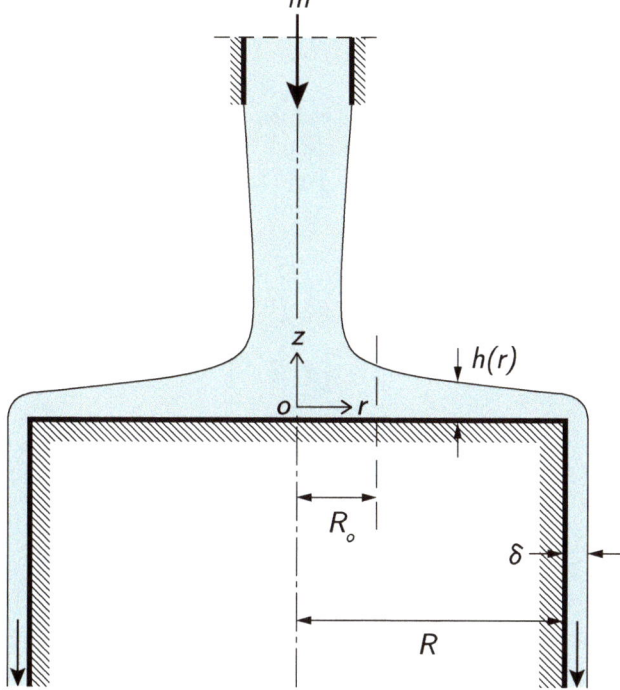

Fig. 5.15 Oil jet dripping on a cylinder

1. Determine the velocity profile in the film.
2. Determine the thickness of the film on the cylinder for $r = R_0 = 0.5$ cm assuming that for $r = R$ the thickness is equal to that of the film forming on the side wall of the cylinder.
3. Determine the value of the shear stress at the wall, τ_w, for $r = R_0$ and $r = R$.

Solution

1. The average velocity of the jet flowing out from the conduit is:

$$\langle v \rangle = \frac{4\dot{m}}{\rho \pi D^2} = 0.04 \text{ m/s} \quad , \tag{5.200}$$

which corresponds to a Reynolds number:

$$Re = \frac{\rho \langle v \rangle D}{\mu} = 5.3 \quad . \tag{5.201}$$

The thickness of the film dripping along the side wall of the cylinder can be calculated through Eq. (4.176):

$$\delta = \sqrt[3]{\frac{3\mu\Gamma}{\rho^2 g}} = \sqrt[3]{\frac{3\mu\dot{m}}{\rho^2 g 2\pi R}} = 1.22 \cdot 10^{-3} \text{ m} \quad . \tag{5.202}$$

The low value of Reynolds number and the small ratio δ/R make it possible to use the lubrication theory approximations. Given the geometry of the problem, cylindrical coordinates are considered and the conservation equations describing the oil flow in the $r - z$ plane are:

$$\frac{1}{r}\frac{\partial}{\partial r}(r v_r) + \frac{\partial v_z}{\partial z} = 0 \quad , \tag{5.203}$$

$$0 = -\frac{\partial P}{\partial r} + \mu\left[\frac{\partial}{\partial r}\left(\frac{1}{r}\frac{\partial}{\partial r}(r v_r)\right) + \frac{\partial^2 v_r}{\partial z^2}\right] \quad , \tag{5.204}$$

$$0 = -\frac{\partial P}{\partial z} \quad , \tag{5.205}$$

having as well applied the symmetry condition $\partial \bullet /\partial\theta = 0$, the steady-state condition ($\partial \bullet /\partial t = 0$) and considered radial flow with $v_\theta = 0$ and $v_z \ll v_r$. A further simplification of Eq. (5.204) is possible if the second derivatives are analyzed in dimensionless form. By setting $\tilde{v}_r = v_r/V$, with V velocity characteristic of the film, $\tilde{r} = r/R$ and $\tilde{z} = z/\delta$, the second derivatives in equation (5.204) can be written as:

$$\frac{\partial}{\partial r}\left(\frac{1}{r}\frac{\partial}{\partial r}(r v_r)\right) + \frac{\partial^2 v_r}{\partial z^2} = \frac{\partial}{\partial\tilde{r}}\left(\frac{1}{\tilde{r}}\frac{\partial}{\partial\tilde{r}}(\tilde{r}\tilde{v}_r)\right)\frac{V}{R^2} + \frac{\partial^2\tilde{v}_r}{\partial\tilde{z}^2}\frac{V}{\delta^2} \quad , \tag{5.206}$$

with $V/R^2 \ll V/\delta^2$. It follows that:

$$\frac{\partial}{\partial r}\left(\frac{1}{r}\frac{\partial}{\partial r}(r v_r)\right) \ll \frac{\partial^2 v_r}{\partial z^2} \quad . \tag{5.207}$$

Integration of equation (5.204), taking into account equation (5.207) and applying the boundary conditions $v_r(r, z = 0) = 0$ and $\tau_{zr}(r, z = h(r)) = 0$ with $h(r)$ film thickness on the cylinder, gives:

$$v_r = \frac{1}{2\mu}\frac{\partial P}{\partial r}\left[z^2 - 2h(r)z\right] \quad , \tag{5.208}$$

with $\partial P/\partial r = dP/dr$ being $P = P(r)$. This expression is valid between $r = R_0$ and $r = R$. The expression of dP/dr can be derived observing that the pressure at each location on the free surface of the film is equal to the atmospheric pressure, p_{atm}. Therefore, according to the equation of statics applied in the z direction, the pressure at a generic point inside the film is $p(r, z) = p_{atm} + \rho g[h(r) - z]$, with

$R_0 \leq r \leq R$. For a fixed z position within the film, the pressure decreases as the coordinate r increases. This because h decreases as r increases. As a consequence:

$$\frac{\partial P}{\partial r} = \frac{\partial p}{\partial r} = \rho g \frac{dh}{dr} \quad . \tag{5.209}$$

Equation (5.209) indicates that the flow of oil towards the periphery of the cylinder is not due to a continuity effect, but rather to a non-uniform distribution of hydrostatic pressure in the radial direction. This is further confirmed by the fact that, since $dh/dr < 0$, the pressure gradient is negative.

2. To calculate the thickness, it is sufficient to obtain the expression for the film flowrate by integrating the velocity profile. In terms of mass flowrate:

$$w = \int_0^{2\pi} \int_0^{h(r)} \rho v_r(r, z) r \, d\theta \, dz = -\frac{2\pi \rho^2 g}{3\mu} r h^3(r) \frac{dh(r)}{dr} \quad . \tag{5.210}$$

Since w is known, Eq. (5.210) can be used to calculate the thickness as a function of r:

$$h^3 dh = -\underbrace{\frac{3\mu w}{2\pi \rho^2 g}}_{const.} \frac{1}{r} dr \rightarrow \int_{h(R_0)}^{h(r)} h^3 dh = -\frac{3\mu w}{2\pi \rho^2 g} \int_{R_0}^r \frac{1}{r} dr \quad . \tag{5.211}$$

After integration:

$$h(r) = \sqrt[4]{h^4(R) + \frac{6\mu w}{\pi \rho^2 g} \ln\left(\frac{R}{r}\right)} \quad , \tag{5.212}$$

with $h(R) = \delta$. At $r = R_0$ the film thickness is $h(R_0) = 5.39 \cdot 10^{-3}$ m.

3. The shear stress on the surface of the cylinder is $\tau_w(r) = \tau_{zr}(r, z = 0)$ with:

$$\tau_{zr} = \mu \left(\frac{\partial v_z}{\partial r} + \frac{\partial v_r}{\partial z}\right) \quad . \tag{5.213}$$

By expressing the two derivatives in dimensionless form, it can be easily verified that $\partial v_z/\partial r << \partial v_r/\partial z$ and, therefore:

$$\tau_{zr}(r, z) = \rho g \frac{dh}{dr} (z - h(r)) \quad , \tag{5.214}$$

where:

$$\frac{dh}{dr} = -\frac{3\mu w}{2\pi \rho^2 g} \frac{1}{r h^3(r)} \quad , \tag{5.215}$$

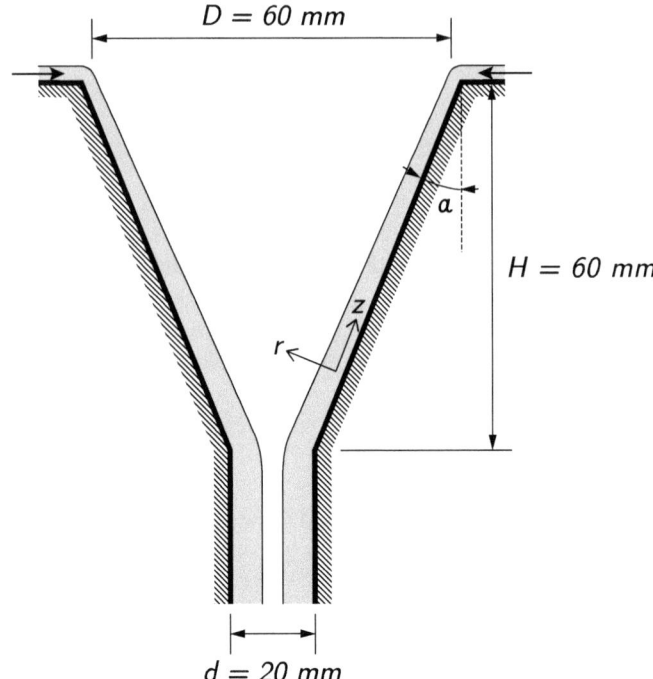

Fig. 5.16 Schematic of a funnel into which a liquid film flows with constant flowrate

using Eq. (5.210). The wall shear stress is:

$$\tau_w(r) = \tau_{z,r}(r, z = 0) = \frac{3\mu w}{2\pi\rho} \frac{1}{r h^2(r)} \quad , \tag{5.216}$$

from which $\tau_w(r = R_0) = 4.93 \ N/m^2$ and $\tau_w(r = R) = 9.62 \ N/m^2$.

$\boxed{8}$ A funnel having the dimensions (expressed in millimeters) shown in Fig. 5.16 is uniformly fed at the top edge by a constant flowrate of 0.05 kg/s of a fluid of density $\rho = 800 \ kg/m^3$ and viscosity $\mu = 10^{-2} Pa \cdot s$. Assuming that the liquid forms a thin laminar film:

1. determine which simplifications can be made;
2. calculate the film thickness in the final section of the funnel.

Solution

1. The problem can be solved using the r, z coordinate system shown in Fig. 5.16.

Because of symmetry considerations, it can be assumed that the velocity distribution does not depend on θ ($\partial \bullet /\partial\theta = 0$), and that the velocity component v_θ is zero. Furthermore, assuming a thin film, and that we are sufficiently far from the bottom opening of the funnel, one can assume $\partial P/\partial r \simeq 0$.

With these assumptions, it is possible to consider the motion of the film in the funnel as equivalent to the motion of a film on an inclined plane, with $\beta = \pi/2 - \alpha$ the inclination angle. We can also assume that, for each position z, the relation between the flowrate and the film thickness is equal to the relation that would apply to the case of constant specific flowrate.

2. The film thickness in the terminal section of the funnel can be calculated using Eq. (4.52):

$$w = \frac{\rho g W \delta^3(z) \cos \alpha}{3\nu} , \qquad (5.217)$$

where W is the spanwise extent of the film:

$$W = 2\pi z \sin \alpha , \qquad (5.218)$$

and:

$$w = \frac{\rho g 2\pi z \sin \alpha \, \delta^3(z) \cos \alpha}{3\nu} . \qquad (5.219)$$

The film thickness as a function of z is:

$$\delta(z) = \sqrt[3]{\frac{3\nu w}{\rho g 2\pi z \cos \alpha \sin \alpha}} . \qquad (5.220)$$

Based on the dimensions shown in the figure, it can be derived that the radius of the outlet section of the funnel is $R = z \sin \alpha = 10$ mm. Furthermore, $\alpha = \tan^{-1}(20/60) \simeq 18°$. For these values, the thickness is $\delta = 1.58$ mm.

Problems

$\boxed{a}$ The axial bearing shown in Fig. 5.17 is fed with oil of density $\rho = 850 \, \text{kg/m}^3$ and viscosity $\mu = 0.02$ Pa $\cdot$ s. Let $Q = 1.26 \cdot 10^{-7} \, \text{m}^3/\text{s}$ be the oil flowrate, $h = 0.2$ mm the (fixed) distance between the two disks and $\mathcal{P} = 10^5$ Pa the pressure outside the gap between the disks. Neglecting inlet effects, determine the force F that the bearing can sustain for the assigned values of h and Q.

$\boxed{b}$ Consider an incompressible Newtonian fluid (viscosity $\mu = 6.0 \cdot 10^{-2}$ Pa $\cdot$ s, density $\rho = 1.6 \cdot 10^3 \, \text{kg/m}^3$) that flows radially in a casting mold and fills the space between two disks as shown in Fig. 5.18. Let the pressure at the inlet be $\mathcal{P} = 1.1 \cdot 10^5 \, Pa$. Using for the velocity distribution v_r the expression calculated in Sect. 5.2.3, Exercise 2, calculate the time required to fill the volume of the casting mold, neglecting the effects of flow redistribution near the inlet and assuming that air outlets are present to keep the air in the casting mold at atmospheric pressure.

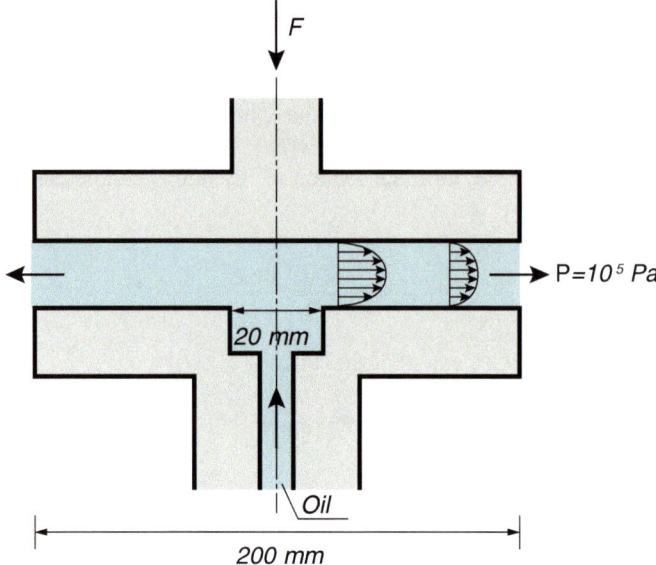

Fig. 5.17 Schematic of axial bearing in which a purely radial flow is established. The velocity profile changes (while maintaining parabolic shape) as the oil flows towards the periphery of the bearing

$\boxed{c}$ A viscous fluid ($\rho = 750$ kg/m^3, $\mu = 0.2$ Pa $\cdot$ s) flows in the truncated-cone pipe with circular cross section shown in Fig. 5.19 The flowrate is $w = 2$ kg/s.

1. Determine the approximate solution of the *N-S* equations for $Re \to 0$.
2. Verify whether the approximate solution found is valid for less restrictive conditions than those given at point 1.

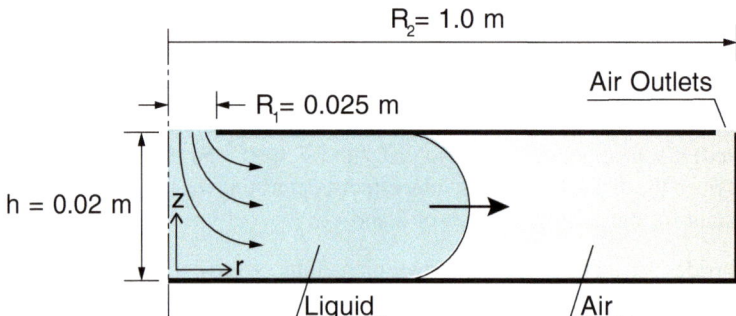

Fig. 5.18 Filling of a casting mold with liquid flowing radially outward, in creeping flow conditions

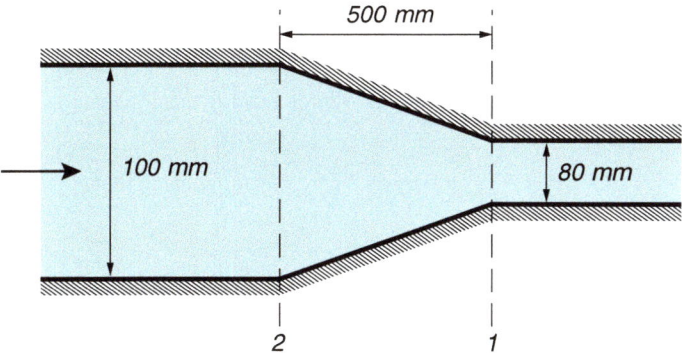

Fig. 5.19 Creeping flow in a truncated-cone pipe

3. Calculate the pressure drop between inlet and outlet in the truncated-cone section of the pipe, using the obtained solution of the *N-S* equations.

$\boxed{d}$ A liquid flows vertically on a solid cone. The flowrate Q is constant and falls uniformly along the lateral surface of the cone, thus forming a film of variable thickness s, which changes with the distance from the vertex of the cone. Given the values of viscosity, μ and density, ρ, of the liquid as well as the half-opening angle of the cone, α, determine the film thickness evolution for laminar motion conditions.

Chapter 6
Approximate Solutions for High Reynolds Number Flows

6.1 Potential Flow

In the previous chapters, the governing equations for Newtonian and incompressible fluids have been applied to internal flows, namely flows in which the fluid moves between solid walls (for example, Couette or Poiseuille motion in a channel or pipe). Many problems of practical interest, however, are characterized by external flows, namely flows over bodies immersed in an unbounded fluid (for example, the flow around cars, airplanes, ships, and submarines).

Considering the general case of flow around an immersed body, like the airfoil shown in Fig. 6.1 for example, three regions can be distinguished: an outer region of *potential flow*, in which viscous effects do not influence the behavior of the fluid and no velocity gradients are present if the incident velocity profile is uniform; a region called *wake*, in which viscous effects are still negligible but the motion of the fluid is affected by the presence of velocity gradients; and a region close to the body, called *boundary layer*, in which the behavior of the fluid is strongly influenced by the action of viscous stresses and by the resulting velocity gradients.

This chapter analyzes the behavior of a fluid under potential flow conditions while Chap. 7 is devoted to the description of the boundary layer.

6.2 Vorticity

The motion of a fluid element can be described as the combination of elementary motions of translation, rotation and deformation. Consider, for simplicity, a two-dimensional fluid element of rectangular shape at time $t = 0$. At time $t + dt$ the element will have undergone a translation to a new position of coordinates $v_x dt$, $v_y dt$,

© The Author(s), under exclusive license to Springer Nature Switzerland AG 2024
A. Soldati and C. Marchioli, *Fluid Mechanics for Mechanical Engineers*,
https://doi.org/10.1007/978-3-031-53950-3_6

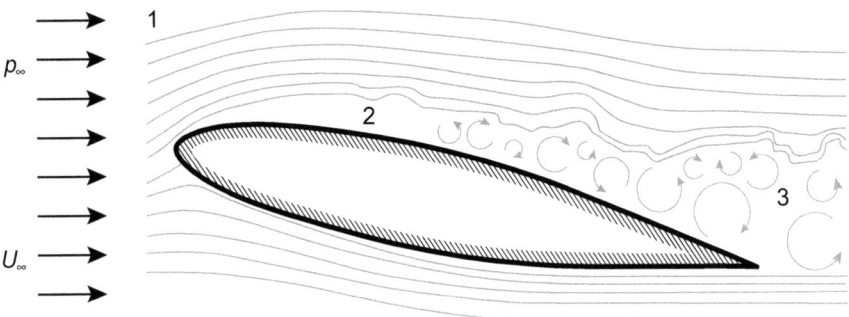

Fig. 6.1 Flow regions around a solid body, in this case an airfoil. Region 1: outer flow (where potential flow conditions apply), Region 2: boundary layer, Region 3: wake. The velocity U_∞ represents the relative velocity between the undisturbed stream and the immersed body. The pressure p_∞ represents the pressure of the undisturbed stream

but also deformation and rotation as shown in Fig. 6.2. In this figure, the angle $d\alpha$ is given by:

$$d\alpha = \tan^{-1}\left[\frac{(A'B')_y}{(A'B')_x}\right] , \qquad (6.1)$$

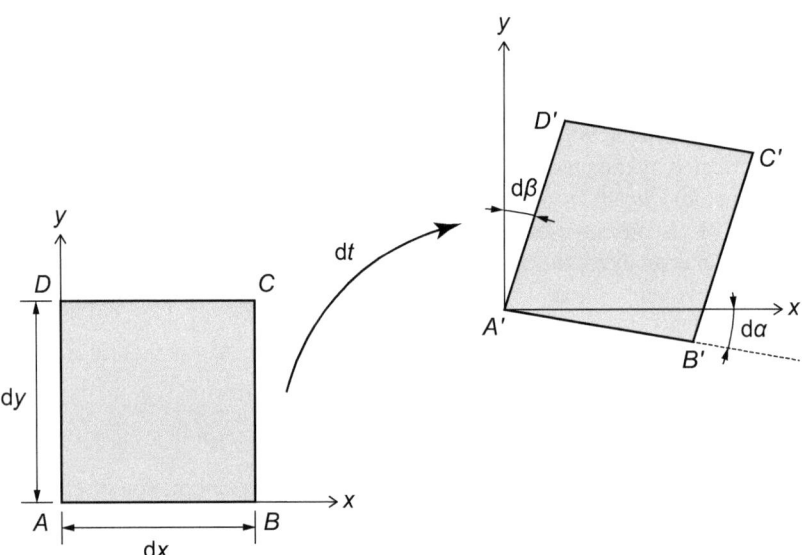

Fig. 6.2 Rotation and deformation of a moving fluid element

where $(A'B')_y$ and $(A'B')_x$ are the x and y components of the segment $A'B'$, which can be expressed as:

$$(A'B')_y = \left(v_y \mid_{x+dx} - v_y \mid_x\right) dt \quad , \tag{6.2}$$

$$(A'B')_x \approx dx \quad . \tag{6.3}$$

This leads to write:

$$d\alpha = \tan^{-1}\left(\frac{\partial v_y}{\partial x} dt\right) \cong \frac{\partial v_y}{\partial x} dt \quad , \tag{6.4}$$

and:

$$\dot{\alpha} = \frac{\partial v_y}{\partial x} \quad . \tag{6.5}$$

Repeating the same procedure for the angle $d\beta$ yields:

$$\dot{\beta} = \frac{\partial v_x}{\partial y} \quad . \tag{6.6}$$

In Fig. 6.2, α represents a counter-clockwise rotation while β represents a clockwise rotation. Therefore, the average counter-clockwise rotation of the fluid element is given by:

$$\frac{1}{2}(\dot{\alpha} - \dot{\beta}) = \frac{1}{2}\left(\frac{\partial v_y}{\partial x} - \frac{\partial v_x}{\partial y}\right) \quad . \tag{6.7}$$

Similarly, the deformation of the fluid element is equal to:

$$\frac{1}{2}(\dot{\alpha} + \dot{\beta}) = \frac{1}{2}\left(\frac{\partial v_y}{\partial x} + \frac{\partial v_x}{\partial y}\right) \quad . \tag{6.8}$$

Note that the deformation given by Eq. (6.8) depends on the tangential stresses τ_{yx} or τ_{xy}, defined by Eq. (3.39).

The *vorticity* of a fluid element is defined as the *curl* of the velocity vector:

$$\omega = \nabla \times \mathbf{v} = \begin{vmatrix} \mathbf{i} & \mathbf{j} & \mathbf{k} \\ \frac{\partial}{\partial x} & \frac{\partial}{\partial y} & \frac{\partial}{\partial z} \\ v_x & v_y & v_z \end{vmatrix} = \tag{6.9}$$

$$= \mathbf{i}\underbrace{\left(\frac{\partial v_z}{\partial y} - \frac{\partial v_y}{\partial z}\right)}_{\omega_x} + \mathbf{j}\underbrace{\left(\frac{\partial v_x}{\partial z} - \frac{\partial v_z}{\partial x}\right)}_{\omega_y} + \mathbf{k}\underbrace{\left(\frac{\partial v_y}{\partial x} - \frac{\partial v_x}{\partial y}\right)}_{\omega_z} \quad . \tag{6.10}$$

When the curl operator is applied to vector $\mathbf{v}$, the i-th component of the resulting vector $\boldsymbol{\omega}$ can be expressed as:

$$\omega_i = \frac{\partial v_k}{\partial x_j} - \frac{\partial v_j}{\partial x_k} \quad . \tag{6.11}$$

This equation, combined with Eq. (6.7), shows that the vorticity is equal to twice the velocity with which the fluid element rotates around its axis.

It should be noted that a fluid element can move along a circular trajectory and have zero vorticity since the rotation occurs around the axis of the element. Similarly a fluid element can move along a trajectory that is rectilinear and have non-zero vorticity. To understand these concepts it is helpful to consider the following examples.

1. Fluid in rigid body rotation with angular velocity Ω. The velocity field is:

$$\mathbf{v} = \Omega \times \mathbf{r} \quad . \tag{6.12}$$

Let z be the axis of rotation. Then, $\Omega = \Omega_z \mathbf{k}$ and $\mathbf{r} = x\mathbf{i} + y\mathbf{j}$, and:

$$\mathbf{v} = -\Omega_z(y\mathbf{i} - x\mathbf{j}) \quad , \tag{6.13}$$

$$\nabla \times \mathbf{v} = \omega_z \mathbf{k} = 2\Omega_z \mathbf{k} \quad , \tag{6.14}$$

and the vorticity is equal to twice the angular velocity at which the fluid rotates, even in the absence of relative motion between the fluid elements.

2. Fluid rotating with the following velocity field:

$$v_\varphi = \frac{k}{r} \quad , \quad v_r = v_z = 0 \quad . \tag{6.15}$$

This is the case of torsional flow already examined in Chap. 4. In this case, the motion is transmitted between fluid elements by the viscous forces. By taking the curl of this flow field in cylindrical coordinates, the vorticity along the z axis can be obtained:

$$r \neq 0 \longrightarrow \omega_z = \frac{1}{r}\frac{\partial}{\partial r}(r v_\varphi) = 0 \quad . \tag{6.16}$$

To understand the difference between cases 1 and 2, it may be helpful to refer to Fig. 6.3.

3. Unidirectional plane flow (Couette or Poiseuille flow):

$$v_x = v_x(y) \qquad\qquad v_y = v_z = 0 \quad . \tag{6.17}$$

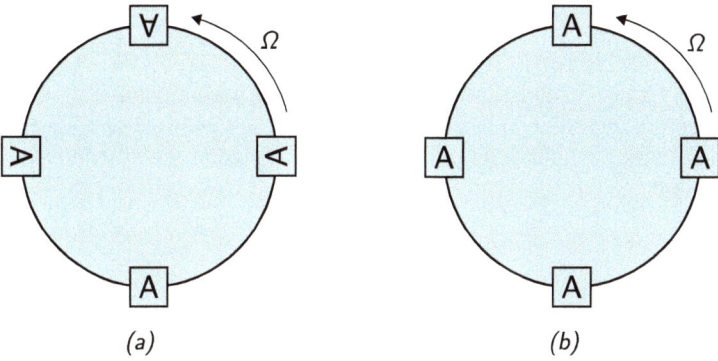

Fig. 6.3 Difference between rotation of a rigid body (**a**), and rotation of a fluid (**b**)

The vorticity w_z is equal to:

$$w_z = -\frac{\partial v_x}{\partial y} \neq 0 \quad . \tag{6.18}$$

In this case, the fluid elements move along a rectilinear trajectory as elementary rotations are compensated by elementary deformations.

6.2.1 Examples

1 Consider the following two-dimensional velocity field:

$$\mathbf{v} = \begin{cases} v_1 = cx_1 \\ v_2 = -cx_2 \end{cases},$$

with c a known constant. Determine the expression of the vorticity, ($\boldsymbol{\omega} = \nabla \times \mathbf{v}$).

Solution The vorticity field has the following components:

$$\boldsymbol{\omega} = \left[c\frac{\partial x_2}{\partial x_3}, c\frac{\partial x_1}{\partial x_3}, -c\left(\frac{\partial x_2}{\partial x_1} + \frac{\partial x_1}{\partial x_2}\right) \right] \quad . \tag{6.19}$$

Since $(\partial x_i)/(\partial x_j) = 0$ for $i \neq j$ with $i, j = 1, 2, 3$, it follows that $\boldsymbol{\omega} = (0, 0, 0)$.

6.3   Vorticity Transport Equation

The vorticity transport equation describes the time evolution of vorticity within a flow and is therefore useful for understanding the mechanisms of rotation and deformation that characterize the fluid during its motion. This equation is easily derived from the *N-S* equations.

6.3.1   Three-Dimensional Steady Flow

When inertial effects prevail, the *N-S* equations for three-dimensional, incompressible, steady flow can be written in the following dimensionless form:

$$(\mathbf{v} \cdot \nabla)\mathbf{v} = -\nabla \mathcal{P} + \frac{1}{Re}\nabla^2\mathbf{v} \quad . \tag{6.20}$$

Applying the curl operator to this equation gives:

$$\nabla \times [(\mathbf{v} \cdot \nabla)\mathbf{v}] = -\nabla \times \nabla \mathcal{P} + \frac{1}{Re}\nabla \times \nabla^2\mathbf{v} \quad . \tag{6.21}$$

The following vector identities can be used to simplify Eq. (6.21):

$$(\mathbf{v} \cdot \nabla)\mathbf{v} = \nabla \left(\frac{1}{2}v^2\right) - \mathbf{v} \times \boldsymbol{\Omega} \quad , \tag{6.22}$$

from which it follows that:

$$\nabla \times [(\mathbf{v} \cdot \nabla)\mathbf{v}] = \nabla \times \left[\nabla \left(\frac{1}{2}v^2\right)\right] - \nabla \times (\mathbf{v} \times \boldsymbol{\Omega}) \quad . \tag{6.23}$$

In this equation, the first term on the right-hand side represents the curl of a gradient and is therefore zero. For the same reason, the term $\nabla \times \nabla \mathcal{P}$ in Eq. (6.21) is zero. The second term on the right-hand side can instead be rewritten as:

$$\nabla \times (\mathbf{v} \times \boldsymbol{\Omega}) = \mathbf{v}(\nabla \cdot \boldsymbol{\Omega}) - \boldsymbol{\Omega}(\nabla \cdot \mathbf{v}) - (\mathbf{v} \cdot \nabla)\boldsymbol{\Omega} + (\boldsymbol{\Omega} \cdot \nabla)\mathbf{v} \quad . \tag{6.24}$$

In this equation, the first two terms on the right-hand side are zero because:

$$\nabla \cdot \boldsymbol{\Omega} = \nabla \cdot (\nabla \times \mathbf{v}) = 0 \quad , \tag{6.25}$$

since the divergence of the *curl* of any vector is zero, and $\nabla \cdot \mathbf{v} = 0$ from the assumption of incompressible fluid. Noting that:

$$\nabla \times \nabla^2 \mathbf{v} = \nabla^2 (\nabla \times \mathbf{v}) = \nabla^2 \mathbf{\Omega} \quad , \tag{6.26}$$

equation (6.21) can be rewritten as:

$$(\mathbf{v} \cdot \nabla)\boldsymbol{\omega} = (\boldsymbol{\omega} \cdot \nabla)\mathbf{v} + \frac{1}{Re}\nabla^2 \boldsymbol{\omega} \quad . \tag{6.27}$$

This equation describes the transport of vorticity by a three-dimensional, incompressible, steady flow. In addition to the convective term, $(\mathbf{v} \cdot \nabla)\boldsymbol{\omega}$, and the diffusive term, $Re^{-1}\nabla^2 \boldsymbol{\omega}$, the equation has an additional term, $(\boldsymbol{\omega} \cdot \nabla)\mathbf{v}$, called *stretching/tilting term*, which accounts for the self-amplification and redistribution of vorticity.

6.3.2   Self-amplification and Distribution of Vorticity

The physical meaning of the stretching/tilting term can be understood by rewriting it in tensor notation, $\omega_j \partial v_i / \partial x_j$, with $i, j = 1, ..., 3$. Two cases may be considered:

Case (1) $i = j$: the term represents a mechanism of self-amplification (*stretching*) of the vorticity associated with a given fluid element along a certain direction of motion. The self-amplification is produced by the velocity gradients in that direction, even in the absence of external forcing. Considering, for example, $i = j = 1$, the stretching/tilting term is $\omega_1 \partial v_1 / \partial x_1$. With reference to Fig. 6.4a, if $\partial v_1 / \partial x_1 > 0$, then the variation of velocity v_1 in the direction x_1 results in a stretching of the fluid element and, by conservation of angular momentum, causes an increase of ω_1, without producing any change in the direction of vorticity. If, on the other hand, $\partial v_1 / \partial x_1 < 0$, then the variation of velocity v_1 in the x_1 direction results in a compression of the fluid element and in a decrease of ω_1. The self-amplification mechanism is analogous to the mechanism by which ice skaters may vary their angular velocity during a rotation.

Case (2) $i \neq j$: the term represents a mechanism of redistribution (*tilting*) of vorticity from one direction of motion to another, caused by the velocity gradients even in the absence of external forcing. Consider, for example, $i = 1$ and $j = 2$. The stretching/tilting term is $\omega_2 \partial v_1 / \partial x_2$. With reference to Fig. 6.4b, if $\partial v_1 / \partial x_2 > 0$, then the variation of velocity v_1 in the direction x_2 results in a deformation of the fluid element associated with the clockwise rotation of the vorticity component ω_2. This causes a decrease of ω_2 and a corresponding redistribution of positive vorticity along the x_1 direction. On the other hand, if $\partial v_1 / \partial x_2 < 0$, then the rotation is counterclockwise and the decrease of ω_2 corresponds to a redistribution of negative vorticity along the x_1 direction.

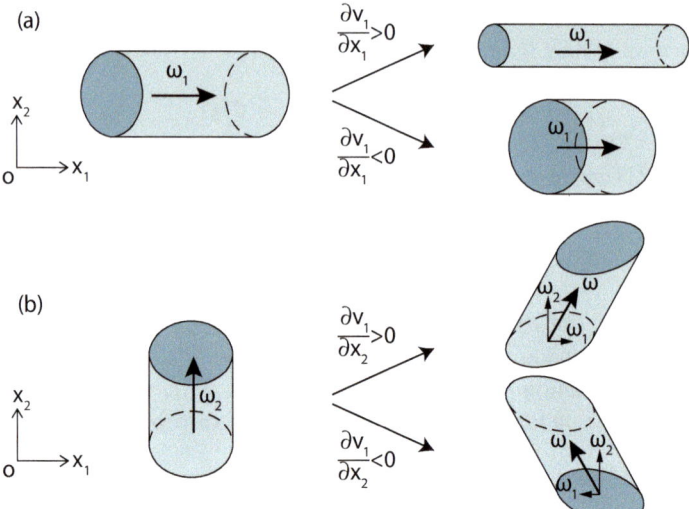

Fig. 6.4 Mechanisms of **a** self-amplification (stretching) and **b** redistribution (tilting) of vorticity

6.3.3 Baroclinicity (Density Variation Effects)

In deriving the vorticity transport equation, the analysis conducted so far has focused on the case of an incompressible flow, with density uniform in space and constant in time. However, many cases of practical interest, such as the motion of large masses of air in the atmosphere or water in oceans and lakes, are characterized by significant fluid density gradients. These gradients are produced by pressure and density changes within the fluid. In cases when the isolines of pressure and the isolines of density are not aligned, the fluid is called baroclinic, instabilities may occur and vorticity can be generated.

Non-zero density gradients imply that, in the vorticity transport equation, the term associated with the pressure gradient is no longer zero. Starting from the *N-S* equations for fluids with variable density and assuming that changes in density do not affect the continuity equation, it can be shown (detailed derivation is omitted) that Eq. (6.27) becomes:

$$(\mathbf{v} \cdot \nabla)\boldsymbol{\omega} = (\boldsymbol{\omega} \cdot \nabla)\mathbf{v} + \frac{1}{Re}\nabla^2\boldsymbol{\omega} + \frac{\nabla\rho \times \nabla\mathcal{P}}{\rho^2} \quad . \tag{6.28}$$

The last term in this equation is called *baroclinicity* and represents a measure of the stratification of a fluid, in other words, the difference in orientation between isolines of pressure and density. As can be seen, this term is zero when the density gradient $\nabla\rho$ is aligned with the pressure gradient $\nabla\mathcal{P}$, namely when the constant-density surfaces (isopycnals) are parallel to the constant-pressure surfaces (isobars). When

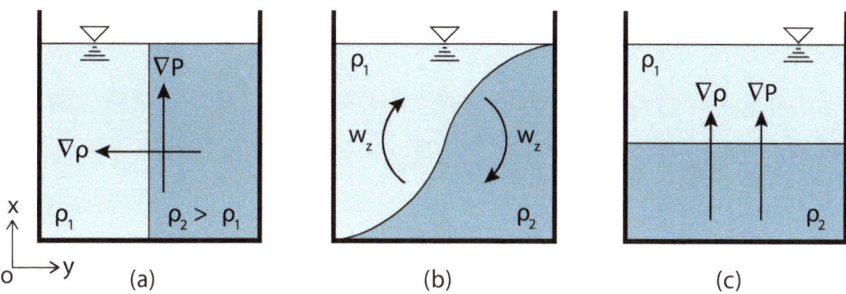

Fig. 6.5 Generation of baroclinic vorticity by density difference

this is not the case, baroclinicity contributes to the generation of vorticity, which is maximum when the two gradients are orthogonal to each other ($\nabla \rho \perp P$). Areas of high atmospheric baroclinicity are characterized by the occurrence of strong fluid dynamic instabilities which are the main mechanism for the formation of mid-latitude cyclones and anticyclones. To understand this mechanism, it is convenient to refer to Fig. 6.5, in which a vessel is filled with two fluids having different densities, initially at rest and separated by a movable wall. Consider, for simplicity, the two-dimensional case.

When the wall is removed, a density gradient is generated at the interface between the two fluids (Fig. 6.5a). This gradient is orthogonal to the static pressure gradient, and this condition leads to the production of baroclinic vorticity within the vessel, resulting in rotation and mixing of the fluids (Fig. 6.5b). The process stops only when the system reaches a new equilibrium configuration in which the two gradients are parallel (Fig. 6.5c).

6.3.4 Two-Dimensional Steady Flow

For a two-dimensional system, the scalar product $\boldsymbol{\Omega} \cdot \nabla$ is zero and the stretching/tilting mechanisms just described are absent. Furthermore, the only non-zero component of the vorticity vector is the one normal to the plane of motion. Equation (6.27) reduces to the scalar equation:

$$(\mathbf{v} \cdot \nabla)\omega = \frac{1}{Re}\nabla^2\omega \ , \tag{6.29}$$

where $\boldsymbol{\Omega}$ has been replaced with its magnitude ω. For a real fluid, when $\mu \to 0$, $Re \to \infty$ and Eq. (6.29) reduces to:

$$(\mathbf{v} \cdot \nabla)\omega = 0 \ , \tag{6.30}$$

which implies that the vectors $\mathbf{v}$ and $\nabla\omega$ are orthogonal.

6.4 Stream Function

A function whose gradient is orthogonal to $\mathbf{v}$ is the stream function $\psi(x, y)$, defined as:

$$v_x = -\frac{\partial \psi}{\partial y} \quad , \quad v_y = \frac{\partial \psi}{\partial x} \quad . \tag{6.31}$$

From this definition, it follows that:

$$\nabla \psi = \frac{\partial \psi}{\partial x}\mathbf{i} + \frac{\partial \psi}{\partial y}\mathbf{j} = v_y\mathbf{i} - v_x\mathbf{j} \quad , \tag{6.32}$$

and the scalar product $\mathbf{v} \cdot \nabla \psi$ is zero. By definition of stream function, ψ, it is easy to shown that Eq. (6.31) satisfies the continuity equation for an incompressible fluid and a two-dimensional flow (or even a three-dimensional flow, as long as the flow exhibits symmetry with respect to at least one flow direction). When the stream function is defined in cylindrical coordinates, i.e. $\psi(r, \theta)$, the following relations hold:

$$v_r = -\frac{1}{r}\frac{\partial \psi}{\partial \theta} \quad , \quad v_\theta = \frac{\partial \psi}{\partial r} \quad . \tag{6.33}$$

6.4.1 Streamlines

The stream function is the mathematical equation that describes the *streamlines* of the flow field, namely the lines of constant values of the stream function, ψ. The streamlines are tangential to the velocity vector at each point, since the condition:

$$\psi(x, y) = \text{constant} \quad , \tag{6.34}$$

implies that:

$$d\psi = \frac{\partial \psi}{\partial x}dx + \frac{\partial \psi}{\partial y}dy = 0 \quad , \tag{6.35}$$

and, for the angular coefficient:

$$\left.\frac{dy}{dx}\right|_{\psi=const.} = -\frac{\partial \psi/\partial x}{\partial \psi/\partial y} = \frac{v_y}{v_x} \quad . \tag{6.36}$$

From this condition, it follows that a fluid cannot move across a streamline. A physical interpretation of the stream function can thus be provided by considering two

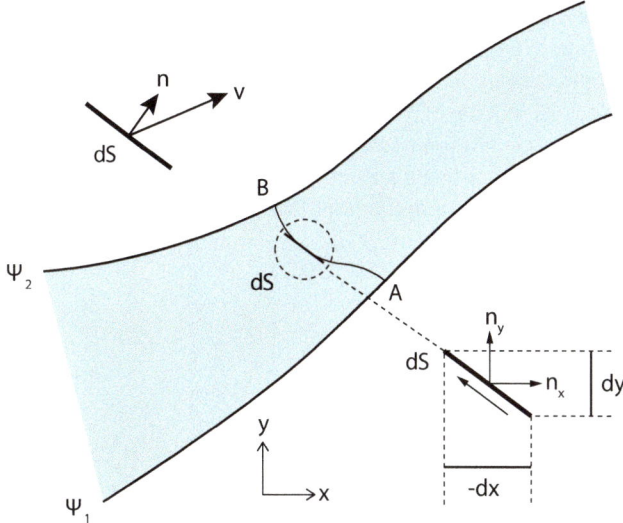

Fig. 6.6 Flowrate between two generic streamlines

generic streamlines $\psi = \psi_1$ and $\psi = \psi_2$, like those shown in Fig. 6.6. The volumetric flowrate Q per unit width W that crosses the generic section $A - B$ is given by the integral:

$$\frac{Q}{W} = \int_A^B \mathbf{v} \cdot d\mathbf{S} = \int_A^B \mathbf{v} \cdot \mathbf{n} dS \quad , \tag{6.37}$$

where $d\mathbf{S} = dS\mathbf{n}$ is an infinitesimal area on section $A - B$ (which can be thought of as a series of subsequent elements $d\mathbf{S}$, each having magnitude dS and normal unit vector $\mathbf{n}$). Note that $d\mathbf{S}$ is assumed to be small enough to neglect local curvature variations and to assume that the velocity $\mathbf{v}$ on $d\mathbf{S}$ is uniform. Equation (6.37) can be rewritten as:

$$\frac{Q}{W} = \int_A^B (v_x n_x + v_y n y) dS = \int_A^B v_x n_x dS + \int_A^B v_y n_y dS \quad , \tag{6.38}$$

where v_x and n_x are the components of $\mathbf{v}$ and $\mathbf{n}$ in the x direction while v_y and n_y are the components of $\mathbf{v}$ and $\mathbf{n}$ in the y direction. The term $n_x dS$ represents the projection of the vector $d\mathbf{S}$ in the x direction and it is therefore possible to set $n_x dS = dy$. Similarly, the term $n_y dS$ represents the projection of the vector $d\mathbf{S}$ in the y direction and it is possible to set $n_y dS = -dx$.

Using Eqs. (6.31) and (6.38) becomes:

$$\frac{Q}{W} = \int_A^B v_x dy - \int_A^B v_y dx = -\int_A^B d\psi = \psi(A) - \psi(B) = \psi_1 - \psi_2 \quad , \tag{6.39}$$

which proves that the difference between the value of the stream function of two distinct streamlines coincides with the flowrate between them. A consequence of this result is that regions of higher fluid velocity are characterized by closer streamlines compared to regions that have lower fluid velocity.

The orthogonality condition between the velocity vector and the gradients of the vorticity as well as the gradients appearing in the definition of the stream function implies that vorticity is only a function of the stream function:

$$\omega = \omega(\psi) \quad . \tag{6.40}$$

In this case, it is possible to write:

$$\nabla\omega = \frac{d\omega}{d\psi}\nabla\psi \quad , \tag{6.41}$$

and the vectors $\nabla\omega$ and $\nabla\psi$ are parallel. Equation (6.40) allows to conclude that the value of ω is constant on each streamline.

6.4.2 Examples

1 A flow is characterized by the following stream function:

$$\psi = 3ax^2y - ay^3 \quad . \tag{6.42}$$

1. Show that the fluid in motion is incompressible.
2. Show that the flow is irrotational.
3. Demonstrate that the value of the velocity at each point is a function of the distance from the origin.

Solution For the given stream function, the velocity components v_x and v_y are:

$$v_x = -\frac{\partial\psi}{\partial y} = -3ax^2 + 3ay^2 \quad , \tag{6.43}$$

$$v_y = \frac{\partial\psi}{\partial x} = 6axy \quad . \tag{6.44}$$

1. The condition of incompressible flow implies that the velocity field must be divergence-free. This condition is verified since:

$$\frac{\partial v_x}{\partial x} + \frac{\partial v_y}{\partial y} = -6ax + 6ax = 0 \quad . \tag{6.45}$$

2. The condition of irrotational flow is satisfied when the vorticity vector is everywhere zero. For the present two-dimensional field, the vorticity components in
the plane of motion are identically zero while the component normal to the plane
of motion is:

$$\omega_z = \frac{\partial v_y}{\partial x} - \frac{\partial v_x}{\partial y} = 6ay - 6ay = 0 \quad . \tag{6.46}$$

Therefore, the field is irrotational.
3. The velocity magnitude, v, is:

$$v = \sqrt{v_x^2 + v_y^2} = 3a\sqrt{x^4 + y^4 + 2x^2y^2} = 3a(x^2 + y^2) \quad ,$$

which is proportional to the square of the distance from the origin.

6.5 Velocity Potential

In two-dimensional flows characterized by non-closed streamlines, $\omega = 0$ in the
entire field if there exists a location, for example at a large distance from an obstacle,
in which $\omega = 0$. Such flows are defined as *irrotational*. Since the *curl* of the gradient
of a scalar function is zero, the irrotational flow condition, $\nabla \times \mathbf{v} = 0$, allows to set:

$$\mathbf{v} = -\nabla\phi \quad , \tag{6.47}$$

where the sign is negative by convention, namely to obtain the velocity vector as
the gradient of a suitable scalar function, ϕ, called *velocity potential*. In Cartesian
coordinates, the function $\phi(x, y)$ is defined as:

$$v_x = -\frac{\partial \phi}{\partial x} \quad , \quad v_y = -\frac{\partial \phi}{\partial y} \quad . \tag{6.48}$$

In cylindrical coordinates, the function $\phi(r, \theta)$ is defined as:

$$v_r = -\frac{\partial \phi}{\partial r} \quad , \quad v_\theta = -\frac{1}{r}\frac{\partial \phi}{\partial \theta} \quad . \tag{6.49}$$

To determine, for an irrotational flow, the expression of ϕ and in turn the resulting
flow field, it is possible to replace the two velocity components given by Eqs. (6.48)
into the continuity equation, thus obtaining the Laplace equation:

$$\nabla^2 \phi = 0 \quad . \tag{6.50}$$

Integrating this equation with the appropriate boundary conditions allows to determine the flow field. The pressure field can be derived from the flow field and the *N-S* equations, which for an ideal fluid and irrotational flow reduce to the Bernoulli equation for a two-dimensional system:

$$\nabla \left[\frac{1}{2} \rho \left(v_x^2 + v_y^2 \right) + p + \rho g h \right] = 0 \quad . \tag{6.51}$$

Alternatively, the problem can be solved exploiting the condition of irrotational flow, expressed by means of the stream function $\psi(x, y)$ as:

$$\omega = \frac{\partial^2 \psi}{\partial x^2} + \frac{\partial^2 \psi}{\partial y^2} = \nabla^2 \psi = 0 \quad . \tag{6.52}$$

Regarding the boundary conditions, it should be noted that, compared with the original problem (which included the viscous terms), the order of the equation degenerates from fourth-order to second-order in ψ. As a consequence, the velocity field can satisfy at most one of the two boundary conditions that are typically set on a wall or on the surface of an immersed object. The first condition is the kinematic condition:

$$\mathbf{v} \cdot \mathbf{n} = 0 \quad , \tag{6.53}$$

on the surface, where $\mathbf{n}$ is the unit vector normal to the surface. The meaning of Eq. (6.53) is that the fluid cannot cross the surface. The second condition is the condition of zero tangential velocity at the wall, namely the no-slip condition at the wall:

$$\mathbf{v} \cdot \mathbf{t} = 0 \quad , \tag{6.54}$$

where $\mathbf{t}$ is the unit vector tangent to the wall. This latter condition is the one that cannot be satisfied by the potential flow theory.

6.5.1 Iso-Potential Lines

The *iso-potential* lines are lines of constant values of the velocity potential ϕ, and are such that their tangent is at each point orthogonal to the velocity vector. Similarly to what has been already derived for the streamlines, it can be observed that the equation:

$$\phi(x, y) = \text{constant} \quad , \tag{6.55}$$

implies:

$$d\phi = \frac{\partial\phi}{\partial x}dx + \frac{\partial\phi}{\partial y}dy = -v_x dx - v_y dy = 0 \quad , \tag{6.56}$$

and, for the angular coefficient:

$$\left.\frac{dy}{dx}\right|_{\phi=const.} = -\frac{\partial\phi/\partial x}{\partial\phi/\partial y} = -\frac{v_x}{v_y} \quad . \tag{6.57}$$

Comparing Eq. (6.57) with Eq. (6.36), it is easy to verify that the product of the angular coefficients is equal to -1, thus confirming that iso-potential lines and streamlines are orthogonal at all points where they intersect.

6.5.2 Complex Velocity Potential

Both the functions ϕ and ψ satisfy the Laplace equation and, as just demonstrated, their tangents are mutually orthogonal at each point they intersect since:

$$v_x = -\frac{\partial\psi}{\partial y} = -\frac{\partial\phi}{\partial x} \quad , \quad v_y = \frac{\partial\psi}{\partial x} = -\frac{\partial\phi}{\partial y} \quad . \tag{6.58}$$

These relations, known as *Cauchy-Riemann* relations, imply analyticity, namely the functions $\psi(x, y)$ and $\phi(x, y)$, are analytic functions and possess all the properties of such functions. In particular, a useful property in the context of the present discussion is that there exists an analytic function, called the *complex potential* $\mathcal{F}$, such that the velocity potential ϕ is the real part of $\mathcal{F}$ and the stream function ψ is its imaginary part:

$$\mathcal{F}(z) = \phi(x, y) + i\,\psi(x, y) \quad , \tag{6.59}$$

where $z = x + i \cdot y$ is the complex variable on which $\mathcal{F}$ depends. It is easy to verify that the derivative with respect to z of the complex potential $\mathcal{F}$ is a function, also analytic, given by the relation:

$$w(z) = \frac{d\mathcal{F}(z)}{dz} = -v_x(x, y) + i\,v_y(x, y) \quad , \tag{6.60}$$

in which $w(z)$ is called the *complex velocity*. The usefulness of the complex potential $\mathcal{F}$ lies in the fact that the flow fields associated with the approximate solution of the *N-S* equations at high Reynolds number can be more simply described in terms of complex potential rather than in terms of velocity potential and stream function.

6.5.3 Examples

1 An inviscid, incompressible fluid flows between two wedge-shaped walls into a small opening, as shown in Fig. 6.7. The opening acts like a sink that removes all the liquid that is entrained. The flow field is described by the velocity potential:

$$\Phi(r, \theta) = -2 \ln r \quad .$$

Assuming that the flow is two-dimensional, determine:

1. the volumetric flowrate per unit width that flows into the opening;
2. the pressure at point 2 if the pressure at point 1 is $p_1 = 30 \cdot 10^3 \ Pa$.

Solution The velocity components are:

$$v_r = -\frac{\partial \Phi}{\partial r} = \frac{2}{r} \quad , \tag{6.61}$$

and:

$$v_\theta = -\frac{1}{r}\frac{\partial \Phi}{\partial \theta} = 0 \quad . \tag{6.62}$$

The volumetric flowrate per unit width is:

$$\frac{Q}{W} = \int_0^{\frac{\pi}{6}} v_r \cdot r \mathrm{d}\theta = \frac{\pi}{3}\frac{m^2}{s} \quad . \tag{6.63}$$

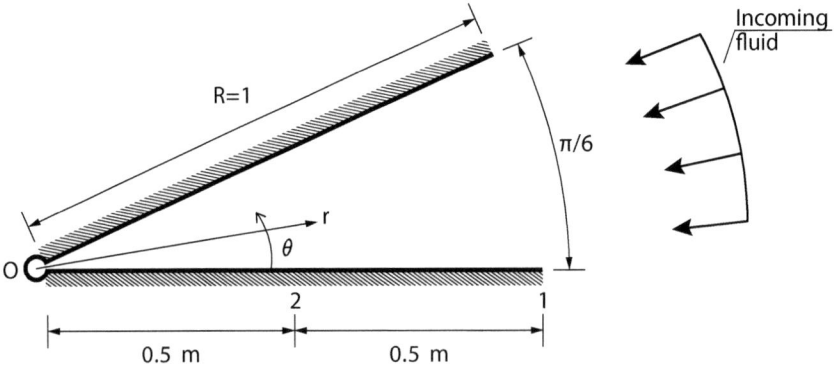

Fig. 6.7 Flow between wedge-shaped walls into a small opening

From the Bernoulli equation:

$$\frac{1}{2}\rho v_i^2 + p_i + \rho g h_i = \text{const.} \quad \rightarrow \quad p_1 + \frac{1}{2}\rho v_1^2 = p_2 + \frac{1}{2}\rho v_2^2 \quad, \tag{6.64}$$

where $h_1 = h_2$. Since $v^2 = v_r^2 + v_\theta^2 = 4/r^2$, it follows that $v_1 = v(r = 1) = 4$, $v_2 = v(r = 0.5) = 16$ and hence:

$$p_2 = p_1 + \frac{1}{2}\rho\left(v_1^2 - v_2^2\right) = 24 \cdot 10^3\, Pa \quad. \tag{6.65}$$

Pressure decreases as the opening, located at the origin O of the frame of reference, is approached since the velocity increases.

6.6 D'Alembert's Paradox

The problem of determining the flow field given any analytic function, $w(z)$, is straightforward. It is much more difficult to determine $w(z)$ for a given flow condition. A classic example is the *flow past a cylinder*. In Fig. 6.8, the streamlines generated by the flow past a cylinder are shown. It is found that, in this problem, the velocity field is described by the complex potential:

$$w(z) = v_\infty\left(z + \frac{R^2}{z}\right) \quad, \tag{6.66}$$

with R radius of the cylinder. To derive ϕ and ψ from Eq. (6.66), it is sufficient to split the potential into its real and imaginary parts:

Fig. 6.8 Streamlines around a cylinder in potential flow conditions

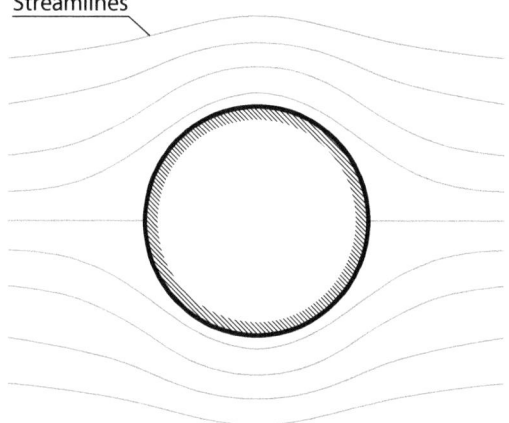

Streamlines

$$w(x, y) = v_\infty x \left(1 + \frac{R^2}{x^2 + y^2} \right) + i v_\infty y \left(1 - \frac{R^2}{x^2 + y^2} \right) \quad . \tag{6.67}$$

It follows that the stream function ψ is given by:

$$\psi(x, y) = v_\infty y \left(1 - \frac{R^2}{x^2 + y^2} \right) \quad , \tag{6.68}$$

while the velocity potential ϕ is given by:

$$\phi(x, y) = v_\infty x \left(1 + \frac{R^2}{x^2 + y^2} \right) \quad . \tag{6.69}$$

In cylindrical coordinates $\phi(r, \theta)$ becomes:

$$\phi(r, \theta) = v_\infty \left(r + \frac{R^2}{r} \right) \cos \theta \quad , \tag{6.70}$$

and the velocity components are:

$$v_r = -v_\infty \left(1 - \frac{R^2}{r^2} \right) \cos \theta \quad , \tag{6.71}$$

$$v_\theta = v_\infty \left(1 + \frac{R^2}{r^2} \right) \sin \theta \quad , \tag{6.72}$$

with $v_r = 0$ and $v_\theta = 2 v_\infty \sin \theta$ on the surface of the cylinder, where $r = R$. From Eq. (6.72), it follows that $v_\theta = 0$ only for $\theta = 0$, and $\theta = \pi$ (stagnation points): in all the other points of the surface, the no-slip condition given by Eq. (6.54) is not satisfied. The square of the velocity magnitude is equal to:

$$v^2 = v_r^2 + v_\theta^2 = v_\infty^2 \left[\left(1 - \frac{R^2}{r^2} \right)^2 + 4 \left(\frac{R}{r} \right)^2 \sin^2 \theta \right] \quad . \tag{6.73}$$

The pressure distribution can be found by applying the Bernoulli equation:

$$P = P_0 - \frac{1}{2} \rho v_\infty^2 \left[1 + \left(1 - \frac{R^2}{r^2} \right)^2 + 4 \left(\frac{R}{r} \right)^2 \sin^2 \theta \right] \quad . \tag{6.74}$$

On the surface of the cylinder:

$$\mathcal{P}^* = \frac{P - P_0}{\frac{1}{2} \rho v_\infty^2} = 1 - 4 \sin^2 \theta \quad . \tag{6.75}$$

It can be observed that the velocity and pressure fields are symmetric. Therefore, no net friction force and pressure force is applied on the cylinder:

$$F_x = \int_0^{2\pi} \tau_{r,\theta}(r = R) \cos\theta R d\theta - \int_0^{2\pi} \mathcal{P} \cos\theta d\theta = 0 \ , \qquad (6.76)$$

with $\tau_{r,\theta}(r = R) = \mu \partial v_\theta / \partial r |_{r=R} = 0$, and:

$$F_y = -\int_0^{2\pi} \tau_{r,\theta}(r = R) \sin\theta R d\theta - \int_0^{2\pi} \mathcal{P} \sin\theta R d\theta = 0 \ , \qquad (6.77)$$

with F_x and F_y forces per unit length of the cylinder.

The analysis of the flow past a cylinder shows how the potential flow theory, although based on seemingly acceptable assumptions for the case at hand, may lead to the erroneous conclusion, known as *D'Alembert's paradox* after the French mathematician, philosopher, and writer Jean Le Rond d'Alembert (1717–1783), that the resultant of the friction and pressure forces on the cylinder is zero. The premise is to use the theory to measure the forces acting on the cylinder for Reynolds numbers sufficiently high to ensure that the flow is dominated by the inertial forces, which balance the pressure forces, and characterized by negligible viscous forces. This is exactly what D'Alembert did: Verify the applicability of the potential flow theory to the case of flow past a stationary cylinder at high Re. The expected result is that the force acting on the cylinder is produced mainly by the pressure (form drag) associated with the blockage effect that the cylinder has on the incident flow. The expected contribution by the shear stresses (friction drag) is only minimal. However, the result obtained by applying the theory is that indeed the friction drag is zero but so is the form drag, in clear contradiction with the experimental measurements! The paradox arises from the fact that, having removed the no-slip assumption, the potential flow theory completely disregards the presence of the boundary layer and the occurrence of friction at the wall. Both the boundary layer and wall friction contribute to significantly decrease the local value of Re in the region where forces are to be measured, that is near the surface of the cylinder (where viscous forces are in fact comparable to inertial forces).

The comparison between theory and experiments, illustrated in Fig. 6.9 for experimental data referring to two different Reynolds numbers, shows good agreement for angles θ, measured from the forward stagnation point on the cylinder, smaller than about $90°$.

According to the Bernoulli equation, the pressure is maximum for $\theta = 0$, where the velocity is zero, and then gradually decreases following the flow along the surface of the cylinder. In this region, a *favorable pressure gradient* $dp/d\theta < 0$ is established and the pressure decreases following the streamlines. For $\theta > 90°$, the pressure tends to increase following the streamlines. In this case, an *adverse pressure gradient* $dp/d\theta > 0$ is established. Under these conditions, the flow may separate from the surface of the cylinder, leading to the formation of two or more stationary vortices, as illustrated in Chap. 7, Fig. 7.8. The formation of these vortices prevents the pressure

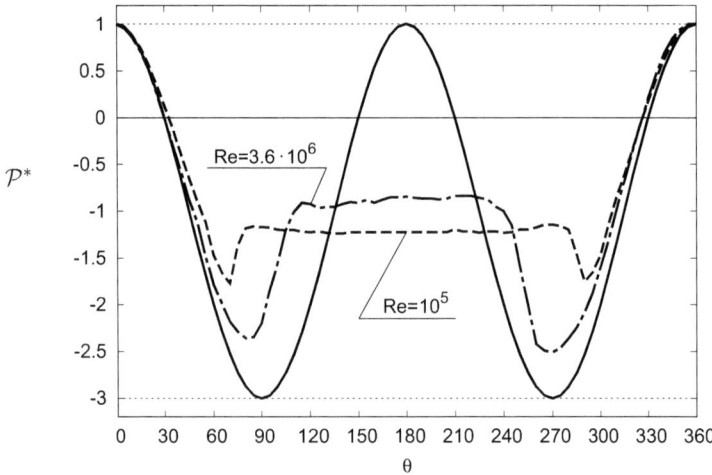

Fig. 6.9 Comparison of experimental data (dashed lines, referring to two different Reynolds numbers) and potential flow theory (solid line, Eq. 6.75) for the dimensionless pressure, $\mathcal{P}^* = 2(\mathcal{P} - \mathcal{P}_0)/\rho v_\infty^2$, as a function of the angle θ

recovery that is predicted by the theory and indeed experiments show how, for $\theta > 90°$, the pressure tends to level off on a value that, for $\theta = 180°$, is much lower than the one expected theoretically. The final result is that, in reality, the cylinder is subject to a significant resistance force. This force increases as the flow Reynolds number increases, leaving no hope on whether the theory can prove to be valid for $Re \to \infty$.

To understand the reasons behind the discrepancy between theoretical analysis and experimental observations, it is first useful to note how flow separation leads to the formation of flow regions of high vorticity. The vorticity equation for a two-dimensional system, Eq. (6.29), has no vorticity generation terms, so the vorticity associated with the flow field can only be injected at the boundary, i.e. at the cylinder surface in the case under examination, and then be transported within the flow field by convection and diffusion. Experiments clearly show that the assumption of negligible viscous effects at the surface of the cylinder is incorrect. However, since this assumption is based on the dimensional analysis of the *N-S* equations, it must be concluded that the dimensional analysis is incorrect. This is indeed the case, since the problem is not characterized by a single characteristic length, which is clearly the diameter of the cylinder, but also by a second length, which represents the extension of the area around the cylinder in which viscous effects are important. This region is called *boundary layer*, and the description of the flow field within this region, provided in Chap. 7, has enabled significant progress in the description of complex flow fields.

6.6.1 Examples

$\boxed{1}$ A chimney 60 m high and with diameter $D = 2$ m, is subject, at heights above 10 m, to the action of wind ($\rho = 1.2$ kg/m³, $\mu = 20 \cdot 10^{-6} Pa \cdot s$), as shown in Fig. 6.10. Knowing that the wind speed is $U_\infty = 20$ m/s, calculate the total force acting on the chimney.

Solution The Reynolds number of the flow past the chimney is:

$$Re = \frac{U_\infty D \rho}{\mu} = 2.4 \cdot 10^6 \quad . \tag{6.78}$$

This value appears to be high enough to apply the potential flow theory. The theory, however, predicts zero net force on the chimney and, therefore, it is not useful for the purposes of the problem.

An alternative way to determine realistically the total force acting on the cylinder is to start from the definition of the drag force:

$$F_D = \frac{1}{2} C_D A_p \rho U_\infty^2 \quad , \tag{6.79}$$

with $A_p = DL$ lateral area of the chimney exposed to the wind. For the Reynolds number calculated before, the diagram reported in Fig. 2.3 can be used to obtain $C_D \simeq 0.8$.

Fig. 6.10 Effect of wind on a chimney

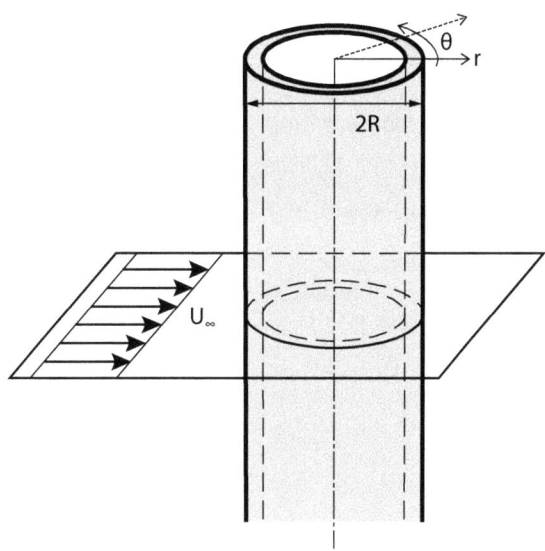

Assuming that the total force coincides with the drag force, it follows that:

$$F_D = \frac{1}{2}\rho U_\infty^2 DLC_D = 2.3 \cdot 10^4 \ N \quad .$$

(6.80)

6.7   Examples of Plane Potential Flows

The Laplace equation (6.50) is a linear partial derivative equation, the solution of which can be obtained as a linear combination of simple solutions. If ϕ_1 and ϕ_2 are two solutions of the Laplace equation, then $\phi_3 = \phi_1 + \phi_2$ is also a solution. This property allows to combine the solutions of simple potential flows to obtain the solution of more complex flows that are, however, of greater practical interest.

The following sections present first several basic velocity potentials that describe simple flows and, subsequently, a series of more complex flows that can be described by suitable combination of the basic potentials. For simplicity, only two-dimensional flows in Cartesian coordinates will be examined.

6.7.1   Uniform Flow

In this flow, the streamlines are straight and parallel to each other, as illustrated in Fig. 6.11. In the simplest case, they are produced by a flow field having velocity components $v_x = U$ and $v_y = 0$. The velocity potential is:

$$\phi = -Ux \quad ,$$

(6.81)

while the stream function is:

Fig. 6.11 Streamlines (dashed lines) and iso-potential lines (solid lines) for uniform potential flow generated by $v_x = U$ e $v_y = 0$

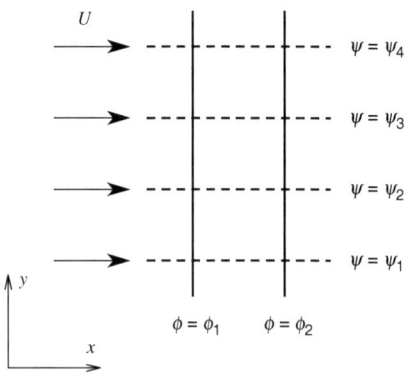

$$\psi = -Uy \quad . \tag{6.82}$$

In the more general case where $v_y \neq 0$, the flow (and therefore the streamlines) forms an angle α with the horizontal direction. The velocity potential and the stream function become:

$$\phi = -U(x \cos \alpha + y \sin \alpha) \quad , \quad \psi = -Uy(x \sin \alpha - y \cos \alpha) \quad . \tag{6.83}$$

6.7.2 Source and Sink

This flow is produced by a fluid flowing radially through the origin perpendicular to the x-y plane of motion, as shown in Fig. 6.12 for the case of a source. The volumetric flowrate per unit length, m, in the direction orthogonal to the plane of motion is:

$$m = (2\pi r)v_r \qquad \rightarrow \qquad v_r = \frac{m}{2\pi r} \quad , \tag{6.84}$$

with $r = \sqrt{x^2 + y^2}$, while $v_\theta = 0$ as the flow is purely radial. Using the relations (6.49), the velocity potential can be derived[2]:

$$\phi = -\frac{m}{2\pi} \ln r \quad . \tag{6.85}$$

Similarly, using relation (6.33), the stream function can be derived as:

$$\psi = -\frac{m}{2\pi}\theta \quad . \tag{6.86}$$

If m is positive, the flow is radially outward, and the flow is considered to be a source flow. If m is negative, the flow is toward the origin, and the flow is considered to be a sink flow. The flowrate m can also be interpreted as the *strength* of the source/sink. The streamlines produced by this flow field are straight lines that converge towards the origin of the flow, where there is a discontinuity due to the fact that mass conservation cannot be satisfied there. The iso-potential lines, on the other hand, are concentric circles centered at the origin of the flow, as shown in Fig. 6.12. Note how, because of the discontinuity associated with the origin, the flow just described does not exist in reality. However, it can be combined with other basic velocity potentials to approximately describe some real flow fields.

---

[2] The relations (6.49) allow to set $d\phi = -v_r dr$ which, once integrated, yields $\phi = -(m/2\pi) \ln r + f(\theta)$, as well as $d\phi = -rv_\theta d\theta$, which, once integrated, yields $\phi = f(r)$, with $f(\theta)$ and $f(r)$ arbitrary integration constants. By choosing $f(\theta) = 0$ and $f(r) = -(m/2\pi) \ln r$, the velocity potential becomes: $\phi = -(m/2\pi) \ln r$.

Fig. 6.12 Streamlines (dashed lines) and iso-potential lines (solid lines) for a source generated by the flow field $v_r = m/2\pi r$ and $v_\theta = 0$

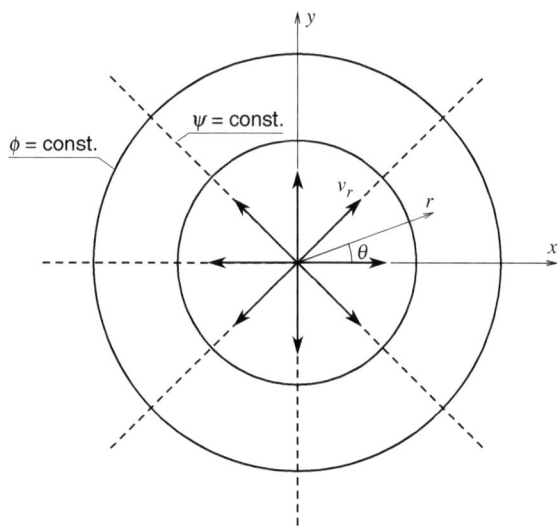

6.7.3   Free (or Irrotational) Vortex

In this flow, the streamlines are concentric circles whose center coincides with the center of the vortex (thus being fully equivalent to the iso-potential lines of the source/sink flow) while the iso-potential lines are straight lines that converge at the center of the vortex (thus being fully equivalent to the streamlines of the source/sink flow), as shown in Fig. 6.13.

The velocity potential and the stream function that characterize the free vortex are:

Fig. 6.13 Streamlines (dashed lines) and iso-potential lines (solid lines) of a free vortex produced by an irrotational flow field $v_r = 0$ and $v_\theta = K/r$

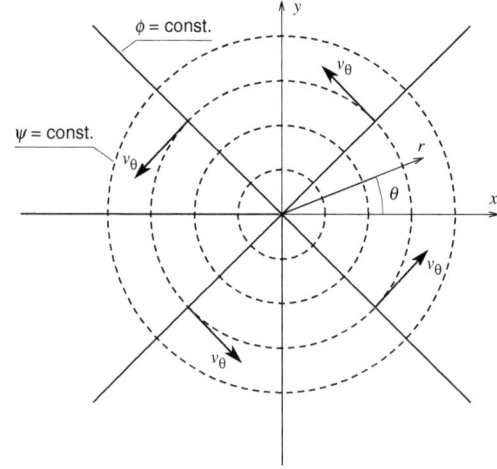

$$\phi = -K\theta \quad , \tag{6.87}$$

while the stream function is:

$$\psi = +K \ln r \quad , \tag{6.88}$$

where K is a constant. The velocity components can be obtained by using either Eqs. (6.33) or (6.49):

$$v_r = -\frac{1}{r}\frac{\partial \psi}{\partial \theta} = -\frac{\partial \phi}{\partial r} = 0 \quad , \tag{6.89}$$

$$v_\theta = \frac{\partial \psi}{\partial r} = -\frac{1}{r}\frac{\partial \phi}{\partial \theta} = \frac{K}{r} \quad . \tag{6.90}$$

Note how this flow field coincides with the one discussed in Sect. 6.2 and given by Eq. (6.15). Also in this type of flow, there is a singularity at the center of the vortex. The free vortex is different from the forced (or rotational) vortex, which is characterized by velocity components $v_r = 0$ and $v_\theta = K_1 r$ (where K_1 is a constant, usually different from K) but cannot be described by a velocity potential. The free vortex represents an approximation of the vortex formed by the swirling motion of the water as it drains from a bathtub, while the forced vortex represents an approximation of the vortex produced by the motion of a liquid contained in a tank that is rotated about its axis with angular velocity Ω. A *real* vortex can be obtained by combining a forced vortex within a certain radial distance r_0 from the origin and a free vortex beyond that distance. The corresponding flow field is:

$$v_\theta = \begin{cases} \Omega r & \text{per } r \le r_0 \quad , \\ \dfrac{K}{r} & \text{per } r > r_0 \quad . \end{cases} \tag{6.91}$$

The velocity potential and the stream function are commonly expressed in terms of the *circulation*, Γ. The circulation is defined as the line integral of the tangential component of the velocity taken around a closed path C in the flow field:

$$\Gamma = \oint_C \mathbf{v} \cdot d\mathbf{S} \quad , \tag{6.92}$$

where $d\mathbf{S}$ is the unit vector tangential to C representing an infinitesimal displacement along the closed line (as shown in Fig. 6.14). For an irrotational flow, the circulation is zero since $\mathbf{v} = \nabla\phi$ and thus $\mathbf{v} \cdot d\mathbf{s} = \nabla\phi \cdot d\mathbf{s} = d\phi$ and:

$$\Gamma = \oint_C d\phi = 0 \quad . \tag{6.93}$$

Fig. 6.14 Notation for the
definition of circulation

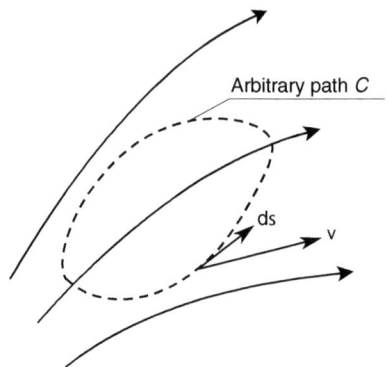

If there are singularities enclosed within the path C, such as the center of a vortex, then the circulation may not be zero. In particular, for a free vortex, the circulation is:

$$\Gamma = \oint_C d\phi = \int_0^{2\pi} K d\theta = 2\pi K \quad , \tag{6.94}$$

with $\Gamma = 0$ only if $K = 0$. Combining Eq. (6.94) with Eqs. (6.89) and (6.90), the velocity potential and the stream function can be obtained as:

$$\phi = -\frac{\Gamma}{2\pi}\theta \quad , \quad \psi = +\frac{\Gamma}{2\pi}\ln r \quad , \tag{6.95}$$

with Γ representing the strength of the vortex.

6.7.4 Dipole and Doublet

When the flow is produced by the combination of a source and a sink of equal strength m, placed at a distance $2a$ from each other, a dipole is formed. With reference to Fig. 6.15, and recalling equation (6.86), the stream function for this combination is:

$$\psi = +\frac{m}{2\pi}(\theta_1 - \theta_2) \quad . \tag{6.96}$$

This expression can be rewritten as:

$$\tan\left(\frac{2\pi\psi}{m}\right) = \tan(\theta_1 - \theta_2) = \frac{\tan\theta_1 - \tan\theta_2}{1 + \tan\theta_1 \tan\theta_2} \quad . \tag{6.97}$$

with:

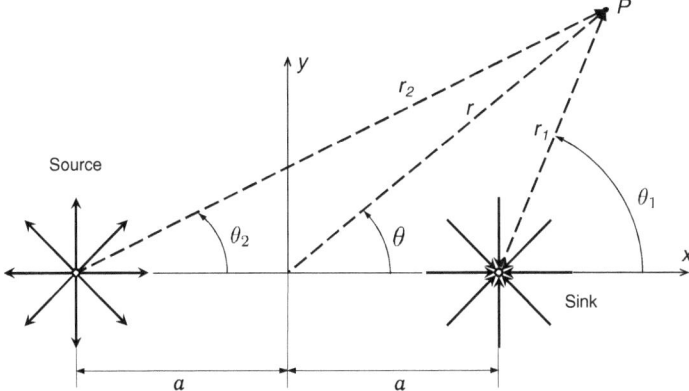

Fig. 6.15 Potential flow produced by the combination of a source and a sink of equal strength

$$\tan \theta_1 = \frac{r \sin \theta}{r \cos \theta - a} \quad , \quad \tan \theta_2 = \frac{r \sin \theta}{r \cos \theta + a} \quad . \tag{6.98}$$

Substituting these expressions into Eq. (6.97) yields:

$$\tan \left(\frac{2\pi\psi}{m} \right) = \frac{2ar \sin \theta}{r^2 - a^2} \quad , \tag{6.99}$$

and, therefore, the stream function is:

$$\psi = \frac{m}{2\pi} \tan^{-1} \left(\frac{2ar \sin \theta}{r^2 - a^2} \right) \quad . \tag{6.100}$$

For sufficiently small values of the distance a, it is possible to write:

$$\psi = \frac{m}{2\pi} \left(\frac{2ar \sin \theta}{r^2 - a^2} \right) = \frac{ma}{\pi} \frac{r}{r^2 - a^2} \sin \theta \quad . \tag{6.101}$$

The doublet is obtained by bringing the source closer to the sink until $a \to 0$, while increasing their intensity until $m \to \infty$ so that the product, ma/π, which represents the strength of the doublet, remains constant. In this case:

$$\lim_{a \to 0} \frac{r}{r^2 - a^2} = \frac{1}{r} \quad , \tag{6.102}$$

and the expression of the stream function reduces to:

$$\psi = \frac{ma}{\pi} \frac{1}{r} \sin \theta \quad . \tag{6.103}$$

Fig. 6.16 Streamlines of a
doublet

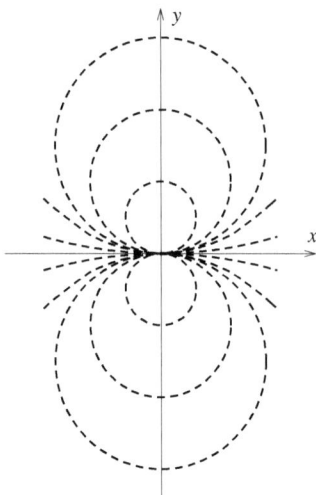

The streamlines are graphically represented in Fig. 6.16. The velocity potential, on
the other hand, is:

$$\phi = -\frac{ma}{\pi}\frac{1}{r}\cos\theta \quad . \tag{6.104}$$

Just as sources and sinks are not physically realistic entities, neither are doublets.
However, the doublet when combined with other basic potential flows provides a
useful representation of some flow fields of practical interest. For example, the combination of a uniform flow and a doublet can be used to represent the flow around a
circular cylinder.

6.7.5 Flow Around a Rankine Oval

The potential flow around an immersed body of oval shape (usually referred to as
Rankine oval) can be obtained by a combination of a source and a sink of equal
intensity with a uniform flow, as illustrated in Fig. 6.17. Thanks to the method of
superposition, the stream function can be obtained as:

$$\psi = \psi_{\text{unif. flow}} + \psi_{\text{source}} + \psi_{\text{sink}} = -Uy + \frac{m}{2\pi}(\theta_1 - \theta_2) \quad , \tag{6.105}$$

with $y = r\sin\theta$. Likewise, for the velocity potential:

$$\phi = \phi_{\text{unif. flow}} + \phi_{\text{source}} + \phi_{\text{sink}} = -Ux + \frac{m}{2\pi}(\ln r_1 - \ln r_2) \quad , \tag{6.106}$$

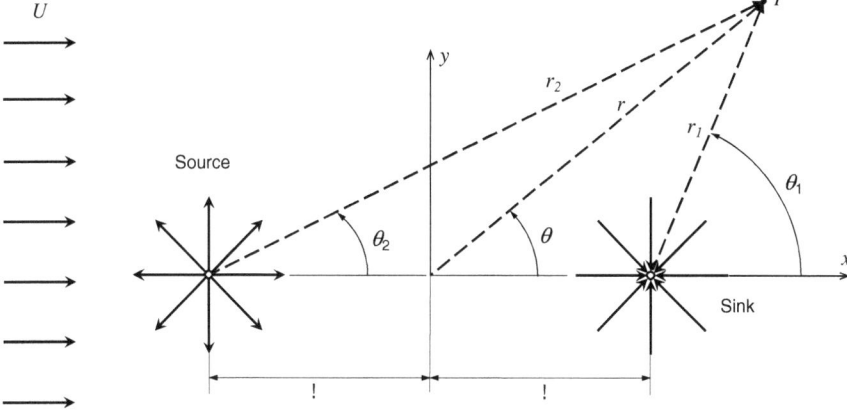

Fig. 6.17 Potential flow produced by combining a source and a sink of equal intensity with a uniform flow

with $x = r \cos \theta$. Recalling equation (6.100), the stream function can be written as:

$$\psi = -Ur \sin \theta + \frac{m}{2\pi} \tan^{-1} \left(\frac{2ar \sin \theta}{r^2 - a^2} \right) , \qquad (6.107)$$

or, in equivalent form:

$$\psi = -Uy + \frac{m}{2\pi} \tan^{-1} \left(\frac{2ay}{x^2 + y^2 - a^2} \right) . \qquad (6.108)$$

The streamlines around a Rankine oval of length $2l$ and height $2h$ are represented graphically in Fig. 6.18. The $\psi = 0$ streamline is a closed line that has exactly the shape of the oval and can therefore be regarded as the boundary of the oval. Indeed, any streamline in an inviscid flow field can be considered as a solid boundary, since the conditions along a solid boundary and a streamline are the same: There is no flow through the boundary or the streamline. Streamlines within this contour have no physical meaning (being inside the body) and therefore are of no practical interest. Note the presence of two stagnation points.

The velocity components at a generic P point outside the oval can be obtained by superposition:

$$v_r = v_{r,\text{unif. flow}} + v_{r,\text{source}} + v_{r,\text{sink}} = U \cos \theta + \frac{m}{2\pi} \frac{1}{r_1} - \frac{m}{2\pi} \frac{1}{r_2} , \qquad (6.109)$$

$$v_\theta = \underbrace{v_{\theta,\text{unif. flow}}}_{} + \underbrace{v_{\theta,\text{source}}}_{=0} + \underbrace{v_{\theta,\text{sink}}}_{=0} = -U \sin \theta , \qquad (6.110)$$

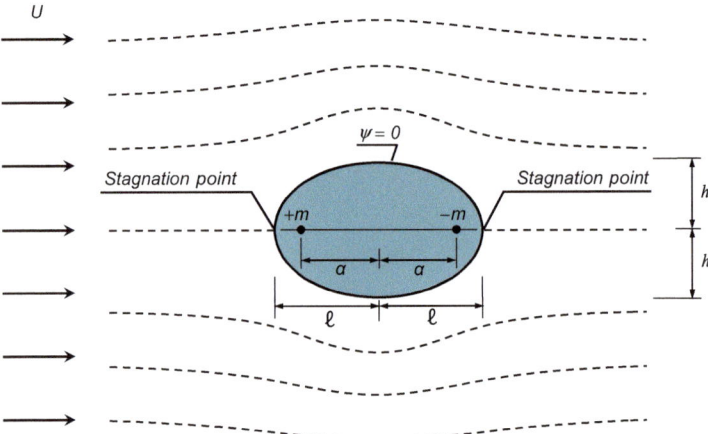

Fig. 6.18 Streamlines around a Rankine oval

with r_1 distance between source and point P and r_2 distance between sink and point P, as shown in Fig. 6.17.

Finally, the dimensions of the oval can be expressed as a function of the parameters U, m and a. In particular, by exploiting the condition $\mathbf{v} = 0$ at one of the stagnation points, e.g. the left one in Fig. 6.18, the following expression is obtained:

$$U = \frac{m}{2\pi}\left(\frac{1}{l-a}\right) - \frac{m}{2\pi}\left(\frac{1}{l+a}\right) \quad , \tag{6.111}$$

where the term on the left-hand side is the velocity induced by the uniform flow at the stagnation point and the two terms on the right-hand side represent the velocity induced at the same point by the source and sink, respectively. Solving with respect to l yields:

$$l = a\left(1 + \frac{m}{\pi U a}\right)^{1/2} \quad . \tag{6.112}$$

The height of the oval can be obtained using Eq. (6.108) and imposing the condition $\psi = 0$ at the point of coordinates $x = 0$ and $y = h$ (or, equivalently $y = -h$). This yields:

$$h = \frac{h^2 - a^2}{2a} \tan\left(\frac{2\pi U h}{m}\right) \quad . \tag{6.113}$$

This relation can be used to obtain the value of h by trial and error. Equations (6.112) and (6.113) show that it is possible to vary the shape of the oval simply by changing the values of U, m and a, i.e. the value of the term Ua/m. High values of Ua/m

correspond to a thin and elongated oval, while small values of Ua/m correspond to a short, bluff oval (with a shape that tends to a circle).

In general, the equations derived for potential flow around a Rankine oval can reproduce with reasonable accuracy the velocity field only outside of the boundary layer and the pressure distribution only in the absence of flow separation (which is usually not observed on the front part of the body). This is because the potential flow theory is unable to reproduce these phenomena.

6.7.6 Flow Around a Rotating Cylinder

The potential flow around a stationary cylinder was examined in Sect. 6.6. Let us now consider a cylinder that is immersed in a uniform flow and rotates with constant angular velocity. The flow field and the pressure distribution around the cylinder can be described by combination of a uniform flow, a doublet and a free vortex. The corresponding stream function and velocity potential are:

$$\psi = U \left(r - \frac{R^2}{r} \right) - \frac{\Gamma}{2\pi} \ln r \quad , \tag{6.114}$$

and:

$$\phi = U \left(r + \frac{R^2}{r} \right) + \frac{\Gamma}{2\pi} \theta \quad , \tag{6.115}$$

where R is the radius of the cylinder and Γ is the circulation of the free vortex. The tangential velocity on the surface of the cylinder $(r = R)$ is:

$$v_{\theta,s} = 2U \sin \theta - \frac{\Gamma}{2\pi R} \quad , \tag{6.116}$$

while $v_{r,s} = 0$, obviously. Equation (6.116) implies that the shear stress applied to the cylinder is zero.

The relations just derived show that it is possible to obtain a virtually infinite variety of different flow fields by simply varying the circulation Γ. For example, changing the position of the stagnation points, which can be obtained by imposing $v_{\theta,s}$ in Eq. (6.116):

$$\theta_s = \sin^{-1} \left(\frac{\Gamma}{4\pi U R} \right) \quad . \tag{6.117}$$

If $\Gamma = 0$ (stationary cylinder), then $\theta_s = 0$ or $\theta_s = \pi$, as already derived in Sect. 6.6 and as shown in Fig. 6.19a. If, on the other hand, $\Gamma/4\pi U R < 1$, then the stagnation points will be at another location (e.g. as shown in Fig. 6.19b), until they perfectly

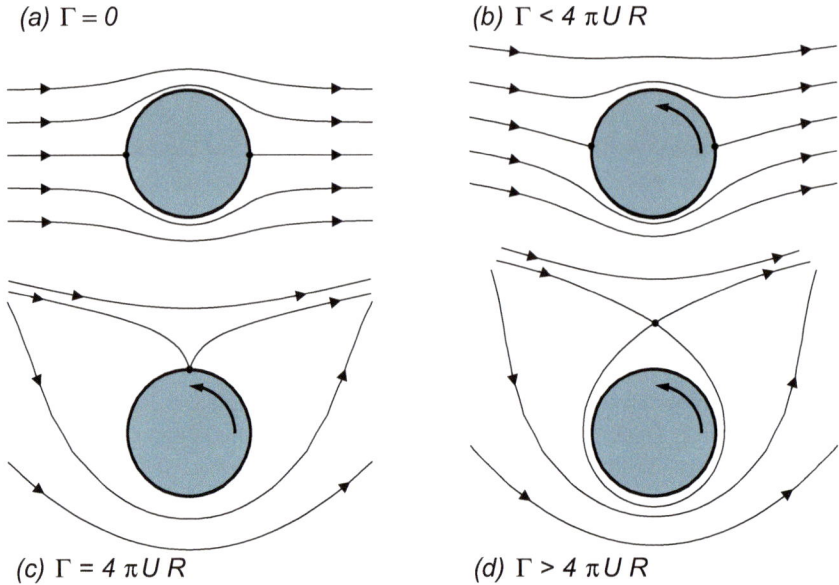

Fig. 6.19 Streamlines and stagnation points for uniform flow around a rotating cylinder

overlap when $\Gamma/4\pi U R = 1$ (or, equivalently, $\theta_s = \pm\pi/2$, Fig. 6.19c). Finally, for $\Gamma/4\pi U R > 1$, the condition (6.117) cannot be satisfied and there will be a single stagnation point located away from the cylinder (Fig. 6.19d).

The pressure distribution on the surface of the cylinder can be obtained by applying the Bernoulli equation, similarly to what has been done to obtain Eq. (6.74). In this case, the equation simplifies as:

$$\mathcal{P}_0 + \frac{1}{2}\rho U^2 = \mathcal{P} - \frac{1}{2}\rho v_{\theta,s}^2 \ , \tag{6.118}$$

and, therefore, the pressure on the surface reads as:

$$\mathcal{P}^* = \frac{\mathcal{P} - \mathcal{P}_0}{\frac{1}{2}\rho v_\infty^2} = 1 - 4\sin^2\theta + \underbrace{\frac{2\Gamma\sin\theta}{\pi U R} - \frac{\Gamma^2}{4\pi^2 U^2 R^2}}_{\substack{\text{Contribution due to} \\ \text{the cylinder rotation}}} \ . \tag{6.119}$$

The behavior of $\mathcal{P}^*$ is represented in Fig. 6.20 for $\Gamma = -\pi U R$. This behavior is compared with that over a stationary cylinder ($\Gamma = 0$). The comparison highlights the asymmetry of the distribution between the upper half-cylinder, where the pressure is larger in magnitude, and the lower half-cylinder, where the pressure is lower in magnitude. Obviously, the flow field between the two half-cylinders is also not symmetric.

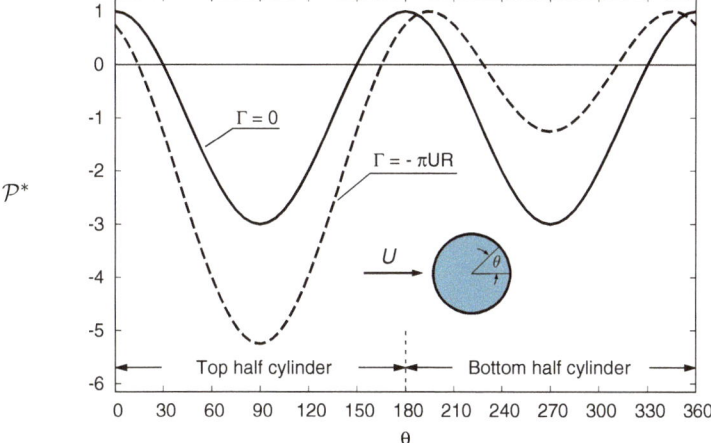

Fig. 6.20 Pressure distribution on the surface of a stationary (solid line) and rotating (dashed line) cylinder immersed in a uniform flow

The force per unit length generated on the cylinder by the pressure distribution along the x direction is equivalent to the drag force, namely to the hydrodynamic resistance exerted by the cylinder in the direction of the uniform flow, and is equal to:

$$F_x = -\int_0^{2\pi} \mathcal{P} \cos \theta R d\theta = 0 \quad , \tag{6.120}$$

as expected from the application of the potential flow theory. The force per unit length generated along the transverse y direction, on the other hand, is:

$$F_y = -\int_0^{2\pi} \mathcal{P} \sin \theta R d\theta = -\rho \Gamma U \quad . \tag{6.121}$$

Therefore, if the cylinder rotates, then it experiences a force called *lift* that acts in the direction perpendicular to that of the undisturbed flow and is proportional to the circulation Γ. This force tends to change the position of the rotating body within the flow thus producing the so-called *Magnus effect*, discovered by the German physicist Heinrich Gustav Magnus (1802–1870). The Magnus effect characterizes the curved trajectory of a baseball or a golf ball thrown in air. The negative sign indicates that, for $U > 0$, the force is directed along the y axis if $\Gamma < 0$ (condition of clockwise cylinder rotation) while it has the opposite direction if $\Gamma > 0$ (counterclockwise cylinder rotation).

Equation (6.121) has been calculated for a rotating cylinder, but the result of the integral gives the lift per unit length that is generated on any two-dimensional body immersed in a uniform flow of an inviscid fluid. The general formulation $F = -\rho \Gamma U$

is referred to as *Kutta-Joukowski theorem*, after the German mathematician Martin Wilhelm Kutta (1867–1944) and the Russian mathematician and engineer Nikolay Yegorovich Joukowski (1847–1921), and is commonly used to determine the lift on airfoils.

Problems

$\boxed{a}$ The velocity $\mathbf{v}$ of a two-dimensional viscous flow of a fluid of density ρ is:

$$\mathbf{v} = \Omega x \mathbf{i} - \Omega y \mathbf{j} \quad . \tag{6.122}$$

Determine whether the flow is irrotational or not. Derive an expression for the viscous stresses $\tau_{xy}, \tau_{xx}, \tau_{yy}$. Derive an expression for $\mathcal{P}(x, y)$ assuming that the value $\mathcal{P}_0$ of the pressure $\mathcal{P}$ at the origin of the axes is known.

$\boxed{b}$ Consider a fluid jet that, under the assumption of potential flow, impinges on a plane, producing the streamlines shown in Fig. 6.21 around a stagnation point located at the origin of the axes. Prove that the velocity potential $\phi = k(x^2 - y^2)$ describes the flow field and determine the velocity components.

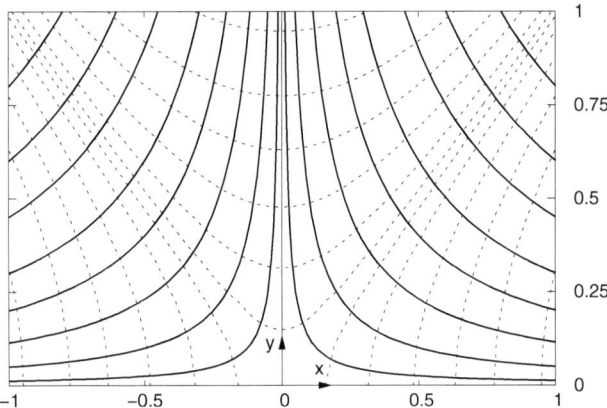

Fig. 6.21 Streamlines (solid) and iso-potential lines (dashed) for potential flow near a stagnation point located on a plane

Chapter 7
Boundary Layers and Self-Similar Solutions

7.1 Flows with Self-Similar Solution

The analysis of the potential flow around immersed bodies at high Reynolds number has led to the identification of two distinct flow regions:

1. An *outer region*, which is not influenced by the no-slip condition on the body surface and is characterized by the macroscopic characteristic scales of the flow field. In this region, the potential flow theory analyzed in Chap. 6 applies.
2. An *inner region*, called boundary layer, which is influenced by viscous effects whatever the flow Reynolds number. The characteristic length of this region, and thus the location of the boundary separating inner and outer region, must be properly identified as it represents, in practice, the thickness of the boundary layer.

The main effect produced by the viscous stresses in the boundary layer region is a change in the velocity of the fluid. Consider, for example, the case of a boundary layer generated by a uniform flow of a viscous fluid on the surface of an infinitely-long flat plate, as shown in Fig. 7.1. With respect to a frame of reference attached to the plate, the velocity gradually changes from zero at the surface of the body to the undisturbed velocity U_∞ of the outer flow. This satisfies the no-slip condition that characterizes real fluids. In the boundary layer, the velocity gradient in the direction normal to the flow determines a rotational flow field (non-zero vorticity) and results in a deformation of the fluid elements along their trajectory, in a way that is illustrated qualitatively in Chap. 6, Fig. 6.2. Outside of the boundary layer, however, the velocity gradients in the direction normal to the flow become relatively small and, therefore, the fluid behaves as if it were inviscid: the flow field is irrotational (zero vorticity) and the fluid elements maintain their shape during the motion.

The thickness of the boundary layer and the specific features of the flow field within it depend on the particular shape of the object. Generally speaking, however, it is always possible to observe an increase in thickness along the downstream direction

© The Author(s), under exclusive license to Springer Nature Switzerland AG 2024   207
A. Soldati and C. Marchioli, *Fluid Mechanics for Mechanical Engineers*,
https://doi.org/10.1007/978-3-031-53950-3_7

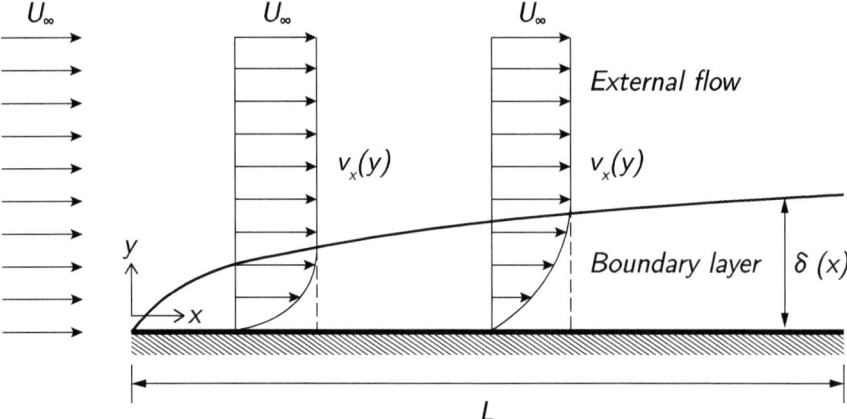

Fig. 7.1 Schematic representation of the formation and growth of the boundary layer produced by a uniform flow of a viscous fluid on an infinitely-long flat plate. The solid line represents qualitatively the separation between the boundary layer and the outer flow region and defines the thickness $\delta(x)$ of the boundary layer. As demonstrated in Sect. 7.3, the thickness increases proportionally to $x^{1/2}$, where x represents a generic position along the plate

starting from the leading edge of the body (located at $x = 0$ for a flat plate, as shown in Fig. 7.1). This results in a reduction of the velocity gradient at the wall and, hence, of the friction generated there. In any case, the thickness is always smaller than the streamwise length of the boundary layer. The flow field, on the other hand, depends on the Reynolds number, defined as $Re_x = U_\infty x/\nu$, where x represents a generic position along the body. The flow field will be laminar below a certain critical value, typically falling within the range $Re_{x,cr} \sim 3.5 \cdot 10^5 \div 10^6$ depending on the shape and the possible roughness of the body but also on the value of U_∞. Above $Re_{x,cr}$, there is a transition from the laminar regime to the turbulent regime: The turbulent boundary layer is characterized by irregular mixing and significant deformation of the fluid elements. Note that mixing is much more pronounced than the one occurring, only at molecular scale, in laminar flow conditions.

Several definitions of boundary layer thickness exist. The simplest and most intuitive definition is the one that sets the thickness as equal to the distance normal to the surface of the body at which the flow velocity has essentially reached its undisturbed value, U_∞. Conventionally, the thickness is defined as $\delta_{99} = y$ with y such that the condition $v_x(y) = 0.99 U_\infty$ is verified, as shown in Fig. 7.2a.

A further definition of thickness, which does not require the adoption of arbitrary values, is based on the velocity defect $U_\infty - v_x(y)$ that exists between the undisturbed flow and the boundary layer. In turn, this velocity defect corresponds to a defect (decrease) in the flowrate. Considering for simplicity a two-dimensional flow, the flowrate around the body in the absence of boundary layer would be:

$$\frac{Q}{W} = \int_0^\infty U_\infty dy \quad , \tag{7.1}$$

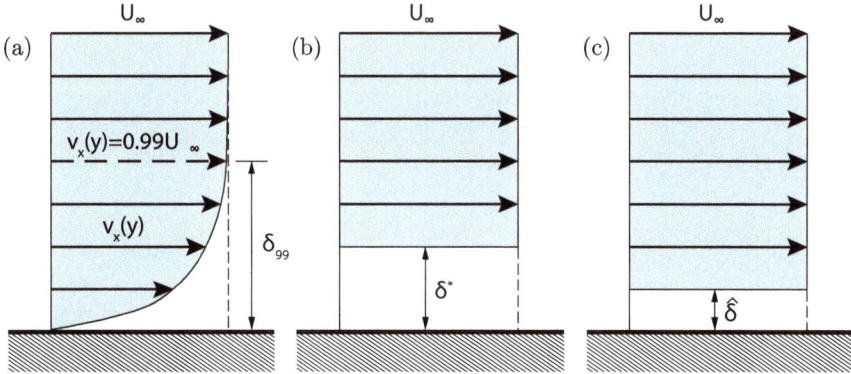

Fig. 7.2 Definitions of boundary layer thickness: **a** δ_{99} thickness; **b** displacement thickness, δ^*; **c** momentum thickness, $\hat{\delta}$

with W characteristic length of the body in the direction normal to the flow. In this case, the velocity profile would be uniform and everywhere equal to U_∞. Because of the boundary layer, the effective flowrate is:

$$\frac{Q_{eff}}{W} = \int_0^\infty v_x(y) dy \quad , \tag{7.2}$$

with $v_x(y) < U_\infty$ inside the boundary layer and $v_x(y) = U_\infty$ outside of it. The corresponding flowrate defect is:

$$\frac{Q}{W} - \frac{Q_{eff}}{W} = \int_0^\infty [U_\infty - v_x(y)] dy \quad . \tag{7.3}$$

The same defect can be produced by virtually displacing the uniform velocity profile away from the surface of the body by a distance δ^*, called the *displacement thickness*, such that the following conditions hold: $v_x = 0$ for $y \leq \delta^*$ and $v_x = U_\infty$ for $y > \delta^*$, as shown in Fig. 7.2b. The deficit in flowrate associated with this profile is:

$$\frac{Q}{W} - \frac{Q_{\delta^*}}{W} = \int_0^\infty U_\infty dy - \underbrace{\left[\int_0^{\delta^*} v_x(y) dy + \int_{\delta^*}^\infty U_\infty dy \right]}_{=0} =$$

$$= \int_0^{\delta^*} U_\infty dy = \delta^* U_\infty \quad , \tag{7.4}$$

Based on the definition given for δ^*, it must be:

$$\frac{Q}{W} - \frac{Q_{\delta^*}}{W} = \frac{Q}{W} - \frac{Q_{eff}}{W} \quad , \tag{7.5}$$

and therefore:

$$\delta^* = \int_0^\infty \left[1 - \frac{v_x(y)}{U_\infty} \right] dy \quad . \tag{7.6}$$

The thickness δ^* compensates for the reduction in flowrate on account of boundary layer formation, which can be now interpreted as a displacement effect: The presence of the boundary layer is equivalent to a displacement of the outer region of potential flow away from the surface of the body and δ^* represents the distance at which the outer region must be displaced in order for the flowrate to be the same as that of the real fluid with the boundary layer. In other words, the effect due to the presence of the boundary layer is *modeled* by adding the displacement thickness to the shape of the body and by treating the entire flow domain as undisturbed.

Another definition of boundary layer thickness can be obtained from the conservation of momentum. In addition to a reduction in flowrate, the velocity defect due to the boundary layer also results in a reduction of momentum compared with the ideal case in which viscous effects are absent. Ideally, this reduction can be obtained by displacing the uniform velocity profile away from the body by a suitable distance $\hat{\delta}$, called the *momentum thickness*, so that:

$$\rho U_\infty \hat{\delta} = \int_0^\infty \rho v_x(y) \left[U_\infty - v_x(y) \right] dy \quad , \tag{7.7}$$

as shown in Fig. 7.2c. Recasting the equation with respect to $\hat{\delta}$ gives:

$$\hat{\delta} = \int_0^\infty \frac{v_x(y)}{U_\infty} \left[1 - \frac{v_x(y)}{U_\infty} \right] dy \quad . \tag{7.8}$$

From a physical point of view, $\hat{\delta}$ is the normal distance at which the outer region of potential must be displaced in order for the momentum to be the same as that of the real fluid with the boundary layer. In this case, the effect due to the presence of the boundary layer is *modeled* by adding the momentum thickness to the shape of the body and by treating the entire flow domain as undisturbed.

These definitions are commonly used in the study of boundary layers and in general it is found that $\delta_{99\%} > \delta^* > \hat{\delta}$. Since the concept of boundary layer is based on the assumption of very small thickness compared to the streamwise extent of the boundary layer, these definitions are valid if and only if $\delta_{99\%} \ll x$, $\delta^* \ll x$ and $\hat{\delta} \ll x$, with x streamwise coordinate along the body. These conditions are verified as long as x is not too close to the leading edge (where $x = 0$).

7.2 Boundary Layer Equations

The fact that the thickness of the boundary layer is always very small compared to its streamwise extent makes it possible to use a Cartesian reference system to describe boundary layers, even when they form on curved surfaces.

Typically, the x coordinate is taken in the flow direction and follows the curvature of the surface exposed to the flow, starting from the leading edge where $x = 0$, while the y coordinate is normal to the surface, as shown in Fig. 7.1. With this coordinate system, and assuming two-dimensional flow, the N-S equations are:

$$\frac{\partial v_x}{\partial x} + \frac{\partial v_y}{\partial y} = 0 \quad , \tag{7.9}$$

$$\rho \left(\frac{\partial v_x}{\partial t} + v_x \frac{\partial v_x}{\partial x} + v_y \frac{\partial v_x}{\partial y} \right) = -\frac{\partial \mathcal{P}}{\partial x} + \mu \left(\frac{\partial^2 v_x}{\partial x^2} + \frac{\partial^2 v_x}{\partial y^2} \right) \quad , \tag{7.10}$$

$$\rho \left(\frac{\partial v_y}{\partial t} + v_x \frac{\partial v_y}{\partial x} + v_y \frac{\partial v_y}{\partial y} \right) = -\frac{\partial \mathcal{P}}{\partial y} + \mu \left(\frac{\partial^2 v_y}{\partial x^2} + \frac{\partial^2 v_y}{\partial y^2} \right) \quad . \tag{7.11}$$

As done for the lubrication theory, these equations can be made dimensionless with respect to the characteristic scales of the problem. The characteristic length scales along x and y are L and δ respectively, where L represents the longitudinal extent of the boundary layer while δ represents its thickness. It is thus possible to define:

$$\tilde{x} = \frac{x}{L} \quad , \quad \tilde{y} = \frac{y}{\delta} \quad . \tag{7.12}$$

For the velocity components, U_∞ can be used as the characteristic velocity along x while a suitably-defined characteristic velocity V must be introduced for the y direction. For time, the ratio L/U_∞ can be used as the characteristic time. For pressure, the characteristic pressure Π can be used, which must also be suitably defined.

The mass conservation equation in dimensionless form reads as:

$$\frac{U_\infty \delta}{VL} \frac{\partial \tilde{v}_x}{\partial \tilde{x}} + \frac{\partial \tilde{v}_y}{\partial \tilde{y}} = 0 \quad . \tag{7.13}$$

Since the two derivatives must have similar order of magnitude, it is possible to define the velocity V as $V = U_\infty \delta/L$. Since $\delta/L \ll 1$, it must be $V/U_\infty \ll 1$. The x component of the N-S equations becomes:

$$\frac{\rho U_\infty \delta^2}{\mu L} \left(\frac{\partial \tilde{v}_x}{\partial \tilde{t}} + \tilde{v}_x \frac{\partial \tilde{v}_x}{\partial \tilde{x}} + \tilde{v}_y \frac{\partial \tilde{v}_x}{\partial \tilde{y}} \right) = -\frac{\Pi \delta^2}{\mu U L} \frac{\partial \tilde{P}}{\partial \tilde{x}} +$$

$$+ \left[\left(\frac{\delta}{L} \right)^2 \frac{\partial^2 \tilde{v}_x}{\partial \tilde{x}^2} + \frac{\partial^2 \tilde{v}_x}{\partial \tilde{y}^2} \right] \quad . \tag{7.14}$$

In this equation, the viscous term:

$$\left(\frac{\delta}{L} \right)^2 \frac{\partial^2 \tilde{v}_x}{\partial \tilde{x}^2} + \frac{\partial^2 \tilde{v}_x}{\partial \tilde{y}^2} \quad ,$$

must have the same order of magnitude as the inertial term:

$$\frac{\rho U_\infty \delta^2}{\mu L} \left(\tilde{v}_x \frac{\partial \tilde{v}_x}{\partial \tilde{x}} + \tilde{v}_y \frac{\partial \tilde{v}_y}{\partial \tilde{y}} \right) \quad ,$$

since the boundary layer is, by definition, the region of the flow field where these two terms are equally important. For this reason, the term $\rho U_\infty \delta^2 / \mu L$ must have order of magnitude equal to one and can be used to define δ as:

$$\delta = \left(\frac{\mu L}{\rho U_\infty} \right)^{1/2} = L Re^{-1/2} \quad , \tag{7.15}$$

with $Re = U_\infty L / \nu$. The characteristic pressure can be defined in such a way that also the coefficient in front of the pressure gradient has order of magnitude equal to one:

$$\Pi = \frac{\mu U_\infty L}{\delta^2} = \rho U_\infty^2 \quad . \tag{7.16}$$

Based on the definition of boundary layer thickness, it is clear from Eq. (7.14) that the term $\partial^2 \tilde{v}_x / \partial \tilde{x}^2$ is negligible with respect to $\partial^2 \tilde{v}_x / \partial \tilde{y}^2$ for $Re \gg 1$, and so Eq. (7.14) becomes:

$$\frac{\partial \tilde{v}_x}{\partial \tilde{t}} + \tilde{v}_x \frac{\partial \tilde{v}_x}{\partial \tilde{x}} + \tilde{v}_y \frac{\partial \tilde{v}_x}{\partial \tilde{y}} = -\frac{\partial \tilde{P}}{\partial \tilde{x}} + \frac{\partial^2 \tilde{v}_x}{\partial \tilde{y}^2} \quad . \tag{7.17}$$

Using the definitions previously obtained for δ and Π, the y component of the N-S equation becomes:

$$\frac{\partial \tilde{P}}{\partial \tilde{y}} = \left(\frac{\delta}{L} \right)^2 \left[\left(\frac{\delta}{L} \right)^2 \frac{\partial^2 \tilde{v}_y}{\partial \tilde{x}^2} + \frac{\partial^2 \tilde{v}_y}{\partial \tilde{y}^2} - \frac{\partial \tilde{v}_y}{\partial \tilde{t}} - \tilde{v}_x \frac{\partial \tilde{v}_y}{\partial \tilde{x}} - \tilde{v}_y \frac{\partial \tilde{v}_y}{\partial \tilde{y}} \right] = 0 \quad . \tag{7.18}$$

From this equation, it follows that $\partial p/\partial y = 0$ and $\mathcal{P} = \mathcal{P}(x)$, which implies that the pressure inside the boundary layer is equal to the pressure in the free stream outside of the boundary layer, at the same x coordinate. Assuming potential flow conditions in the free stream, the Bernoulli equation reads as:

$$\mathcal{P} + \frac{1}{2}\rho U_\infty^2(x) = \text{constant} \quad , \tag{7.19}$$

$$\frac{d\mathcal{P}}{dx} = -\rho U_\infty \frac{dU_\infty}{dx} \quad . \tag{7.20}$$

With this assumption, the boundary layer equations in dimensional form are:

$$\frac{\partial v_x}{\partial x} + \frac{\partial v_y}{\partial y} = 0 \quad , \tag{7.21}$$

$$\rho \left(\frac{\partial v_x}{\partial t} + v_x \frac{\partial v_x}{\partial x} + v_y \frac{\partial v_x}{\partial y} \right) = \rho U_\infty \frac{dU_\infty}{dx} + \mu \frac{\partial^2 v_x}{\partial y^2} \quad . \tag{7.22}$$

These equations can be integrated with the appropriate boundary conditions, once the flow field outside of the boundary layer has been determined. The term $\partial v_x/\partial t$ will appear only in the case of unsteady boundary layers.

7.3 Boundary Layer on a Flat Plate

The boundary layer equations can be readily applied to the case of incompressible flow over a thin flat plate, since $dU_\infty/dx = 0$ and $\mathcal{P} = \text{constant}$ away from the wall. In 2D, the equations become:

$$\frac{\partial v_x}{\partial x} + \frac{\partial v_y}{\partial y} = 0 \quad , \tag{7.23}$$

$$\rho \left(v_x \frac{\partial v_x}{\partial x} + v_y \frac{\partial v_x}{\partial y} \right) = \mu \frac{\partial^2 v_x}{\partial y^2} \quad , \tag{7.24}$$

and the boundary conditions are the following:

$$y = 0 \qquad v_x, \ v_y = 0 \quad , \tag{7.25}$$

$$y \to \infty \qquad v_x \to U_\infty \quad , \tag{7.26}$$

$$x \le 0 \qquad v_x = U_\infty \quad . \tag{7.27}$$

To solve the problem, which was first treated by H. Blasius (1908) in his doctoral thesis, the assumption is made that the velocity profile has the same shape everywhere in the boundary layer, i.e. is self-similar. Based on this assumption, the validity of which shall be verified a posteriori, the following relation holds:

$$\frac{v_x(x, y)}{U_\infty} = \phi\left[\frac{y}{\delta(x)}\right] \quad , \tag{7.28}$$

where $\delta(x)$ is a function of x that can be interpreted as the thickness of the boundary layer at position x along the wall. To determine the function $\delta(x)$, it is convenient to introduce the so-called similarity variable:

$$\eta = \frac{y}{\delta(x)} \quad , \tag{7.29}$$

defined so that the velocity profile obeys the following relation:

$$\hat{v}_x(\hat{\eta}) = \hat{v}_x[\hat{\eta}(x_1, y_1)] = \hat{v}_x[\hat{\eta}(x_2, y_2)] \quad , \tag{7.30}$$

where (x_1, y_1) and (x_2, y_2) are the coordinates of two generic points with the same value of velocity $v_x = \hat{v}_x$ (and, consequently, the same value $\eta = \hat{\eta}$) within the boundary layer.

If the velocity field is only a function of η, it must be $v_x/U_\infty = \phi(\eta)$, and the system of partial differential Eqs. (7.23) and (7.24) can be transformed into a system of total differential equations. Considering Eq. (7.29), the derivatives with respect to x and y can be transformed into derivatives with respect to the variable η as follows:

$$\frac{\partial}{\partial y} = \frac{d}{d\eta}\frac{\partial\eta}{\partial y} = \frac{1}{\delta(x)}\frac{d}{d\eta} \quad , \tag{7.31}$$

$$\frac{\partial}{\partial x} = \frac{d}{d\eta}\frac{\partial\eta}{\partial x} = -\frac{y}{\delta(x)^2}\frac{d\delta}{dx}\frac{d}{d\eta} = -\frac{\eta}{\delta(x)}\frac{d\delta}{dx}\frac{d}{d\eta} \quad . \tag{7.32}$$

Therefore, the continuity equation becomes:

$$-\frac{\eta}{\delta(x)}\frac{d\delta(x)}{dx}\phi'U_\infty + \frac{1}{\delta(x)}\frac{dv_y}{d\eta} = 0 \quad , \tag{7.33}$$

from which v_y can be obtained as:

$$v_y = \frac{d\delta(x)}{dx}U_\infty\int_0^\eta \eta\phi'd\eta \quad , \tag{7.34}$$

where:

$$\int_0^\eta \eta \phi' d\eta = \eta\phi - \int_0^\eta \phi d\eta \quad . \tag{7.35}$$

Introducing the new variable, $f(\eta)$, defined as:

$$f(\eta) = \int_0^\eta \phi(\eta) d\eta \quad , \tag{7.36}$$

it follows that $f'(\eta) = \phi(\eta)$ and:

$$\frac{v_x}{U_\infty} = f'(\eta) \quad . \tag{7.37}$$

Equation (7.34) becomes:

$$\frac{v_y}{U_\infty} = \frac{d\delta(x)}{dx} \left[\eta f'(\eta) - f(\eta) \right] \quad . \tag{7.38}$$

To highlight the physical meaning of the function $f(\eta)$, it is convenient to observe that the stream function $\psi(x, y)$ can be expressed as:

$$\psi(x, y) = -U_\infty \delta(x) f(\eta) \quad . \tag{7.39}$$

Indeed, relations (7.37) and (7.38) follow immediately from Eq. (7.39) recalling that, by definition:

$$v_x = -\frac{\partial \psi}{\partial y} \quad , \quad v_y = \frac{\partial \psi}{\partial x} \quad . \tag{7.40}$$

With these values of v_x and v_y and making the coordinate change given by Eq. (7.29), the N-S equation becomes:

$$\rho U_\infty \left[-f' \frac{\eta}{\delta(x)} \frac{d\delta(x)}{dx} f'' + \frac{d\delta(x)}{dx} (\eta f' - f) \frac{f''}{\delta(x)} \right] = \frac{\mu}{\delta(x)^2} f''' \quad , \tag{7.41}$$

from which:

$$\frac{\rho U_\infty}{\mu} \delta(x) \frac{d\delta(x)}{dx} ff'' + f''' = 0 \quad . \tag{7.42}$$

In order for f to be only a function of η, the coefficient of the first term must not depend on x:

$$\frac{\rho U_\infty}{\mu} \delta(x) \frac{d\delta(x)}{dx} = \text{constant} \quad . \tag{7.43}$$

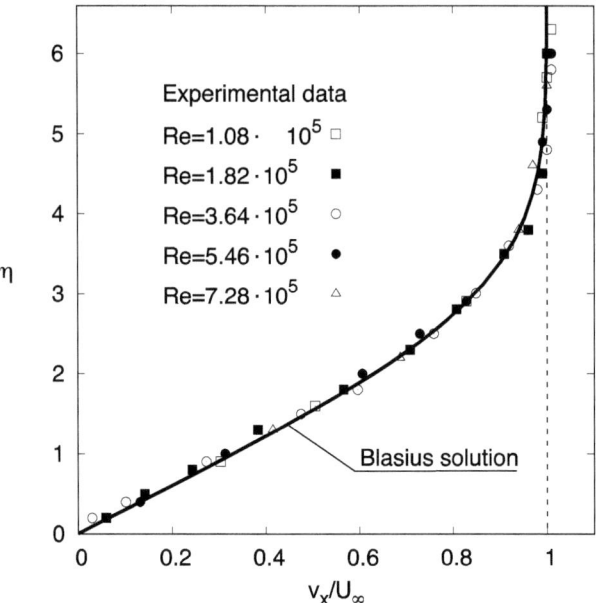

Fig. 7.3 Dimensionless velocity profile, v_x/U_∞, in the boundary layer as a function of the similarity variable, $\eta = y/\delta(x)$. The Blasius solution (solid line) accurately reproduces the experimental measurements (symbols)

Integration of Eq. (7.43) yields:

$$\frac{1}{2}\delta^2(x) = \text{constant}\ \frac{\mu}{\rho U_\infty}x \quad , \tag{7.44}$$

since $\delta(0) = 0$. The solution of the problem is independent of the value assigned to the constant that appears in Eq. (7.44). Here, the value $1/2$ is assumed. Therefore, Eqs. (7.42) and (7.44) can be written in the following form:

$$ff'' + 2f''' = 0 \quad , \tag{7.45}$$

$$\delta(x) = \sqrt{\frac{\mu x}{\rho U_\infty}} \quad . \tag{7.46}$$

Equation (7.45) is also known as the Blasius equation. The boundary conditions of the original problem can be recast as follows:

$$f(0) = f'(0) = 0 \ \text{ for } \ \eta = 0 \quad , \tag{7.47}$$

$$f' = 1 \ \text{ for } \ \eta \to \infty \quad . \tag{7.48}$$

It may seem that the condition $v_x = U_\infty$ for $x \leq 0$ was not used. However, having set $\delta = 0$ at $x = 0$, it must be $\eta = y/\delta(0) \to \infty$ and so this condition is taken into account by Eq. (7.48).

With the assigned boundary conditions, the differential equation for $f(\eta)$ can be solved numerically and the result of the numerical integration (Blasius solution) is shown in Fig. 7.3. Also numerically, it is possible to derive the following expression for the shear stress exerted on the plate:

$$\tau_w = \mu \left.\frac{\partial v_x}{\partial y}\right|_{y=0} = 0.332\, \mu\, U_\infty \sqrt{\frac{\rho U_\infty}{\mu x}} \ . \tag{7.49}$$

The local friction coefficient, measured at a given position x along the plate, is:

$$C_{f,x} = \frac{\tau_w}{\frac{1}{2}\rho U_\infty^2} = 0.664 \sqrt{\frac{\mu}{\rho U_\infty x}} = 0.664\, Re_x^{-1/2} \ , \tag{7.50}$$

where $Re_x = \rho U_\infty x / \mu$. This Reynolds number can be interpreted as the dimensionless form of the x position. The displacement thickness of the boundary layer, on the other hand, is:

$$\delta^* = \int_0^\infty \left[1 - \frac{v_x(y)}{U_\infty}\right] dy = 1.72 \sqrt{\frac{\mu x}{U_\infty}} = 1.72 \frac{\mu}{U_\infty} \sqrt{Re_x} \ . \tag{7.51}$$

Equation (7.51) shows that the thickness increases proportionally to $x^{1/2}$ while the shear stress at the wall decreases proportionally to $x^{-1/2}$, in agreement with the observation that the slope of the velocity profile within the boundary layer decreases as x increases.

7.3.1 Examples

1 Consider the flow at zero incidence of a stream of water ($\rho = 10^3$ kg/m^3, $\mu = 1 \cdot 10^{-3}$Pa $\cdot$ s) moving with velocity $U_\infty = 2$ m/s over a square flat plate ($L = W = 1$ m), as shown in Fig. 7.4. Assuming that the boundary layer that forms on the plate is laminar and neglecting wake effects, calculate the force exerted on the plate.

Solution The shear stress at the wall ($y = 0$) is given by Eq. (7.49):

$$\tau_w = \mu \left.\frac{\partial v_x}{\partial y}\right|_{y=0} = 0.332\, \mu\, U_\infty \sqrt{\frac{\rho U_\infty}{\mu x}} \ . \tag{7.52}$$

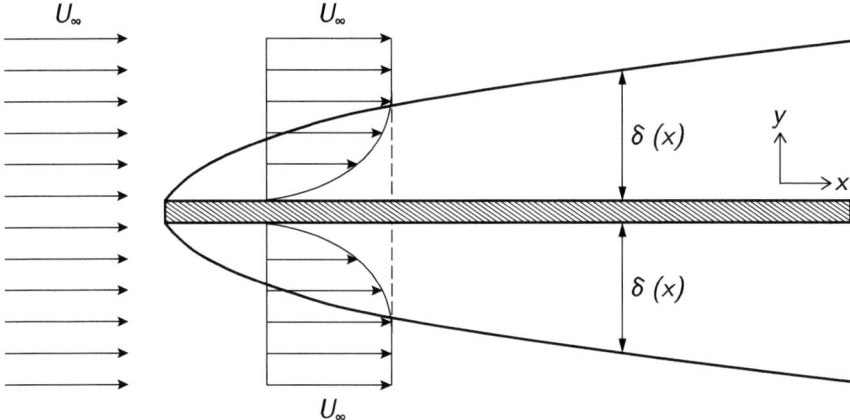

Fig. 7.4 Boundary layer on a flat plate due to a uniform flow at zero incidence

The force exerted on the wall is:

$$F = \int_A \tau_w \, dA \quad , \tag{7.53}$$

where A is the area of the wall. It follows that the force per unit width of the plate is:

$$\frac{F}{W} = \int_0^L \tau_w \, dx = \int_0^L 0.332 \, \mu \, U_\infty \sqrt{\frac{\rho U_\infty}{\mu x}} \, dx =$$

$$= 0.332 \, \sqrt{\mu \rho} \, U_\infty^{3/2} \int_0^L \frac{dx}{\sqrt{x}} = 0.332 \, \sqrt{\mu \rho} \, U_\infty^{3/2} \, 2 \sqrt{x} \, \big|_0^L \quad . \tag{7.54}$$

This is the force on just one face of the plate. Therefore, the total force is:

$$F_t = 4 \cdot 0.332 \, \sqrt{\mu \rho} \, U_\infty^{3/2} \, \sqrt{x} \, \big|_0^L = 3.75 \, N/m \quad . \tag{7.55}$$

$\boxed{2}$ Consider the zero-incidence flow, characterized by a free stream velocity U_∞, over the flat plate shown in Fig. 7.5. Let the plate be porous so that the fluid can flow through it with velocity $-v_0$. Derive the expression of the velocity profile and the expression of the wall shear.

Solution The solution to the problem is sought among those for which $\mathcal{P} = $ constant and $v_x = v_x(y)$. Namely, a solution to the balance equations is sought for which the value of the suction velocity v_0 is such that the tangential velocity component is independent of x. By assuming steady flow and uniform fluid density, the continuity equation (for the two-dimensional problem) becomes:

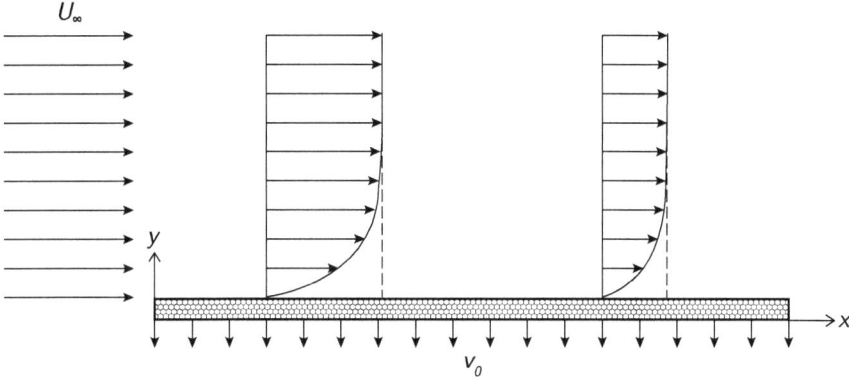

Fig. 7.5 Schematic representation of boundary layer with suction

$$\frac{\partial v_y}{\partial y} = 0 \quad , \tag{7.56}$$

while the *N-S* equations read as:

$$v_y \frac{dv_x}{dy} = \nu \frac{d^2 v_x}{dy^2} \quad ,$$

$$v_x \frac{\partial v_y}{\partial x} + v_y \frac{\partial v_y}{\partial y} = \nu \left(\frac{\partial^2 v_y}{\partial x^2} + \frac{\partial^2 v_y}{\partial y^2} \right) \quad . \tag{7.57}$$

The boundary conditions are:

$$\begin{aligned}
y = 0 &\rightarrow v_x(0) &&= 0 \quad , \\
y = 0 &\rightarrow v_y(x, 0) &&= -v_0 \quad , \\
y = \infty &\rightarrow v_x(y) &&= U_\infty \quad .
\end{aligned} \tag{7.58}$$

Integration of the continuity equation yields a constant value for the wall-normal velocity component: $v_y(x, y) = -v_0$. Using this result, the momentum conservation equations can be recast as:

$$- v_0 \frac{dv_x}{dy} = \nu \frac{d^2 v_x}{dy^2} \quad . \tag{7.59}$$

Equation (7.59) can be integrated with respect to y to obtain:

$$\nu \frac{d^2 v_x}{dy^2} + v_0 \frac{dv_x}{dy} = 0 \longrightarrow \frac{dv_x}{dy} + \frac{v_0}{\nu} v_x + C_1 = 0 \quad . \tag{7.60}$$

This equation admits the particular solution $v_{xp} = -C_1 \nu / v_0$. Hence, the complete solution is:

$$v_x(y) = C_2 e^{-v_0 y / \nu} - C_1 \frac{\nu}{v_0} \quad . \tag{7.61}$$

The two remaining boundary conditions yield:

$$v_x(y = 0) = C_2 - C_1 \frac{\nu}{v_0} = 0 \quad \rightarrow \quad C_2 = C_1 \frac{\nu}{v_0} \quad ,$$

$$v_x(y \rightarrow \infty) = -C_1 \frac{\nu}{v_0} = U_\infty \rightarrow C_1 = -U_\infty \frac{v_0}{\nu} \rightarrow C_2 = -U_\infty \quad , \tag{7.62}$$

and the velocity field is:

$$
\begin{aligned}
v_x(y) &= U_\infty \left(1 - e^{-v_0 y / \nu} \right) \quad , \\
v_y(x, y) &= -v_0 \quad .
\end{aligned}
\tag{7.63}
$$

These expressions verify the *N-S* equations. Note that the velocity profile does not depend on x.

The tangential shear stress at the wall is:

$$\tau_{xy} = \tau_0 = \mu \left(\frac{dv_x}{dy} \right) = \mu U_\infty \frac{v_0}{\nu} = \rho U_\infty v_0 \quad . \tag{7.64}$$

Note that τ_0 does not depend on the fluid viscosity.

7.4 Boundary Layer on a Wall Suddenly Set into Motion

Another case in which the self-similarity assumption can be applied is provided by the boundary layer that develops over time rather than along a spatial coordinate. This problem, which was first solved by G.G. Stokes (1856) in his famous treatment of the pendulum and is often referred to as the *first Stokes problem* in the literature, is illustrated in Fig. 7.6, where the flow field near a wall suddenly set in motion at time $t = 0$ and at constant velocity U_∞, is shown.

Simple considerations allow to conclude that the solution of the problem has the form $v_x = v_x(y, t)$, $v_y = v_z = 0$.

The three scalar components of the *N-S* equation become:

$$\rho \frac{\partial v_x}{\partial t} = \mu \frac{\partial^2 v_x}{\partial y^2} \quad , \tag{7.65}$$

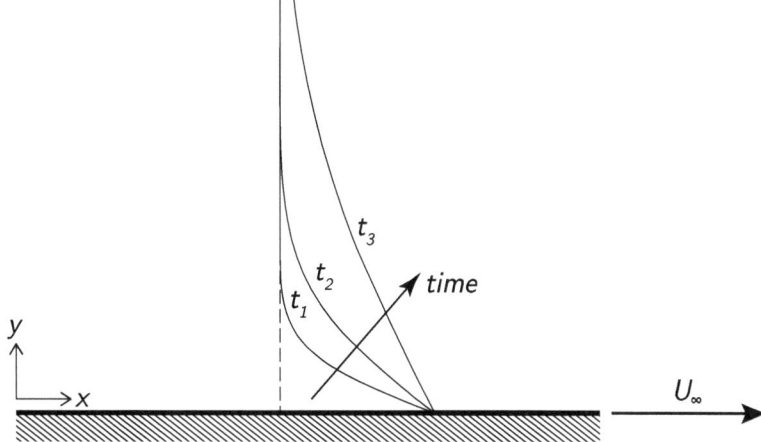

Fig. 7.6 Time evolution of the velocity profile generated by a wall suddenly set into motion

$$\frac{\partial \mathcal{P}}{\partial y} = 0 \ , \tag{7.66}$$

$$\frac{\partial \mathcal{P}}{\partial z} = 0 \ . \tag{7.67}$$

Equation (7.65) shows how, in this problem, there is no convective transport of momentum along the y direction, but only transport by diffusion. In this respect, the problem is analogous to the process of heat transfer by conduction (determined by a temperature gradient and governed by Fourier's law) or the process of mass transfer (determined by a concentration gradient and governed by Fick's law). Equations (7.66) and (7.67) imply $\mathcal{P} =$ constant and it can be assumed that the solution for v_x has the form:

$$\frac{v_x(y, t)}{U_\infty} = \phi \left[\frac{y}{\delta(t)} \right] . \tag{7.68}$$

It is possible to define the similarity variable $\eta = y/\delta(t)$ so that the velocity profile obeys the following relation:

$$\hat{v}_x(\hat{\eta}) = \hat{v}_x[\hat{\eta}(y_1, t_1)] = \hat{v}_x[\hat{\eta}(y_2, t_2)] \ , \tag{7.69}$$

where $\hat{v}_x$ is the velocity measured at distance y_1 from the wall at time t_1 but also at distance y_2 at a later time t_2 chosen such that $\hat{\eta}(y_1, t_1) = \hat{\eta}(y_2, t_2)$. Proceeding as in the case of boundary layer on a flat plate and replacing the variable x with t, the following equation can be written:

$$-\rho\frac{\eta}{\delta(t)}\frac{d\delta(t)}{dt}U_\infty\phi' = \frac{\mu}{\delta(t)^2}\phi''U_\infty \quad . \tag{7.70}$$

Rearraning the equation, it can be recast as:

$$\frac{\rho}{\mu}\delta(t)\frac{d\delta(t)}{dt}\eta\phi' + \phi'' = 0 \quad , \tag{7.71}$$

with:

$$\frac{\rho}{\mu}\delta(t)\frac{d\delta(t)}{dt} = \text{constant} = 2 \quad . \tag{7.72}$$

For this value of the constant, the following expressions are obtained for $\delta(t)$, η and $\phi(\eta)$:

$$\delta(t) = 2\sqrt{\frac{\mu}{\rho}t} \quad , \tag{7.73}$$

$$\eta = \frac{y}{\sqrt{4\mu t/\rho}} \quad , \tag{7.74}$$

$$2\eta\phi' + \phi'' = 0 \quad . \tag{7.75}$$

Equation (7.75) can be solved analytically with the following boundary conditions:

$$\eta = 0 \qquad \phi = 1 \quad , \tag{7.76}$$

$$\eta \to \infty \qquad \phi = 0 \quad . \tag{7.77}$$

By integrating twice with respect to η, the analytical solution of Eq. (7.75) can be derived, which has the following general form:

$$\phi = C_1\int_0^\eta e^{-\eta^2}d\eta + C_2 \quad . \tag{7.78}$$

The integration constants can be obtained from the boundary conditions. In particular, the condition $\phi(\eta = 0) = 1$ yields $C_2 = 1$, while the condition $\phi(\eta \to \infty) = 0$ yields:

$$\phi = C_1\int_0^\infty e^{-\eta^2}d\eta + 1 = 0 \qquad \to \qquad C_1 = -\frac{1}{\int_0^\infty e^{-\eta^2}d\eta} \quad . \tag{7.79}$$

Fig. 7.7 Dimensionless velocity profile generated in a fluid by a wall suddenly set into motion as a function of the variable $\eta = y/\sqrt{4\mu t/\rho}$

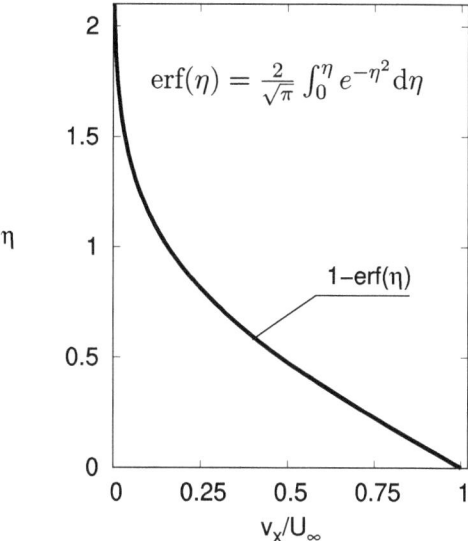

$$\text{erf}(\eta) = \frac{2}{\sqrt{\pi}} \int_0^\eta e^{-\eta^2} d\eta$$

1−erf(η)

v_x/U_∞

Since $\int_0^\infty e^{-\eta^2} d\eta = \sqrt{\pi}/2$, it follows that $C_1 = -\frac{2}{\sqrt{\pi}}$ and:

$$\phi = 1 - \frac{2}{\sqrt{\pi}} \int_0^\eta e^{-\eta^2} d\eta \quad . \tag{7.80}$$

The integral on the right-hand side of Eq. (7.80) is the *error function*, also denoted as erf(η), and is a tabulated function of η. Figure 7.7 shows the solution in dimensionless form.

7.4.1 Examples

$\boxed{1}$ Derive the expression of the velocity profile and shear stress for the boundary layer on a flat plate that moves with uniform velocity U_∞ for $t < 0$ and instantaneously stopped at $t = 0$.

Solution Let us denote by x the direction along which the wall moves and by y the direction normal to the wall. The velocity components are $v_x = v_x(y, t)$ and $v_y = v_z = 0$. The continuity equation is identically satisfied, while the N-S equations become:

$$\frac{\partial v_x}{\partial t} = \nu \frac{\partial^2 v_x}{\partial y^2} \quad , \tag{7.81}$$

$$\frac{\partial P}{\partial y} = 0 \quad , \tag{7.82}$$

$$\frac{\partial P}{\partial z} = 0 \quad . \tag{7.83}$$

Equations (7.82) and (7.83) imply that $\mathcal{P} = \text{const}$. Assuming that the solution for v_x has the form:

$$v_x(y, t) = U_\infty \left[\frac{y}{\delta(t)} \right] = U_\infty \phi(\eta) \quad , \tag{7.84}$$

with $\eta = y/\delta(t)$, it is possible to rewrite the terms in Eq. (7.81) as:

$$\frac{\partial v_x}{\partial t} = -U_\infty \frac{\eta}{\delta(t)} \frac{d\delta}{dt} \frac{d\phi}{d\eta} \quad , \tag{7.85}$$

and as:

$$\frac{\partial^2 v_x}{\partial y^2} = U_\infty \frac{1}{\delta^2(t)} \frac{d^2\phi}{d\eta^2} \quad . \tag{7.86}$$

Therefore, Eq. (7.81) becomes:

$$\eta \, \delta(t) \frac{d\delta}{dt} \frac{d\phi}{d\eta} + \nu \frac{d^2\phi}{d\eta^2} = 0 \quad . \tag{7.87}$$

Setting $\phi' = d\phi/d\eta$ and $\phi'' = d^2\phi/d\eta^2$, and being $\delta(t) = 2\sqrt{\nu t}$, Eq. (7.87) becomes:

$$2 \, \eta \, \phi' + \phi'' = 0 \quad . \tag{7.88}$$

The boundary conditions are:

$$v_x(y \to +\infty, t) = U_\infty \quad \longrightarrow \quad \phi(\eta \to +\infty) = 1 \quad , \tag{7.89}$$

$$v_x(y \to 0, t) = 0 \quad \longrightarrow \quad \phi(\eta \to 0) = 0 \quad . \tag{7.90}$$

Integration of Eq. (7.88) with these conditions gives the following solution:

$$\phi = C_1 \int_0^\eta e^{-\eta^2} d\eta + C_2 \quad , \tag{7.91}$$

with:

$$C_1 = \frac{1}{\int_0^\infty e^{-\eta^2} d\eta} = \frac{2}{\sqrt{\pi}} \quad , \quad C_2 = 0 \quad . \tag{7.92}$$

Therefore, the velocity profile in dimensionless form is:

$$\frac{v_x(\eta)}{U_\infty} = \frac{2}{\sqrt{\pi}} \int_0^\eta e^{-\eta^2} d\eta \quad . \tag{7.93}$$

The shear stress is:

$$\tau_{xy} = \mu \frac{dv_x}{dy} = \mu \frac{dv_x}{d\eta} \frac{d\eta}{dy} = \mu \frac{dv_x}{d\eta} \frac{1}{\delta(t)} = \mu U_\infty \frac{d\phi}{d\eta} \frac{1}{\delta(t)} \quad , \tag{7.94}$$

or, equivalently (recalling that $\delta(t) = 2\sqrt{\nu t}$ and $d\phi/d\eta = 2e^{-\eta^2}/\sqrt{\pi}$):

$$\tau_{xy}(\eta) = \frac{\mu U_\infty}{\sqrt{\pi \nu t}} e^{-\eta^2} \quad . \tag{7.95}$$

7.5 Separation of the Boundary Layer

In general, it is not possible to predict the forces acting on a body immersed in a moving fluid by means of the potential flow theory, even when the theory is coupled with the description of the boundary layer. This because conditions favoring the separation of the boundary layer may be achieved even at low Reynolds number. To understand this phenomenon, it is useful to consider the case illustrated in Fig. 7.8, which represents the flow field around a cylinder in the presence of boundary layer separation and formation of a couple of stationary, couter/rotating vortices in the rear part of the cylinder. The velocity profiles generated by such a flow field are shown in Fig. 7.9 as well as in Fig. 7.10, which refers to the case of negligible curvature of the cylinder surface. This assumption is justified by the fact that the boundary layer thickness is much smaller than the cylinder radius: $\delta << R$.

In the forward portion of the cylinder, which is directly exposed to the incoming flow and is indicated qualitatively by point A in Fig. 7.10, the velocity profiles vary from zero at the surface of the cylinder (no-slip condition) to a relatively large value within a short distance from the surface, their slope being $\partial v_x/\partial y > 0$. The slope is maximum at the surface of the cylinder and gradually decreases as the distance from the surface increases, vanishing for $y \geq \delta$: This implies $\partial^2 v_x/\partial y^2 < 0$ (condition of locally concave profile). Recall now that, in a two-dimensional system, the vorticity equation is given by:

$$\mathbf{v} \cdot \nabla \omega = \frac{1}{Re} \nabla^2 \omega \quad , \tag{7.96}$$

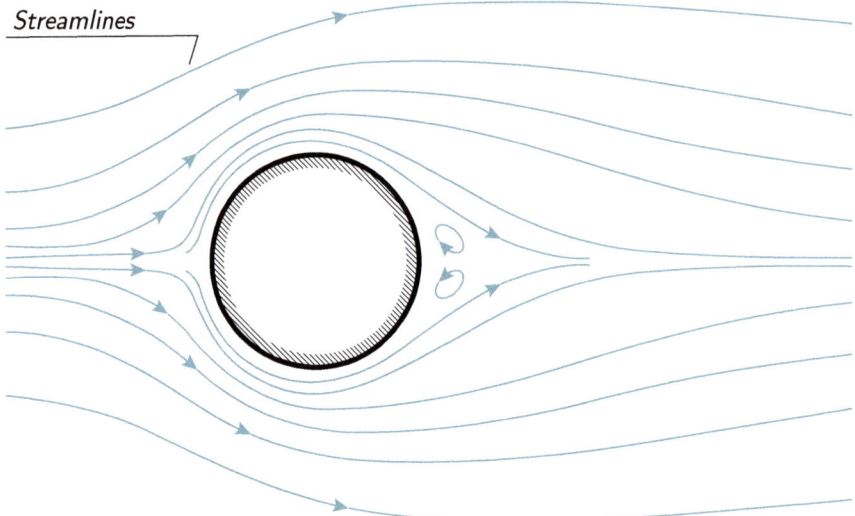

Fig. 7.8 Illustration of the boundary layer separation in the rear portion of a cylinder. As a result of separation, two stationary counter-rotating vortices are formed in the wake of the cylinder

with:

$$\omega = \frac{\partial v_y}{\partial x} - \frac{\partial v_x}{\partial y} \quad , \tag{7.97}$$

by definition. At the wall, $v_x = v_y = 0$, and therefore:

$$\left.\frac{\partial v_y}{\partial x}\right|_{y=0} = 0 \quad , \quad \left.\frac{\partial v_x}{\partial x}\right|_{y=0} = 0 \quad . \tag{7.98}$$

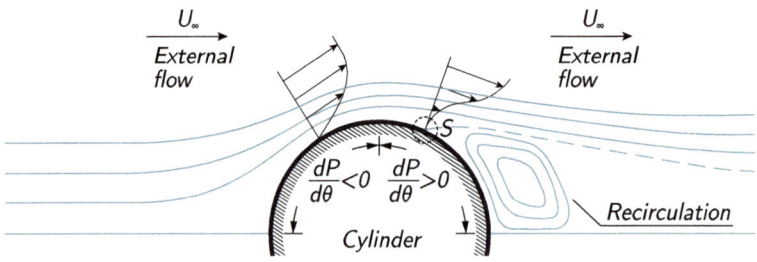

Fig. 7.9 Schematic illustration of boundary layer separation in the wake of a cylinder. The different velocity profiles and the formation of an adverse pressure gradient are shown. The velocity profiles are useful to understand the structure of the flow field around the cylinder. The velocity of the incoming fluid is tangent to the cylinder only in the boundary layer region, while being always parallel to U_∞ in the outer flow region

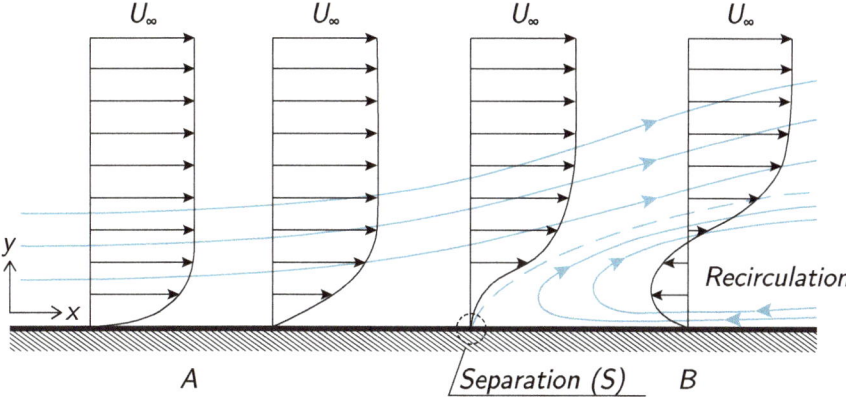

Fig. 7.10 Schematic representation of boundary layer separation. The different velocity profiles are also shown. The change in concavity is associated with the formation of an adverse pressure gradient

From the continuity equation, it follows that $(\partial v_y / \partial y)|_{y=0} = 0$. Therefore, according to Eq. (7.97), the following relations apply at the wall:

$$
\omega|_{y=0} = - \left. \frac{\partial v_x}{\partial y} \right|_{y=0} = - \frac{\tau_{yx}|_{y=0}}{\mu} \quad , \quad \left. \frac{\partial \omega}{\partial y} \right|_{y=0} = - \left. \frac{\partial^2 v_x}{\partial y^2} \right|_{y=0} . \tag{7.99}
$$

Equation (7.99) indicates that in the forward, flow-facing portion of the cylinder: $\omega|_{y=0} < 0$, $\tau_{yx}|_{y=0} > 0$ and $(\partial \omega / \partial y)|_{y=0} > 0$. Moreover, considering the boundary layer equation in the form:

$$
\rho \left(v_x \frac{\partial v_x}{\partial x} + v_y \frac{\partial v_x}{\partial y} \right) = -\frac{\partial P}{\partial x} + \mu \left(\frac{\partial^2 v_x}{\partial y^2} \right) \quad , \tag{7.100}
$$

at the wall, the following relation holds:

$$
\left. \left(-\frac{\partial P}{\partial x} + \mu \frac{\partial^2 v_x}{\partial y^2} \right) \right|_{y=0} = 0 \quad , \tag{7.101}
$$

and thus, from Eq. (7.99):

$$
\left. \frac{\partial P}{\partial x} \right|_{y=0} = \mu \left. \frac{\partial^2 v_x}{\partial y^2} \right|_{y=0} \quad \rightarrow \quad \left. \frac{\partial P}{\partial x} \right|_{y=0} = -\mu \left. \frac{\partial \omega}{\partial y} \right|_{y=0} \quad . \tag{7.102}
$$

Therefore, in the forward portion of the cylinder, there is a *favorable* pressure gradient, $(\partial P / \partial x)|_{y=0} < 0$, in the streamwise direction. This gradient corresponds to a decrease of pressure along the surface of the cylinder, which generates a minimum of pressure before the fluid is able to flow past the entire surface of the cylinder. At the location where the minimum occurs, the pressure gradient is $(\partial P/\partial x)|_{y=0} = 0$, and the second derivative of velocity is $\partial^2 v_x/\partial y^2 = 0$: This condition corresponds to a change of concavity in the velocity profile. As a consequence, at points located downstream of the minimum, there is a change of sign of both the second derivative, $\partial^2 v_x/\partial y^2 > 0$ (condition of locally convex profile), and the pressure gradient, $(\partial P/\partial x)|_{y=0} > 0$, according to Eq. (7.102). The pressure gradient in the flow direction becomes unfavorable (or *adverse*).

The change in concavity of the velocity profiles, associated with the change from a favorable to an adverse pressure gradient, is shown in Fig. 7.10 for the case of a boundary layer on a flat plate, which also mimics the case of a boundary layer on a cylinder with small curvature or on an airfoil. An important consequence of the change in concavity is that the slope, $\partial v_x/\partial y$, of the velocity profile gradually decreases. The point S in Figs. 7.9 and 7.10 represents the position on the surface of the cylinder at which the slope becomes zero, $(\partial v_x/\partial y)|_{y=0} = 0$, and separation occurs. At points located downstream of S, there is a change of sign of the first derivative, $\partial v_x/\partial y > 0$, which corresponds to an inversion of the slope of the velocity profile, namely to a recirculation of fluid near the surface of the cylinder (see point B in Fig. 7.10). The vorticity and the shear stress also change sign, according to Eq. (7.99).

Summarising, with reference to Fig. 7.10:

At point A:

$$\frac{\partial P}{\partial x} < 0 \quad , \quad \frac{\partial v_x}{\partial y} > 0 \quad , \quad \tau_{yx} > 0 \quad , \quad \omega < 0 \quad , \quad \frac{\partial \omega}{\partial y} > 0 \quad . \quad (7.103)$$

At point S:

$$\frac{\partial P}{\partial x} > 0 \quad , \quad \frac{\partial v_x}{\partial y} = 0 \quad , \quad \tau_{yx} = 0 \quad , \quad \omega = 0 \quad , \quad \frac{\partial \omega}{\partial y} < 0 \quad . \quad (7.104)$$

At point B:

$$\frac{\partial P}{\partial x} > 0 \quad , \quad \frac{\partial v_x}{\partial y} < 0 \quad , \quad \tau_{yx} < 0 \quad , \quad \omega > 0 \quad , \quad \frac{\partial \omega}{\partial y} < 0 \quad . \quad (7.105)$$

Note that, downstream of the separation point S, the wall region is characterized by positive vorticity and negative vorticity derivative. Such a region is necessary to introduce positive vorticity into a flow where vorticity is negative. Indeed, as indicated by the lack of a source term in Eq. (7.96), vorticity can only be generated at the walls in the flow under consideration.

The analysis carried out so far shows that the presence of an adverse pressure gradient is a necessary condition for triggering the separation of the boundary layer. This condition alone, however, is not sufficient. To have separation, the contribution of the viscous effects acting on any fluid element that moves within the boundary layer must be taken into account. Because of these effects, the fluid element loses part of its kinetic energy (or, equivalently, momentum): if this loss is high enough to prevents the element from completing its motion in the region where $\partial P/\partial x > 0$, then the conditions required to have separation (at point S) are met. A fluid element moving in the region outside the boundary layer, on the other hand, is not subject to viscous effects, albeit being subject to the same pressure field (according to Eq. (7.18), the pressure in the boundary layer is equal to the pressure in the free stream at the same x coordinate).

Recalling Eq. (7.22), it is possible to verify that a favorable pressure gradient implies $dU_\infty/dx > 0$. This condition corresponds to an acceleration in the outer region of the flow that promotes the adhesion of the fluid to the surface of the cylinder through the generation of high shear stress at the wall. In this case, the thickness of the boundary layer is reduced and convective effects dominate over diffusive effects, which would tend to thicken the boundary layer. In contrast, an adverse pressure gradient implies $dU_\infty/dx < 0$, a condition that corresponds to deceleration in the outer region of the flow and favors the thickening of the boundary layer as well as the detachment of fluid from the surface, due to the predominance of viscous diffusion effects.

7.6 Plane Free Jet

In analogy to the boundary layer on a flat wall, the problem of a fluid jet injected into a quiescent fluid of equal physical properties through a long, thin gap can be solved using a similar procedure.

For simplicity, let us assume that the gap issuing the jet is infinitesimally thin and that the jet is issued at infinite velocity to ensure that the initial flowrate and momentum of the jet have a finite value. As illustrated in Fig. 7.11, x is the coordinate along the jet axis and y is the coordinate parallel to the wall. The velocity components in x and y are v_x and v_y, respectively. The quiescent fluid is at uniform pressure, and so is the fluid within the jet. Since the momentum per unit volume is ρv_x and is transported downstream with velocity v_x, the total momentum flux, $M(x)$, per unit width of the gap is:

$$M(x) = \int_{-\infty}^{\infty} \rho v_x^2(x, y)\mathrm{d}y \quad .$$

(7.106)

Assuming a negligible pressure gradient along x, the total momentum along x must remain constant, namely: $M(x) = M_0 = \text{constant}$.

The equations describing the velocity field of a two-dimensional laminar jet are the continuity equation:

$$\frac{\partial v_x}{\partial x} + \frac{\partial v_y}{\partial y} = 0 \quad ,$$

(7.107)

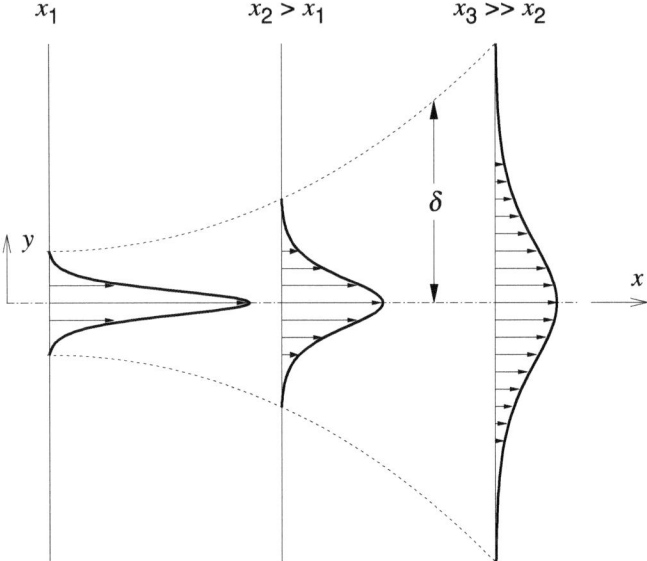

Fig. 7.11 Spatial development of a plane free jet. The evolution of the velocity profiles within the jet is shown. The dashed line indicates the jet width, δ

and the *N-S* equation along the x direction:

$$v_x \frac{\partial v_x}{\partial x} + v_y \frac{\partial v_x}{\partial y} = \nu \frac{\partial^2 v_x}{\partial y^2} \quad . \tag{7.108}$$

The difference with respect to the boundary layer solution lies in the different boundary conditions that must be adopted:

$$v_x \to 0 \quad \text{for} \quad y \to \pm\infty \ ,$$

$$\frac{\partial v_x}{\partial y} = 0 \text{ and } v_y = 0 \quad \text{for} \quad y = 0 \quad . \tag{7.109}$$

These conditions stem from the fact that the ambient fluid into which the jet is issued is quiescent and from the fact that the jet is symmetrical, respectively.

To solve the system of Eqs. (7.107) and (7.108), it is assumed that the velocity profiles are self-similar and that there exists a function $\delta(x)$, representing the width of the jet, such that the velocity field can be represented as a function of the variable $\eta = y/\delta(x)$ (see Fig. 7.11).

In this problem, there is no characteristic velocity that can be used to make the *N-S* equations non-dimensional. At most, it is possible to represent the x component of the velocity in the form:

$$\frac{v_x(x, y)}{U(x)} = \phi(\eta) \quad , \tag{7.110}$$

where $U(x)$ is the maximum velocity of the jet at distance x from its origin. Equation (7.110) states that the $v_x(x, y)$ profiles along y, normalized by $U(x)$, are similar and, therefore, they can be represented as a function of the similarity variable η. Equation (7.106) allows to determine a relation between $U(x)$ and $\delta(x)$, once written in the form:

$$M_0 = \int_{-\infty}^{+\infty} \rho U(x)^2 \phi(\eta)^2 \delta(x) d\eta \quad , \tag{7.111}$$

where $dy = \delta(x)d\eta$. In order for M_0 to be independent from x, it must be:

$$U(x)^2 \delta(x) = \text{constant} \quad . \tag{7.112}$$

By substituting Eq. (7.110) into the continuity equation and recalling Eqs. (7.31) and (7.32), it is possible to find an expression for $v_y(x, y)$:

$$v_y(x, y) = \frac{dU(x)}{dx} \delta(x)(f - 2\eta f') \quad , \tag{7.113}$$

where $f(\eta) = \int_0^{\eta} \phi(\eta)d\eta$. The function $f(\eta)$ is directly related to the stream function $\psi(x, y)$ by the relation:

$$\psi(x, y) = -\delta(x)U(x)f(\eta) \quad . \tag{7.114}$$

While the y component of the velocity, $v_y(x, y) = \partial\psi/\partial x$, is defined by the expression (7.113), the x-component reads as follows:

$$v_x(x, y) = -\frac{\partial\psi}{\partial y} = U(x)f'(\eta) \quad . \tag{7.115}$$

By substituting expressions (7.113) and (7.115) into Eq. (7.108), and skipping some steps of the derivation, the following equation for $f(\eta)$ is obtained:

$$-\frac{\delta^2(x)}{\nu}\frac{dU}{dx}(f'^2 + f \cdot f'') + f''' = 0 \quad . \tag{7.116}$$

In this equation, f depends solely on η if and only if:

$$-\frac{\delta^2(x)}{\nu}\frac{dU}{dx} = \text{constant} \quad . \tag{7.117}$$

The value of the constant is arbitrary and, in this case, it is convenient to set the constant equal to 2. Equation (7.117) can be easily integrated together with Eq. (7.112), and the solution can be written as:

$$U(x) = \alpha x^{-1/3} \quad , \quad \delta(x) = \beta x^{2/3} \quad , \tag{7.118}$$

where the two constants α and β must satisfy the relation:

$$\alpha\beta^2 = 6\nu \quad . \tag{7.119}$$

It is now necessary to integrate Eq. (7.116). The integration yields:

$$f(\eta) = \tanh(\eta) \quad . \tag{7.120}$$

The unknown constants α and β can be obtained from Eq. (7.119) and from the conservation of momentum, Eq. (7.111), which becomes:

$$M_0 = \rho U(x)^2 \delta(x) \int_{-\infty}^{+\infty} \sinh^{-4}(\eta)\mathrm{d}\eta = \frac{48\nu^2\rho}{\beta^3} \quad . \tag{7.121}$$

From this equation, it follows that:

$$\beta = \left(\frac{48\nu^2\rho}{M_0}\right)^{1/3} \quad , \quad \alpha = \left(\frac{3M_0^2}{32\rho^2\nu}\right)^{1/3} \quad . \tag{7.122}$$

The dimensionless jet velocity, v_x/U, is shown in Fig. 7.12.
    The jet flowrate per unit width is:

$$Q = \int_{-\infty}^{+\infty} v_x\mathrm{d}y = \left(\frac{36M_0\nu}{\rho}\right)^{1/3} x^{1/3} \quad . \tag{7.123}$$

Interestingly, the flowrate increases along the x coordinate. This is due to the entrainment of the initially quiescent fluid by the jet.

Problems

a The water in a deep lake (density ρ, viscosity μ) is initially at rest. Suddenly, a horizontal wind begins to blow with constant velocity v, setting the upper water layers into motion without altering the flatness of the free surface. Describe the development of the velocity profile in the lake, from the free surface, where the velocity is v, to the deeper layers, where the velocity of water remains for all practical purposes equal to zero.

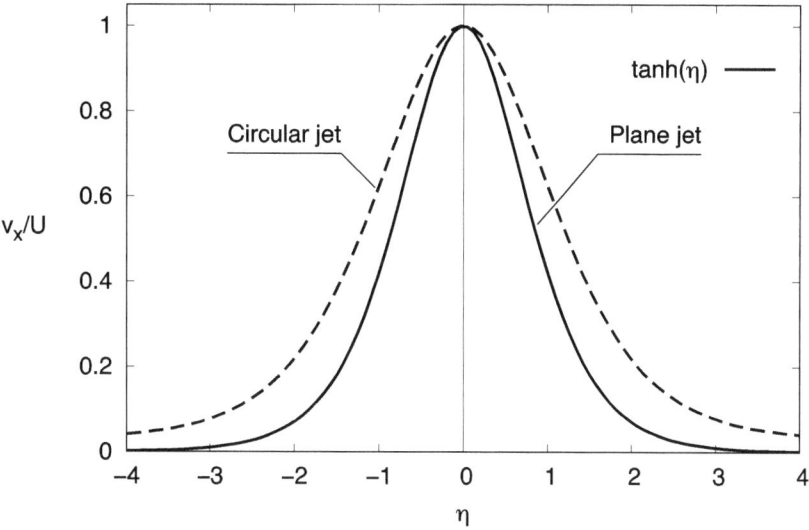

Fig. 7.12 Axial velocity distribution in a plane free jet (solid line), compared with the velocity distribution in a circular jet (dashed line)

$\boxed{b}$ A fluid jet issuing from a submerged orifice is injected into a quiescent fluid that has the same physical properties. Assume that the size of the orifice is infinitesimal and that the same assumptions used to calculate the velocity profile of a plane jet can be applied (Sect. 7.6).

1. Derive an expression for the dimensionless velocity profile shown in Fig. 7.12.
2. Prove that the volumetric flowrate can be expressed as $Q = 8\pi\nu z$, where z is the axial coordinate.

Chapter 8
Introduction to Turbulent Flows

8.1 Laminar and Turbulent Flows

The turbulent flow of a fluid is characterized by temporal and spatial fluctuations of its macroscopic properties, that is velocity, pressure and possibly temperature (in the case of non-isothermal flows), while preserving constant mean values over time in the case of steady flow in a broad sense. Turbulence can be described as the result of instability phenomena that occur in the flow. Consider, for example, the flow of a fluid in a pipe. The Reynolds number based on the pipe diameter and on the average fluid velocity is the parameter for determining whether the flow is laminar or turbulent. Below the critical value $Re = 2100$, the flow is laminar and stable: even if a perturbation of arbitrary amplitude (e.g. produced by vibrations of the pipe) is applied, laminar flow conditions are re-established after a transient associated with the perturbation. Above the critical Reynolds number, on the other hand, the effect of a perturbation is amplified and the flow becomes turbulent. The turbulent regime is strongly diffusive, as it promotes a much stronger mixing between fluid elements than molecular diffusion.

To understand the difference between turbulent diffusion and molecular diffusion, a useful example is provided by the transport of cigarette smoke inside a room. If the transport occurred by molecular diffusion only, then the time taken to cover a certain distance L could be estimated as $T \simeq L^2/\nu$, with ν the kinematic viscosity of air, providing a measure of momentum diffusivity. Assuming $L = 4\ m$ (size of a room, for example), the time would be $T \simeq 10^6\ s$, which is about 12 days. The estimate is reduced only slightly using, instead of ν, the diffusivity of smoke in air, $k = \nu \cdot Sc$ with Sc the Schmidt number (representing the ratio of momentum diffusivity to mass diffusivity, and equal to 0.7 in the present example). The estimated values obtained for T are significantly higher than those suggested by empirical experience, which suggests that only a few seconds are needed to sense the presence of smoke in the room. The reason for this discrepancy is due to the fact that the previous estimates do not take into account turbulent diffusion.

© The Author(s), under exclusive license to Springer Nature Switzerland AG 2024   235
A. Soldati and C. Marchioli, *Fluid Mechanics for Mechanical Engineers*,
https://doi.org/10.1007/978-3-031-53950-3_8

The turbulent regime is also strongly dissipative, as it promotes mixing between fluids elements carrying different momentum and produces an increase in the internal energy of the fluid at the expense of kinetic energy. In addition, the turbulent regime is strongly rotational since velocity fluctuations generate vorticity fluctuations. The turbulent flow field is three-dimensional and time-dependent and, in general, no simplifications can be made that would allow to solve the *N-S* equations in closed form, given the strong non linearity that characterizes the convective terms. The only way to obtain the flow field is via direct numerical solution of the equations, which is however feasible only for relatively low values of the Reynolds number and simple geometries. This because the computational cost of the direct numerical solution is proportional to the third power of the Reynolds number and quickly becomes too expensive for existing computers. An alternative strategy that is commonly used for solving the problem, consists of developing simplified methods for the description of the turbulent flow field. Some of these methods are described in this chapter.

8.2 Reynolds Procedure

The most common procedure available for the analysis of turbulent flow fields is based on a time-averaging procedure, originally proposed by Osborne Reynolds, which decomposes the flow field in time-averaged terms and fluctuating terms. Assume that an instantaneous velocity measurement is available at a point in a turbulent flow: This measurement is of no practical use because it provides no information about what the velocity will be at the same point at later times. Instead, it is useful to know the time-averaged value of the velocity at that point. The method proposed by Reynolds is presented in the following sections.

8.2.1 Time Averages

To apply the Reynolds procedure, some definitions must be introduced first. The average value of a variable ξ, of which N values are known, is defined as:

$$\bar{\xi} = \frac{1}{N} \sum_{i=1}^{N} \xi_i \quad , \tag{8.1}$$

or, for a continuous-time variable:

$$\bar{\xi}(t) = \frac{1}{2T} \int_{t-T}^{t+T} \xi(\tau) \mathrm{d}\tau \quad , \tag{8.2}$$

where it is assumed that even the mean value of ξ can be a function of time. This means that the value of T must be large enough to eliminate the effect of turbulent fluctuations on the mean value, but smaller than the timescales that characterize the macroscopic time evolution of the flow field. Obviously, the time variations of the mean value must be slower than the fluctuations due to turbulence. If $\overline{\xi}$ does not vary over time, the process is defined as steady.

In the particular case of a flow field, the time-averaged value of the x component, v_x, of the velocity vector will be:

$$\overline{v}_x(x, y, z) = \lim_{T \to \infty} \frac{1}{2T} \int_{-T}^{T} v_x(x, y, z, t)dt \quad . \tag{8.3}$$

Having introduced these concepts, it is now possible to use the following decomposition of the fluid velocity, written here for the x component:

$$v_x(x, y, z, t) = v_x'(x, y, z, t) + \overline{v}_x(x, y, z) \quad , \tag{8.4}$$

where $v_x'(x, y, z, t)$ is defined as the velocity fluctuation of the flow field. It follows from the definition that the time average of v_x' is equal to zero. The introduction of this decomposition into the N-S equations leads to the appearance of terms such as:

$$\overline{v}_y \frac{\partial v_x'}{\partial y} \quad , \tag{8.5}$$

the average value of which is zero since differentiation and integration can be interchanged in order:

$$\overline{\overline{v}_y \frac{\partial v_x'}{\partial y}} = \lim_{T \to \infty} \frac{1}{2T} \int_{-T}^{T} \overline{v}_y \frac{\partial v_x'}{\partial y} dt =$$

$$= \overline{v}_y \lim_{T \to \infty} \frac{1}{2T} \int_{-T}^{T} \frac{\partial v_x'}{\partial y} dt = \overline{v}_y \frac{\partial}{\partial y} \lim_{T \to \infty} \frac{1}{2T} \int_{-T}^{T} v_x' dt = 0 \quad . \tag{8.6}$$

8.2.2 Time-Averaged Continuity Equation

Consider the continuity equation for an incompressible fluid:

$$\frac{\partial v_x}{\partial x} + \frac{\partial v_y}{\partial y} + \frac{\partial v_z}{\partial z} = 0 \quad . \tag{8.7}$$

Introducing the decomposition presented in Eq. (8.4), the equation can be recast as:

$$\frac{\partial \bar{v}_x}{\partial x} + \frac{\partial v'_x}{\partial x} + \frac{\partial \bar{v}_y}{\partial y} + \frac{\partial v'_y}{\partial y} + \frac{\partial \bar{v}_z}{\partial z} + \frac{\partial v'_z}{\partial z} = 0 \quad . \tag{8.8}$$

From the evaluation of the time-averaged value of this equation, it follows that:

$$\frac{\partial \bar{v}_x}{\partial x} + \frac{\partial \bar{v}_y}{\partial y} + \frac{\partial \bar{v}_z}{\partial z} = 0 \quad . \tag{8.9}$$

Subtracting Eq. (8.9) from Eq. (8.8) gives:

$$\frac{\partial v'_x}{\partial x} + \frac{\partial v'_y}{\partial y} + \frac{\partial v'_z}{\partial z} = 0 \quad . \tag{8.10}$$

8.2.3   Time-Avaraged Navier-Stokes Equations

Consider the x component of the *N-S* equations. The application of the decomposition (8.4) and the time-averaging operation yield:

$$\rho \left(\frac{\partial \bar{v}_x}{\partial t} + \overline{v_x \frac{\partial v_x}{\partial x}} + \overline{v_y \frac{\partial v_x}{\partial y}} + \overline{v_z \frac{\partial v_x}{\partial z}} \right) =$$

$$= -\frac{\overline{\partial \mathcal{P}}}{\partial x} + \mu \left[\overline{\frac{\partial^2 v_x}{\partial x^2}} + \overline{\frac{\partial^2 v_x}{\partial y^2}} + \overline{\frac{\partial^2 v_x}{\partial z^2}} \right] \quad . \tag{8.11}$$

The term $\partial \bar{v}_x / \partial t$ is zero assuming a steady flow field. Furthermore, by substituting the velocity components with the expressions in which the averaged quantities appear, the non-linear terms can be rewritten as:

$$\rho \left[\overline{(\bar{v}_x + v'_x) \frac{\partial}{\partial x}(\bar{v}_x + v'_x)} + \overline{(\bar{v}_y + v'_y) \frac{\partial}{\partial y}(\bar{v}_x + v'_x)} + \overline{(\bar{v}_z + v'_z) \frac{\partial}{\partial z}(\bar{v}_x + v'_x)} \right] \quad .$$

It is possible to express the averaged terms, like $\overline{(\bar{v}_x + v'_x)\, \partial(\bar{v}_x + v'_x)/\partial x}$, in the following way:

$$\overline{(\bar{v}_x + v'_x) \frac{\partial}{\partial x}(\bar{v}_x + v'_x)} = \overline{\bar{v}_x \frac{\partial \bar{v}_x}{\partial x}} + \overline{v'_x \frac{\partial \bar{v}_x}{\partial x}} + \overline{\bar{v}_x \frac{\partial v'_x}{\partial x}} + \overline{v'_x \frac{\partial v'_x}{\partial x}} =$$

$$= \bar{v}_x \frac{\partial \bar{v}_x}{\partial x} + \overline{v'_x \frac{\partial v'_x}{\partial x}} \quad , \tag{8.12}$$

since the terms in which only the fluctuating component of velocity or its time-averaged derivative appear, are equal to zero. Therefore, the left-hand side of Eq. (8.11) becomes:

$$\rho \left[\bar{v}_x \frac{\partial \bar{v}_x}{\partial x} + \overline{v'_x \frac{\partial v'_x}{\partial x}} + \bar{v}_y \frac{\partial \bar{v}_x}{\partial y} + \overline{v'_y \frac{\partial v'_x}{\partial y}} + \bar{v}_z \frac{\partial \bar{v}_x}{\partial z} + \overline{v'_z \frac{\partial v'_x}{\partial z}} \right] ,$$

where, for each term, e.g. $\overline{v'_x \, \partial v'_x / \partial x}$, the following equality holds:

$$\overline{v'_x \frac{\partial v'_x}{\partial x}} = \frac{\partial}{\partial x}(\overline{v'_x v'_x}) - \overline{v'_x \left(\frac{\partial v'_x}{\partial x} \right)} . \tag{8.13}$$

By substituting this equality into the momentum conservation equation and considering that Eq. (8.10) allows to write:

$$\overline{v'_x \frac{\partial v'_x}{\partial x}} + \overline{v'_x \frac{\partial v'_y}{\partial y}} + \overline{v'_x \frac{\partial v'_z}{\partial z}} = \overline{v'_x \left(\frac{\partial v'_x}{\partial x} + \frac{\partial v'_y}{\partial y} + \frac{\partial v'_z}{\partial z} \right)} = 0 , \tag{8.14}$$

the time-averaged x component of the N-S equation is obtained as:

$$\rho \left[\bar{v}_x \frac{\partial \bar{v}_x}{\partial x} + \bar{v}_y \frac{\partial \bar{v}_x}{\partial y} + \bar{v}_z \frac{\partial \bar{v}_x}{\partial z} \right] = -\frac{\partial \mathcal{P}}{\partial x} +$$

$$+ \mu \left[\frac{\partial^2 \bar{v}_x}{\partial x^2} + \frac{\partial^2 \bar{v}_x}{\partial y^2} + \frac{\partial^2 \bar{v}_x}{\partial z^2} \right] - \rho \left[\frac{\partial \overline{(v'_x v'_x)}}{\partial x} + \frac{\partial \overline{(v'_x v'_y)}}{\partial y} + \frac{\partial \overline{(v'_x v'_z)}}{\partial z} \right] . \tag{8.15}$$

The y and z components of the momentum conservation equation can be derived in a similar way.

Denoting by $i = 1, 2, 3$ the generic coordinate and using the convention that a repeated index implies a summation operation:

$$a_j \frac{\partial b_i}{\partial x_j} \equiv \sum_j a_j \frac{\partial b_i}{\partial x_j} , \tag{8.16}$$

the i-th component of the Reynolds equations for steady-state flow can be written in compact form as:

$$\rho \bar{v}_j \frac{\partial \bar{v}_i}{\partial x_j} = -\frac{\partial \mathcal{P}}{\partial x_i} + \mu \left[\frac{\partial^2 \bar{v}_i}{\partial x_j^2} \right] - \rho \left[\frac{\partial \overline{(v'_i v'_j)}}{\partial x_j} \right] . \tag{8.17}$$

In these equations, known as *RANS* (Reynolds-Averaged Navier-Stokes) equations, terms of the type $\partial \overline{(v'_i v'_j)}/\partial x_j$ are called *Reynolds stresses*.

8.2.4 Reynolds Stresses

To understand the physical meaning of the Reynolds stresses, it is useful to consider the time-averaged velocity profile shown in Fig. 8.1. Also shown are two planes parallel to the $x - z$ plane, labelled 1 and 2 respectively. Consider a fluid element that, due to a positive fluctuation $v'_y > 0$, moves from plane 1 to plane 2. This fluid element will have velocity along x lower than the average, i.e. $v'_x < 0$ since the velocity of the fluid elements on plane 1 is on average smaller than the velocity of the fluid elements on plane 2. It follows that, on average:

$$\overline{\rho v'_x v'_y} < 0 \quad . \tag{8.18}$$

Bringing the fluid element to the average velocity of plane 2 results in a lower average velocity on that plane. Now consider an element initially located on plane 2 that, due to a negative fluctuation $v'_y < 0$, moves to plane 1. It is possible to repeat the same reasoning as before, obtaining: $\overline{\rho v'_x v'_y} < 0$.

The expression for the total shear stress can be introduced into the RANS equations as follows:

$$\tau_{xy} = \mu \frac{d\bar{v}_x}{dy} - \overline{\rho v'_x v'_y} \quad . \tag{8.19}$$

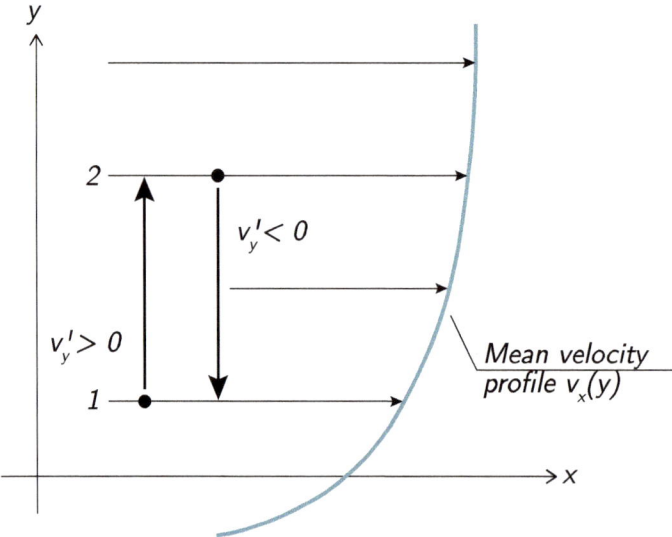

Fig. 8.1 Schematic representation of the Reynolds stresses in a turbulent velocity profile and explanation of their negative sign

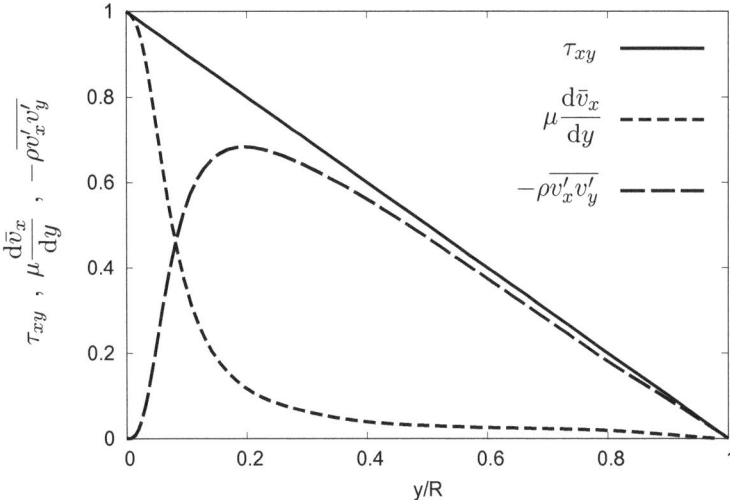

Fig. 8.2 Wall-normal behavior of the Reynolds stresses, viscous stresses and total stresses for turbulent flow in a pipe of radius R

Figure 8.2 shows the behavior of each term appearing in Eq. (8.19), made dimensionless with respect to the wall shear stress, for the case of turbulent flow in a pipe of radius R. The curves are drawn as a function of the dimensionless distance from the wall, y/R.

8.2.5 Turbulent (or Eddy) Viscosity

To solve the time-averaged *N-S* equations using the Reynolds procedure, a model for the Reynolds stresses is required. The simplest model, proposed in 1877 by the French mathematician and physicist Frenchman Joseph Boussinesq (1842–1929), assumes for the turbulent stresses a form that is analogous to the stress tensor for a Newtonian fluid and laminar flow conditions:

$$\tau_{xy}^t = \mu^t \left(\frac{\partial \bar{v}_x}{\partial y} + \frac{\partial \bar{v}_y}{\partial x} \right) \quad , \tag{8.20}$$

where μ^t is called *turbulent viscosity* (or *eddy viscosity*) and has the same dimensions as the dynamic viscosity. The equations that can be obtained using the definition (8.20) differ from the usual ones for the viscous terms alone, which are now expressed as the sum of the usual terms that depend on the molecular viscosity of the fluid (a well-defined physical property that is constant and uniform for a Newtonian fluid) and of the terms that depend on the turbulent viscosity. In general, the turbulent viscosity depends on the particular position within the flow field and on the specific flow configuration. Therefore, it is not a physical property of the fluid but, rather, a parameter that must be properly evaluated to solve the so-called closure problem.[1] The fact that the value of μ^t cannot be constant inside the flow and must depend on position becomes clear by considering that the first derivative of the velocity profile does not vanish at the wall, where the Reynolds stresses must be zero because of the no-slip condition (the velocity at the wall is zero).

8.2.6 Prandtl's Mixing Length Model

The simplest model for determining the turbulent viscosity is based on the concept of mixing length, proposed in 1933 by the German engineer and physicist Ludwig von Prandtl (1875–1953). By analogy with the concept of mean free path in the kinetic theory of gases, Prandtl proposed the following expression for the turbulent viscosity:

$$\mu^t = \rho l_m^2 \left| \frac{d\bar{v}_x}{dy} \right| . \tag{8.21}$$

In this equation, l_m is the mixing length and is the analogue of the mean free path in the kinetic theory of gases, defined as average distance traveled by a moving molecule between successive collisions with other molecules. In a turbulent flow, the mixing length corresponds to the distance beyond which a mass of fluid moving coherently,[2] loses such coherence.

The expression proposed by Prandtl can be derived qualitatively by referring to the situation illustrated in Fig. 8.1, which can be interpreted as the case of a turbulent boundary layer near a plane wall. Consider a small fluid element initially at position 1, and denote by y_1 the distance from the wall and by $\bar{v}_x(y_1)$ the corresponding average velocity. Assume now that the element moves away from the wall as a result of a displacement equal to l_m, which brings the element at position 2, where the average velocity $\bar{v}_x(y_2) = \bar{v}_x(y_1 + l_m)$ is larger than $\bar{v}_x(y_1)$. The displacement results in a negative velocity fluctuation that has order of magnitude:

$$\Delta \bar{v}_x = \bar{v}_x(y_1 + l_m) - \bar{v}_x(y_1) . \tag{8.22}$$

---

[1] The turbulent viscosity μ^t is a parameter that allows to derive the time-averaged effect of the Reynolds stresses by knowing the time-averaged velocity profile.

[2] A mass of fluid moves coherently when its velocity at a given instant of time is a function of the velocity at the previous times. For example, a vortex created by wind in the atmosphere moves coherently, since it moves through space while maintaining its structure over a certain amount of time.

In the limit of sufficiently small displacement l_m, it is possible to set:

$$\Delta \bar{v}_x^- \simeq l_m \frac{d\bar{v}_x}{dy}\bigg|_{y_1+l_m} \quad . \tag{8.23}$$

Similarly, if the small element moves closer to the wall as a result of a displacement l_m, then it reaches a distance $y_1 - l_m$ where $v_x(y_1 - l_m) < v_x(y_1)$ and a positive velocity fluctuation is produced:

$$\Delta \bar{v}_x^+ \simeq l_m \frac{d\bar{v}_x}{dy}\bigg|_{y_1} \quad , \tag{8.24}$$

with order of magnitude:

$$\Delta v_x^- \simeq l_m \frac{d\bar{v}_x}{dy}\bigg|_{y_1+l_m} \quad . \tag{8.25}$$

Assuming that the velocity fluctuation in the x direction is given by the average of $\Delta \bar{v}_x^-$ and $\Delta \bar{v}_x^+$:

$$v_x' \simeq \frac{\Delta v_x^- + \Delta v_x^+}{2} = l_m \frac{d\bar{v}_x}{dy}\bigg|_{y_1+l_m} \quad , \tag{8.26}$$

and that the velocity fluctuation in the wall-normal direction has the same order of magnitude ($v_y' \sim v_x'$), it is found that:

$$\overline{v_x' v_y'} \simeq l_m^2 \left(\frac{d\bar{v}_x}{dy}\right)^2 \quad , \tag{8.27}$$

or, more precisely:

$$\overline{v_x' v_y'} \simeq l_m^2 \left|\frac{d\bar{v}_x}{dy}\right| \frac{d\bar{v}_x}{dy} \quad , \tag{8.28}$$

taking into account the sign of $\overline{v_x' v_y'}$ for each of the two displacements considered. As it must be:

$$\rho \overline{v_x' v_y'} = -\mu^t \frac{d\bar{v}_x}{dy} \quad , \tag{8.29}$$

it follows that the turbulent viscosity can be expressed as:

$$\mu^t = \rho l_m^2 \left|\frac{d\bar{v}_x}{dy}\right| \quad . \tag{8.30}$$

Equation (8.30) shows that the turbulent viscosity is positive, in agreement with the fact that it should generate diffusion, but its value is still unknown since the mixing length has not been specified. The mixing length is a quantity that depends on the specific flow configuration. Two expressions, proposed respectively by Prandtl and by the American physicist and engineer Edward Reginald Van Driest (1913–2005), will be reviewed in the next section.

The main merit of Prandtl's model is its simplicity, as it does not require the use of differential equations. The main limitations are that the model was developed for the specific case of near-wall flow and that it predicts $\mu^t = 0$ whenever $d\bar{v}_x/dy = 0$. Therefore, the model is not applicable to flows with recirculations or to flows where $\mu^t \neq 0$ even if $d\bar{v}_x/dy = 0$, as in the case of flow in channels or axisymmetric pipes or in the center of a jet. To overcome these limitations, numerous alternative formulations of the original Prandtl's theory (not discussed in this textbook) have been proposed to derive suitable expressions of l_m for the specific flow under consideration.

8.3 Turbulent Pipe Flow

The time-averaged momentum conservation equations written in cylindrical coordinates lead to the following relation:

$$\frac{d\overline{\mathcal{P}}}{dz} = \frac{1}{r}\frac{d}{dr}r(\overline{\tau}_{rz} + \tau^t_{rz}) \quad . \tag{8.31}$$

A force balance on a pipe section of axial length Δz yields:

$$\pi R^2 \left(-\frac{d\overline{\mathcal{P}}}{dz}\right)\Delta z = 2\pi R \tau_w \Delta z \quad , \tag{8.32}$$

where τ_w is the average shear stress at the wall. The combination of these two relations gives:

$$\overline{\tau}_{rz} + \tau^t_{rz} = -\frac{r}{R}\tau_w \quad . \tag{8.33}$$

Using Eq. (8.21), the stresses on the left-hand side can be expressed as:

$$\overline{\tau}_{rz} = \mu\frac{d\bar{v}_z}{dr} \quad , \tag{8.34}$$

$$\tau^t_{rz} = \rho l_m^2 \left|\frac{d\bar{v}_z}{dr}\right|\frac{d\bar{v}_z}{dr} \tag{8.35}$$

By substituting Eqs. (8.34) and (8.35) into Eq. (8.33) and defining $y = R - r$, Eq. (8.3) becomes:

$$\rho l_m^2 \left(\frac{d\bar{v}_z}{dy}\right)^2 + \mu \frac{d\bar{v}_z}{dy} - \left(1 - \frac{y}{R}\right) \tau_w = 0 \ . \tag{8.36}$$

Equation (8.36) can be made dimensionless introducing characteristic parameters like the *friction velocity*:

$$v^* = \sqrt{\frac{\tau_w}{\rho}} \ , \tag{8.37}$$

which is a velocity scale obtained from the wall shear stress, τ_w, and the term $\mu/\rho v^*$. Using these parameters, the dimensionless form of Eq. (8.36) reads as:

$$l_+^2 \left(\frac{dv_+}{dy_+}\right)^2 + \frac{dv_+}{dy_+} - \left(1 - \frac{y_+}{R_+}\right) = 0 \ , \tag{8.38}$$

where $l_+ = l_m (\rho v^*)/\mu$ and $v_+ = v_z/v^*$. From Eq. (8.38), dv_+/dy_+ can be derived as:

$$\frac{dv_+}{dy_+} = \frac{-1 + \sqrt{1 + 4l_+^2 (1 - y_+/R_+)}}{2l_+^2} \ . \tag{8.39}$$

This equation must satisfy the boundary condition $v_+ = 0$ at $y_+ = 0$ and can be integrated once the function $l_+ = l_+(y_+)$ is given. In his model, Prandtl proposed to define:

$$l_+ = k \ y_+ \ . \tag{8.40}$$

However, this expression is not correct near the wall, where l_+ tends rapidly to zero. It is thus convenient to assume for l_+ the expression proposed by Van Driest:

$$l_+ = k y_+ \left(1 - e^{-y_+/A}\right) \ , \tag{8.41}$$

where $A = 36$. With the help of Eqs. (8.40) or (8.41) for l_+, Eq. (8.39) can be integrated analytically at least in some regions of the flow field. Near the wall, $y_+/R_+ \to 0$ and the function $v_+ = v_+(y_+)$ that is obtained is independent of R_+ and valid in the vicinity of a wall regardless of the specific flow geometry. It can be shown experimentally that this result is correct and the resulting velocity profile is referred to as *universal*.

The integral of Eq. (8.39) for $y_+/R_+ \to 0$:

$$v_+ = \int_0^{y_+} \frac{-1 + \sqrt{1 + 4l_+^2}}{2l_+^2} dy_+ = 2 \int_0^{y_+} \frac{dy_+}{1 + \sqrt{1 + 4l_+^2}} \quad , \tag{8.42}$$

can be easily calculated for two cases:

Viscous sublayer, $l_+ \to 0$: Near the wall, l_+ tends rapidly to zero and Eq. (8.42) gives:

$$v_+ = y_+ \quad . \tag{8.43}$$

The viscous sublayer represents the portion of the boundary layer that is closest to the wall, where the flow is dominated by the viscous terms (that is by the viscous stresses). Equation (8.43) is valid for $y_+ < 5$. This value, therefore, corresponds to the thickness of the viscous sublayer.

Inertial sublayer, $4l_+^2 \gg 1$: The inertial sublayer represents the portion of the boundary layer in which the flow is dominated by the inertial terms (that is by the Reynolds stresses). In this case, the integral of Eq. (8.42) can be written as:

$$v_+ - v_+^v = \int_{y_+^v}^{y_+} \frac{dy_+}{l_+} \quad , \tag{8.44}$$

where the coordinate y_+^v represents the distance from the wall at which viscous effects become negligible. For $l_+ = ky_+$, Eq. (8.44) yields:

$$v_+ = \frac{1}{k} \ln y_+ + b \quad , \tag{8.45}$$

where:

$$b = v_+^v - \frac{1}{k} \ln y_+^v \quad . \tag{8.46}$$

This logarithmic dependence was determined using simple dimensional considerations. It was assumed that, near the wall, the velocity v_z is a function of only three parameters v^*, y and ν:

$$v_z = f(v^*, y, \nu) \quad . \tag{8.47}$$

This leads to a dimensionless relation of the type:

$$v_+ = f(y_+) \quad . \tag{8.48}$$

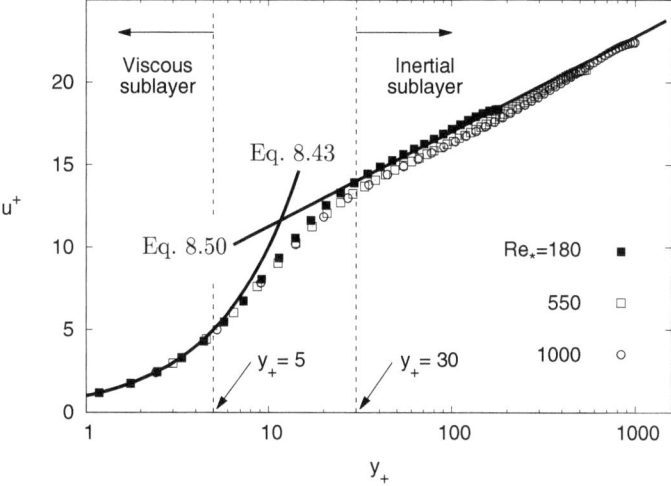

Fig. 8.3 Experimental (■) and numerical (□, ○) velocity profiles compared with Eqs. (8.43) and (8.50). Equation (8.43) is valid within the viscous sublayer ($y_+ < 5$), while Eq. (8.50) is valid within the inertial sublayer ($y_+ > 30$). The intermediate layer ($5 < y_+ < 30$) is characterized by the presence of both viscous and inertial effects, and by velocity values that cannot be reproduced by analytically-derived expressions

Viscosity is important only near the wall. At some distance from the wall, however, if the absolute value of the velocity still depends on ν, then the derivative $\partial v_z / \partial y$ must be independent of viscosity. Namely:

$$\frac{\partial v_z}{\partial y} \sim \frac{v^*}{y} \quad . \tag{8.49}$$

The integral of Eq. (8.49) has precisely the form of Eq. (8.45). Figure 8.3 shows the velocity profile (averaged over time) that characterizes a wall-bounded turbulent flow (such as the flow in a pipe or a channel). The lines in this figure correspond to Eqs. (8.43) and (8.45), while the symbols correspond to data obtained either experimentally or by direct numerical simulation of the *N-S* equations.

Experimental measurements indicate that $k = 0.4$ and $b = 5.5$. The value of the constant b corresponds to a distance $y_+ = 11.6$ from the wall, assuming that the velocity profile in the viscous sublayer, $v_+ = y_+$, can be extended to this height. With these values for k and b, Eq. (8.45) becomes:

$$v_+ = 2.5 \ln y_+ + 5.5 \quad . \tag{8.50}$$

As shown in Fig. 8.3, this equation reproduces very accurately the experimental and numerical results for $y_+ > 30$.

The region of the boundary layer in the range $5 < y_+ < 30$ is characterized by the presence of both viscous and inertial effects. In this region, the velocity profile cannot be obtained analytically but only by means of experimental measurements or sophisticated numerical simulations.

Away from the wall, in the core region of the pipe, the integral of Eq. (8.39) depends on R_+ and can be written as:

$$v_+ = v_{+,max} + \int_{y_+}^{R_+} \frac{1 - \sqrt{1 + 4l_+^2 (1 - y_+/R_+)}}{2l_+^2} dy_+ \quad , \tag{8.51}$$

where $v_{+,max}$ is the maximum velocity, which occurs at $y_+ = R_+$. Equation (8.51) can be immediately simplified for $y^+ \gg 1$ assuming $l_+ = ky_+$:

$$v_+ = v_{+,max} - \int_{y_+}^{R_+} \frac{(1 - y_+/R_+)^{1/2}}{ky_+} dy_+ \quad . \tag{8.52}$$

The knowledge of the velocity profile allows to calculate the friction factor. Recalling the definition of average velocity over the cross section of the pipe, it is possible to write:

$$v^* = \sqrt{\frac{f}{2}} \langle v_z \rangle = \sqrt{\frac{f}{2}} \left(\frac{1}{\pi R^2} \int_0^R 2\pi r v_z dr \right) \quad , \tag{8.53}$$

and:

$$\sqrt{\frac{2}{f}} = \frac{2}{R^2} \int_0^R r v_+ dr = \frac{2}{R^2} \int_o^R (R - y) v_+ dy \quad . \tag{8.54}$$

To integrate this equation, in principle, it would be necessary to substitute v_+ with the velocity profiles previously determined for the different regions in which the pipe section can be divided. However, the logarithmic profile (8.45) provides a good approximation for the entire section. Using this profiles yields:

$$\sqrt{\frac{2}{f}} = \frac{2}{R^2} \int_0^R (R - y)(2.5 \ln y_+ + 5.5) dy =$$

$$= 2 \int_0^1 (1 - \xi) \left(2.5 \ln \xi + 5.5 + 2.5 \ln Re \sqrt{\frac{f}{8}} \right) d\xi \quad , \tag{8.55}$$

where $\xi = y/R$ and, by definition $R_+ = Re \sqrt{f/8}$. Equation (8.55) allows to derive the following expression for the friction factor:

Fig. 8.4 Comparison of velocity profiles for laminar (i) and turbulent (ii) flow in a pipe. The laminar profile (i) and the turbulent profile (ii) refer to flows with the same mean velocity (Reynolds number $Re = 10^4$). The laminar profile (i) and the turbulent profile (iii) refer to flows with the same pressure gradient

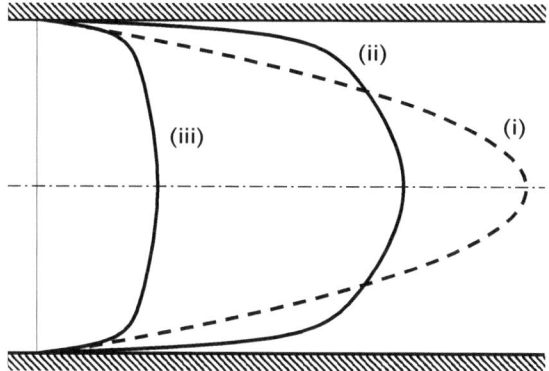

$$\frac{1}{\sqrt{f}} = 1.8 \ln(Re\sqrt{f}) - 0.6 \quad , \tag{8.56}$$

which is very similar to the von Kármán-Nikuradse equation introduced in Chap. 2. Turbulent velocity profiles are often represented in the form:

$$\frac{v_z}{v_{z,max}} = \left(\frac{y}{R}\right)^m \quad , \tag{8.57}$$

namely in terms of the average velocity $\langle v_z \rangle$:

$$\frac{v_z}{\langle v_z \rangle} = \frac{(m+1)(m+2)}{2} \left(\frac{y}{R}\right)^m \quad . \tag{8.58}$$

In these equations, the value of m varies in the range $[1/10 \div 1/6]$. Using this expression it is also possible to calculate the friction factor. For $m = 1/7$, the Blasius equation is retrieved.

At this point of the discussion, it is interesting to compare the velocity profiles that are obtained for laminar and turbulent flows. In Fig. 8.4, it can be seen that the turbulent velocity profile is flatter than the laminar one and that dissipation is higher in a turbulent flow.

8.3.1 Examples

$\boxed{1}$ A fluid with viscosity $1.83 \, \text{Pa} \cdot \text{s}$ and density $1300 \, \text{kg/m}^3$ flows in a horizontal pipe of radius $50 \, \text{mm}$, driven by a pressure gradient. For which value of the pressure gradient will the transition from laminar to turbulent regime occur?

Solution In laminar flow conditions, the pressure gradient can be expressed as:

$$\frac{dP}{dz} = 2\rho v^2 \frac{f}{D} = 2\frac{\rho v^2}{D}\frac{16}{Re} = 32\rho v^2 \frac{1}{D^3} Re \quad . \tag{8.59}$$

Since the transition from laminar flow to turbulent flow in a pipe begins when $Re = 2100$, the critical pressure gradient is:

$$\frac{dP}{dz} = 32\rho v^2 \frac{1}{D^3} 2100 = 1.73 \cdot 10^5 \, \text{Pa/m} \quad . \tag{8.60}$$

2 A drop of water ($\rho_p = 1000 \, \text{kg/m}^3$) with diameter $D_p = 2 \cdot 10^{-5}$m is generated at the wall of a pipe with diameter $D = 20$ mm in which air flows at 10 m/s ($\rho = 1.2 \, \text{kg/m}^3$, $\mu = 15 \cdot 10^{-6}$Pa $\cdot$ s). The drop has an initial velocity $v_r = 1$ m/s and the radial one is the only non-zero velocity component of the drop. Neglect the effects of gravity and determine:

1. the maximum distance traveled by the droplet in the radial direction,
2. the time required to cross the viscous sublayer ($y_+ = 5$),
3. the axial distance traveled by the drop for the time obtained at point 2, assuming that the drop has always the same axial velocity of the air.

Assume that the drop moves in the Stokes regime.

Solution Considering that the pipe is vertical, the x coordinate is parallel to the axis of the pipe, and the y coordinate has its origin at the wall, the system of equations that must be solved to determine the motion of the water drop, together with the boundary conditions, is:

$$\begin{aligned}
\ddot{x} &= (U_f - \dot{x})/\tau_p \quad , \\
\ddot{y} &= -\dot{y}/\tau_p \quad , \\
\dot{x}(0) &= 0 \, , \quad x(0) = 0 \, , \\
\dot{y}(0) &= v_i \, , \quad y(0) = 0 \, ,
\end{aligned} \tag{8.61}$$

where the dot denotes a time derivative and $\tau_p = \rho_p D_p^2/18 \, \mu = 1.481 \cdot 10^{-3}$ is the response time of the drop. For the horizontal component, y, the equation of motion is:

$$\ddot{y} = -\frac{\dot{y}}{\tau_p} \quad \rightarrow \quad \dot{y} = -\frac{y}{\tau_p} + A \quad \rightarrow \quad \frac{dy}{A - y/\tau_p} = dt \quad .$$

Therefore:

$$t = -\tau_p \ln(y - A\tau_p) + B' \quad \rightarrow \quad y(t) = Be^{-t/\tau_p} + A\tau_p \quad . \tag{8.62}$$

The boundary conditions yield $A = v_i$ and $B = -\tau_p v_i$ and:

$$y(t) = v_i \tau_p (1 - e^{-t/\tau_p}) \quad,$$
$$\dot{y}(t) = v_i e^{-t/\tau_p} \quad.$$

(8.63)

1. The maximum distance that the drop can cover in the radial direction is:

$$y_{max} = \lim_{t \to \infty} y(t) = v_i \tau_p = 1.48 \cdot 10^{-3} \text{ m} \quad.$$

(8.64)

2. The viscous sublayer has a thickness of $y_+ = 5$. Since:

$$y_+ = y \frac{u^*}{\nu} = \frac{y}{\nu} \sqrt{\frac{\tau_w}{\rho}} \quad,$$

(8.65)

and the wall shear stress is $\tau_w = 0.039 \, \rho \, v_f^2 \, Re^{-1/4} = 0.416 \text{ N/m}^2$, it follows that $y_+ = 5$ corresponds to a dimensional distance $y = 1.06 \cdot 10^{-4}$ m. The time required to cover this distance is:

$$t(y_+ = 5) = -\tau_p \ln\left(1 - \frac{y}{v_i \tau_p}\right) = 1.1 \cdot 10^{-4} \text{ s} \quad.$$

(8.66)

3. In the viscous sublayer, the velocity profile depends on the wall coordinate, and $v_+ = yv^*/\nu$. The friction velocity is:

$$v^* = \sqrt{\frac{\tau_w}{\rho}} = 0.59 \text{ m/s} \quad.$$

(8.67)

Since the axial velocity of the drop is the same as that of the fluid, the distance covered will be equal to:

$$x = \int_0^{1.1 \cdot 10^{-4}} v_+ v^* dt = \int_0^{1.1 \cdot 10^{-4}} y \frac{v^{*2}}{\nu} dt = \frac{v^{*2}}{\nu} \int_0^{1.1 \cdot 10^{-4}} v_i \tau_p \left(e^{-t/\tau_p}\right) dt =$$

$$= \frac{v^{*2}}{\nu} 1.1 \cdot 10^{-4} v_i \tau_p \left[\tau_p e^{-t/\tau_p} + t - \tau_p\right]_0^{1.1 \cdot 10^{-4}} = 9.25 \text{ mm} \quad.$$

(8.68)

⎡3⎤ Consider a pipe of radius $R = 25$ mm in which water ($\mu = 1 \cdot 10^{-3} \text{Pa} \cdot \text{s}; \rho = 1 \cdot 10^3 \text{kg/m}^3$) flows with an average velocity $\bar{v}_z = 1.8$ m/s. The flow is turbulent and the following expressions can be used to approximate the velocity profile:

$$\bar{v}_z = \bar{v}_{z,max} \left(1 - \frac{r}{R}\right)^{\frac{1}{7}} \quad.$$

(8.69)

Calculate:

1. the total shear stress for $r/R = 0.9$,
2. the fraction of the total shear stress due to turbulent fluctuations at $r/R = 0.9$.

Solution The wall shear stress can be calculated using the Blasius equation for smooth pipes:

$$\tau_w = 0.039 \rho \bar{v}_z^2 Re^{-1/4} \quad . \tag{8.70}$$

From the data of the problem, it follows that $Re = 9 \cdot 10^4$ and $\tau_w = 7.29\ \text{N/m}^2$. Since the total shear stress decreases linearly from the wall to the center of the pipe, its expression as a function of the pipe radius will be:

$$\tau\left(\frac{r}{R}\right) = \tau_w\left(\frac{r}{R}\right) \quad . \tag{8.71}$$

Therefore, for $r/R = 0.9$, Eq. (8.71) gives $\tau(0.9) = 6.56\ \text{N/m}^2$. The total shear stress is also given by the following expression:

$$\tau_w = \mu\frac{d\bar{v}_z}{dr} - \rho\overline{v_z' v_r'} = \mu\frac{d\bar{v}_z}{dr} + \mu^t\frac{d\bar{v}_z}{dr} \quad . \tag{8.72}$$

From the knowledge of the velocity profile:

$$\bar{v}_z = \bar{v}_{z,max}\left(1 - \frac{r}{R}\right)^{\frac{1}{7}} \quad , \tag{8.73}$$

it is easy to determine the laminar part of the shear stress. A simple integration gives $\langle\bar{v}_z\rangle = 0.81\bar{v}_{z,max} \rightarrow \bar{v}_{z,max} = 1.8/0.81 = 2.22\ \text{m/s}$. The laminar part of the shear stress is then given by:

$$\tau^l = \mu\frac{d\bar{v}_z}{dr} = \frac{\mu\ \bar{v}_{z,max}}{7}\frac{1}{R}\left(1 - \frac{r}{R}\right)^{-\frac{6}{7}} \quad . \tag{8.74}$$

Therefore, for $r/R = 0.9$, Eq. (8.74) gives $\tau^l(0.9) = 0.0914\ \text{N/m}^2$ and the amount of shear stress due to turbulent fluctuations is:

$$\tau^t(0.9) = \tau(0.9) - \tau^l(0.9) = 6.56 - 0.0914 = 6.46\ \text{N/m}^2 \quad . \tag{8.75}$$

4 Consider a turbulent flow of water (density $\rho = 1000\ \text{kg/m}^3$, viscosity $\mu = 10^{-3}\text{Pa}\cdot\text{s}$) in a pipe of diameter $D = 1\ \text{m}$.

1. Given the velocity profile in terms of wall variables:

$$v_{z,+}(y_+) = 2.5 \ \ln y_+ + 5.5 \quad , \tag{8.76}$$

derive an expression for the mixing length l^+ as a function of y_+ for $y_+ > 30$.
2. Knowing that the pressure gradient is $dP/dx = 8 \cdot 10^3$ Pa/m, calculate the value of the mixing length at a distance $y = D/4$ from the wall.

Solution

1. The shear stress at a given distance y from the wall can be expressed as:

$$\tau(y) = \tau_w \left(1 - \frac{y}{R}\right) = \mu' \frac{dv}{dy} + \mu \frac{dv}{dy} = \rho l^2 \left(\frac{dv}{dy}\right)^2 + \mu \frac{dv}{dy} \quad , \tag{8.77}$$

and, in dimensionless form, as:

$$1 - \frac{y_+}{R_+} = l_+^2 \left(\frac{dv_+}{dy_+}\right)^2 + \frac{dv_+}{dy_+} \quad . \tag{8.78}$$

The derivative of the velocity profile is $(dv_+)/(dy_+) = 2.5/y_+$. Substitution of this derivative into Eq. (8.78) gives:

$$1 - \frac{y_+}{R_+} = l_+^2 \left(\frac{2.5}{y_+}\right)^2 + \frac{2.5}{y_+} \quad . \tag{8.79}$$

The solution of this equation provides the expression of l_+ as a function of y_+:

$$l_+ = \frac{y_+}{2.5} \sqrt{\left(1 - \frac{y_+}{R_+}\right) - \frac{2.5}{y_+}} \quad . \tag{8.80}$$

2. The shear stress at the wall is linked to the pressure gradient by the relation: $\tau_w = dp/dx \cdot D/4$. Therefore, $\tau_w = 2000 \ Pa$. The friction velocity is $v^* = \sqrt{\tau_w/\rho} = 1.41$ m/s and the dimensionless length is $l^* = 7.07 \cdot 10^{-7}$. Therefore, the dimensionless mixing length:

$$l_+(y = D/4) = \frac{y_+}{2.5} \sqrt{\left(1 - \frac{1}{2}\right) - \frac{2.5}{y_+}} = 5 \cdot 10^{-8} \quad , \tag{8.81}$$

which corresponds to a dimensional value $l = 0.071$ m.

5 For a Reynolds number equal to $4 \cdot 10^4$, the velocity profile of a fluid flowing in a smooth pipe can be approximated by the equation:

$$\frac{v_z}{v_{z,max}} = 1 - 0.351 \left(\frac{r}{R}\right)^2 - 0.649 \left(\frac{r}{R}\right)^{66} \quad . \tag{8.82}$$

Calculate the friction factor at the wall.

Solution Assuming that the given profile provides a good approximation of the flow field near the wall, the magnitude of τ_w can be calculated as:

$$|\tau_w| = -\mu \frac{\partial v_z}{\partial r}\bigg|_{r/R=1} \quad . \tag{8.83}$$

Substituting the expression of the velocity profile into this equation, the previous equation can be rewritten as:

$$|\tau_w| = \mu \left[0.702 \left(\frac{r}{R}\right) + 42.834 \left(\frac{r}{R}\right)^{65}\right] \frac{v_{z,max}}{R} = 43.536 \frac{\mu v_{z,max}}{R} \quad . \tag{8.84}$$

The average value of velocity is given by:

$$\bar{v}_z = \frac{v_{z,max}}{\pi R^2} \int_0^R 2\pi r \left[1 - 0.351 \left(\frac{r}{R}\right)^2 - 0.649 \left(\frac{r}{R}\right)^{66}\right] dr \quad , \tag{8.85}$$

and, substituting $r/R = x$ and $dr = R dx$, integration yields:

$$\bar{v}_z = 0.8054 \, v_{z,max} \quad . \tag{8.86}$$

It turns out that the shear stress can be expressed as a function of the mean velocity as:

$$|\tau_w| = \mu \, 43.536 \frac{1}{0.8054} \frac{\bar{v}_z}{R} \quad . \tag{8.87}$$

Comparing this expression with the one that links τ_w with the friction factor:

$$|\tau_w| = \frac{1}{2} f \rho \bar{v}_z^2 \quad , \tag{8.88}$$

allows to write:

$$f = 2 \cdot 54.0543 \frac{\mu}{\rho \bar{v}_z R} = 108.11 \frac{\mu}{\rho \bar{v}_z R} \quad . \tag{8.89}$$

Therefore, the value of the friction factor at the wall is:

$$f = 4 \cdot 54.0543 \, Re^{-1} = 5.4 \cdot 10^{-3} \quad . \tag{8.90}$$

6 Derive the transport equation of the mean kinetic energy for a Poiseuille flow in a plane channel, considering both laminar and turbulent flow conditions. For simplicity, consider the case of two-dimensional flow.

Solution For the laminar flow regime, referring to Fig. 4.2 and considering the following assumptions:

$$v_y = v_z = 0 \quad \text{unidirectional flow along } x,$$
$$\partial \bullet /\partial z = 0 \quad \text{two-dimensional flow} , \tag{8.91}$$
$$\partial \bullet /\partial t = 0 \quad \text{steady flow} ,$$

the continuity equation and the *N-S* equations read as:

$$\frac{\partial v_x}{\partial x} = 0 , \tag{8.92}$$

$$0 = -\frac{\partial \mathcal{P}}{\partial x} + \mu \frac{\partial^2 v_x}{\partial y^2} . \tag{8.93}$$

To obtain the energy equation, it suffices to multiply Eq. (8.93) by v_x:

$$0 = -v_x \frac{\partial \mathcal{P}}{\partial x} + \mu v_x \frac{\partial^2 v_x}{\partial y^2} . \tag{8.94}$$

Observing that $\partial(\tau_{yx} v_x)/\partial y = \tau_{yx}(\partial v_x/\partial y) + v_x(\partial \tau_{yx}/\partial y)$ and recalling that $\tau_{yx} = \mu \partial v_x/\partial y$, it is possible to rewrite Eq. (8.94) as:

$$0 = -\frac{\partial (\mathcal{P} v_x)}{\partial x} + \frac{\partial (\tau_{yx} v_x)}{\partial y} - \mu \left(\frac{\partial v_x}{\partial y}\right)^2 . \tag{8.95}$$

The first term on the right-hand side represents the work of the pressure forces: since the flow is driven by a pressure gradient, this is the only term representing a *gain* of energy for the flow under consideration. When integrated over the entire section, the term provides a positive contribution. The second term represents the work of the viscous forces. The integral of this term over the cross section is zero, meaning that the term represents a *redistribution* of energy within the flow. The third term represents viscous dissipation and, once integrated over the cross section, leads to a negative contribution.

For a laminar flow, the analytical expression of the velocity profile is known, and it is therefore possible to derive the differential energy transport equation. Since the average velocity can be expressed as:

$$v_x = \frac{3}{2} \langle v_x \rangle \left[1 - \left(\frac{y}{a}\right)^2\right] , \tag{8.96}$$

where $\langle v_x \rangle = Q/A$ and $a = H/2$ is the half-height of the channel, the following terms can be derived:

$$\tau_{yx} = -\frac{3\mu\langle v_x \rangle}{a^2}y \quad , \tag{8.97}$$

$$\frac{\partial \mathcal{P}}{\partial x} = -\frac{3\mu\langle v_x \rangle}{a^2} \quad , \tag{8.98}$$

Substitution of these terms into Eq. (8.95) gives:

$$0 = \frac{9\mu\langle v_x \rangle^2}{2a^2}\left[\underbrace{\left(1 - \frac{y^2}{a^2}\right)}_{1} + \underbrace{\left(\frac{3y^2}{a^2} - 1\right)}_{2} - \underbrace{\left(\frac{2y^2}{a^2}\right)}_{3}\right] \quad . \tag{8.99}$$

Figure 8.5 shows the wall-normal behavior of the terms denoted as 1, 2 and 3 in Eq. (8.99). It can be seen that the work done by the pressure gradient (term 1) is maximum at the channel center and zero at the wall. The work done by the viscous forces (term 2) produces a loss of energy in the center of the channel and a gain of energy near the wall, which is balanced by the loss produced by viscous dissipation (term 3). This latter term is maximum at the wall and decreases to zero at the centerline.

To derive the transport equation of the mean kinetic energy for turbulent flow, it is possible to proceed in a similar way, considering however the *RANS* equations as a starting point (in addition to the continuity equation). Given that the flow is steady and, on average, unidirectional ($\bar{v}_y = \bar{v}_z = 0$), the equations read as:

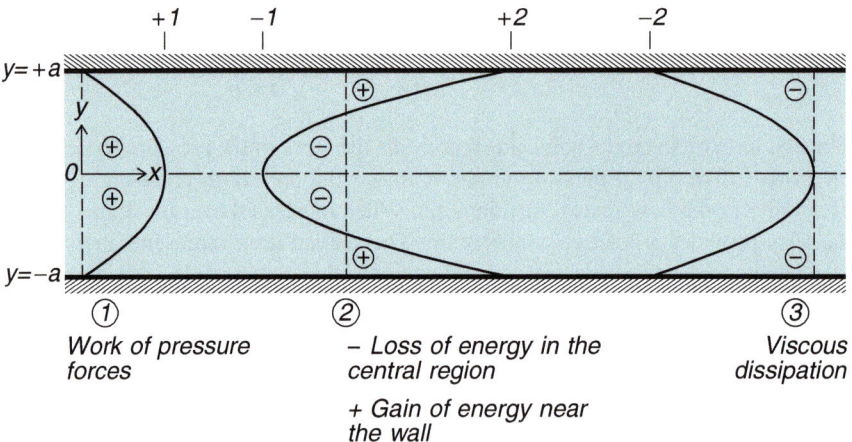

Fig. 8.5 Behavior of the terms appearing in the transport equation of the mean kinetic energy for laminar Poiseuille flow in a plane channel

$$\frac{\partial \overline{v}_x}{\partial x} = 0 \ , \tag{8.100}$$

$$0 = -\frac{\partial \overline{P}}{\partial x} + \frac{\partial}{\partial y}\left[\mu\frac{\partial \overline{v}_x}{\partial y} - \rho\overline{v'_x v'_y}\right] = -\frac{\partial \overline{P}}{\partial x} + \frac{\partial \overline{\tau}_{xy}}{\partial y} \ . \tag{8.101}$$

Multiplying Eq. (8.101) by $\overline{v}_x$ and using the product rules for derivatives, the transport equation can be obtained as:

$$\underbrace{\frac{\partial\left(\overline{P}\overline{v}_x\right)}{\partial x}}_{1} = \underbrace{\frac{\partial}{\partial y}\left[\overline{v}_x\left(\mu\frac{\partial \overline{v}_x}{\partial y} - \rho\overline{v'_x v'_y}\right)\right]}_{2} + \underbrace{\left[\underbrace{-\mu\left(\frac{\partial \overline{v}_x}{\partial y}\right)^2}_{a} + \underbrace{\rho\overline{v'_x v'_y}\frac{\partial \overline{v}_x}{\partial y}}_{b}\right]}_{3} \ .$$

The term labelled as 1 in Eq. (8.102) represents the work done by pressure. As for the laminar case, this is the only source of energy since the pressure gradient drives the flow. The term labelled as 2 represents the work done by shear stresses, and represents a redistribution of energy (its integral over the cross section is equal to zero). The term labelled 3 is the dissipative term and consists of a dissipation term, labelled as 3a, that is due to the principal deformation, and an energy loss term, labelled as 3b, that is due to the production of turbulent velocity fluctuations.

Figure 8.6 shows the wall-normal behavior of these terms, highlighting in particular the contribution of term 3b (dashed line) and showing that term 3a is the only one that does not vanish at the wall.

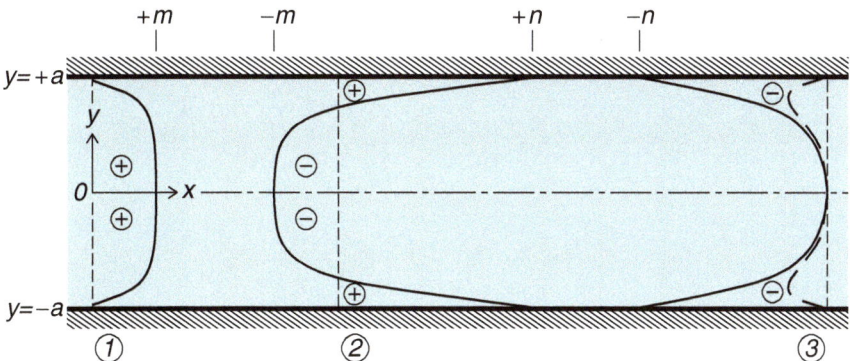

Fig. 8.6 Behavior of the terms appearing in the transport equation of the mean kinetic energy transport equation for turbulent Poiseuille flow in a plane channel

8.4   Turbulent Boundary Layer

The turbulent boundary layer can be studied using the so-called *integral method*, an approximate method proposed by von Kármán together with the German mathematician Karl Pohlhausen (1892–1980). This method is very useful when, as in the turbulent case, it is not possible to obtain an exact solution for the boundary layer on a flat plate from the *N-S* equations. The method requires to integrate the *N-S* equations, starting from the form already derived in Sect. 7.2 for two-dimensional flow, along the direction normal to the plate (y direction) from the plate wall ($y = 0$) to a point far away from it (ideally placed at $y \to \infty$). For the x component of the *N-S* equations:

$$\int_0^\infty \rho \left(v_x \frac{\partial v_x}{\partial x} + v_y \frac{\partial v_x}{\partial y} \right) dy = \int_0^\infty \left[-\frac{\partial \mathcal{P}}{\partial x} + \mu \left(\frac{\partial^2 v_x}{\partial x^2} + \frac{\partial^2 v_x}{\partial y^2} \right) \right] dy \ , \quad (8.102)$$

which, for incompressible and Newtonian fluid, can be recast as:

$$\rho \left(\int_0^\infty v_x \frac{\partial v_x}{\partial x} dy + \int_0^\infty v_y \frac{\partial v_x}{\partial y} dy \right) =$$

$$= -\int_0^\infty \frac{\partial \mathcal{P}}{\partial x} dy + \mu \left(\int_0^\infty \frac{\partial^2 v_x}{\partial x^2} dy + \int_0^\infty \frac{\partial^2 v_x}{\partial y^2} dy \right) \ . \quad (8.103)$$

Integrating by parts and exploiting the continuity equation, the second term on the left-hand side of Eq. (8.103) can be rewritten as:

$$\int_0^\infty v_y \frac{\partial v_x}{\partial y} dy = (v_x \cdot v_y) \Big|_0^\infty - \int_0^\infty v_x \frac{\partial v_y}{\partial y} dy =$$

$$= \underbrace{v_x(\infty)}_{=U_\infty} \cdot v_y(\infty) - \underbrace{v_x(0)}_{=0} \cdot v_y(0) - \int_0^\infty v_x \left(-\frac{\partial v_x}{\partial x} \right) dy =$$

$$= U_\infty \cdot v_y(\infty) + \int_0^\infty v_x \frac{\partial v_x}{\partial x} dy \ , \quad (8.104)$$

with:

$$v_y(\infty) = -\int_0^\infty \frac{\partial v_x}{\partial x} dy \ , \quad (8.105)$$

from the integration of the continuity equation. The last integral on the right-hand side of Eq. (8.103) can be recast as:

$$\int_0^\infty \frac{\partial^2 v_x}{\partial y^2} dy = \frac{\partial v_x}{\partial y}\bigg|_0^\infty = \underbrace{\frac{\partial v_x}{\partial y}\bigg|_{y\to\infty}}_{=0} - \frac{\partial v_x}{\partial y}\bigg|_{y=0} = -\frac{1}{\mu}\tau_w \quad, \quad (8.106)$$

with τ_w the wall shear stress. Substituting into Eq. (8.103) and recalling that $\partial \mathcal{P}/\partial x = -\rho U_\infty dU_\infty/dx$ yields:

$$\rho\left(\int_0^\infty v_x \frac{\partial v_x}{\partial x} dy - \int_0^\infty U_\infty \frac{\partial v_x}{\partial x} dy + \int_0^\infty v_x \frac{\partial v_x}{\partial x} dy\right) =$$

$$= \rho\int_0^\infty U_\infty \frac{dU_\infty}{dx} dy - \tau_w \quad . \quad (8.107)$$

Rearranging the integrals, the following expression for τ_w can be obtained:

$$\tau_w = \rho\int_0^\infty \left[\frac{\partial}{\partial x}(v_x \cdot U_\infty - v_x^2)\right] dy + \int_0^\infty \frac{\partial U_\infty}{\partial x}(U_\infty - v_x) dy =$$

$$= \rho\frac{\partial}{\partial x}\left[U_\infty^2 \underbrace{\int_0^\infty \frac{v_x}{U_\infty}\left(1 - \frac{v_x}{U_\infty}\right) dy}_{\text{Momentum thickness, } \hat{\delta}}\right] + \frac{dU_\infty}{dx}U_\infty \underbrace{\int_0^\infty \left(1 - \frac{v_x}{U_\infty}\right) dy}_{\text{Displacement thickness, } \delta^*}, \quad (8.108)$$

or, in a more compact form:

$$\frac{\tau_w}{\rho} = \frac{d(U_\infty^2 \hat{\delta})}{dx} + U_\infty \frac{dU_\infty}{dx}\delta^* \quad . \quad (8.109)$$

This equation is known as the von Kármán-Pohlhausen *momentum-integral equation* and can be used to describe the boundary layer generated by the two-dimensional flow of a Newtonian and incompressible fluid. This equation is applicable to both the laminar and the turbulent cases, provided that an appropriate choice for the expression of τ_w is made.

If the flow is characterized by a zero pressure gradient, then $dU_\infty/dx = 0$ and the equation reduces to:

$$\frac{\tau_w}{\rho} = U_\infty^2 \frac{d\hat{\delta}}{dx} \quad . \quad (8.110)$$

For a turbulent boundary layer, one possibility to solve this equation is to use equations (8.43) and (8.50), that is (dropping the subscript $+$ for simplicity of notation and indicating the flow direction with x instead of z):

$$\bar{v}_x = \begin{cases} y & \text{for } y \le \delta_{sv} \quad, \\ 2.5 \ln y + 5.5 & \text{for } y > \delta_{sv} \quad, \end{cases} \quad (8.111)$$

with δ_{sv} the thickness of the viscous sublayer. The resulting integral, however, would be difficult to calculate. A better choice is to use the power law:

$$\frac{\overline{v}_x}{U_\infty} = \left[\frac{y}{\delta(x)}\right]^m \quad , \tag{8.112}$$

with $\delta(x)$ thickness of the turbulent boundary layer and $m = 1/7$. Note that this formulation of the power law for turbulent boundary layer is analogous to the formulation written for the turbulent flow in a pipe, Eq. (8.57), with U_∞ replacing $\langle v_z \rangle$ and $\delta(x)$ replacing R. This law provides an excellent approximation for the logarithmic part of the velocity profile, but not for its linear part, namely for $y\delta_{sv}$. In spite of this, it is possible to assume that the power law is valid up to the wall ($y = 0$), so that the integral can be calculated. This assumption introduces an error in the integration, but this error is typically small since $\delta_{sv} << \delta(x)$. Therefore, Eq. (8.110) becomes:

$$\frac{\tau_w}{\rho} = \frac{d}{dx} \int_0^\delta U_\infty \left(\frac{y}{\delta}\right)^{1/7} \left[U_\infty - U_\infty \left(\frac{y}{\delta}\right)\right] = \frac{7}{72}\frac{d}{dx}\left[U_\infty^2 \delta(x)\right] \quad , \tag{8.113}$$

and, since U_∞ is constant from the assumption of zero pressure gradient:

$$\tau_w = \frac{7}{72}\rho U_\infty^2 \frac{d\delta(x)}{dx} \quad . \tag{8.114}$$

It is now necessary to derive an additional expression for τ_w so that Eq. (8.114) can be used to derive an expression for the thickness $\delta(x)$. To this aim, from Eq. (8.57), it is possible to set:

$$\frac{\overline{v}_x}{\langle \overline{v}_x \rangle} = \frac{(m+1)(m+2)}{2}\left[\frac{y}{\delta(x)}\right]^m \quad . \tag{8.115}$$

Recalling that:

$$\overline{v}_{x,+} = \frac{\overline{v}_x}{v^*} = \frac{\overline{v}_x}{\sqrt{\tau_w/\rho}} = \frac{\overline{v}_x}{\langle \overline{v}_x \rangle}\sqrt{\frac{2}{f}} \quad , \tag{8.116}$$

with:

$$\tau_w = \frac{1}{2}\rho\langle \overline{v}_x \rangle^2 f \quad , \tag{8.117}$$

Equation (8.115) becomes:

$$\overline{v}_{x,+}\sqrt{\frac{f}{2}} = \frac{(m+1)(m+2)}{2}\left[\frac{y}{\delta(x)}\right]^m . \qquad (8.118)$$

Since $y_+ = yv^*/\nu$ and $\delta_+ = \delta(x)v^*/\nu$, this equation can be rewritten as:

$$\overline{v}_{x,+}\sqrt{\frac{f}{2}} = \frac{(m+1)(m+2)}{2}\left(\frac{y_+}{\delta_+}\right)^m . \qquad (8.119)$$

Equation (8.119) must hold also in the viscous sublayer and, therefore, it must satisfy the boundary condition:

$$\overline{v}_{x,+} = y_+ \quad \text{for} \quad y_+ = \delta_{sv,+} , \qquad (8.120)$$

with $\delta_{sv,+}$ dimensionless thickness of the viscous sublayer in wall units. Equation (8.120) is analogous to the law of the wall derived for turbulent pipe flow. Imposing this condition and noting that δ_+ can be written as $\delta_+ = Re\sqrt{f/2}$ with $Re = \langle\overline{v}_x\rangle\delta/\nu$, after some algebra that is omitted here, it is possible to rewrite Eq. (8.119) as:

$$f = 2\left[\frac{(m+1)(m+2)}{2}\right]^{\frac{2}{m+1}}\left[\frac{(\delta_{sv,+})^{\frac{2(m-1)}{m+1}}}{Re^{\frac{2m}{m+1}}}\right] . \qquad (8.121)$$

For $m = 1/7$, this equation yields:

$$f = 3.5\left(\delta_{sv,+}\right)^{-3/2} Re^{-1/4} , \qquad (8.122)$$

which becomes the Blasius equation if $3.5\left(\delta_{sv,+}\right)^{-3/2} = 0.079$ if $\delta_{sv,+} \simeq 12.6$. This value is very close to the distance from the wall $y_+ = 11.6$ where the law of the wall and the logarithmic profile cross each other.

The boundary condition (8.120) can be then rewritten as:

$$\overline{v}_{x,+} = 12.6 = \frac{\overline{v}_x}{v^*} \quad \text{for} \quad y_+ = 12.6 = \frac{yv^*}{\nu} , \qquad (8.123)$$

and the power law (8.112) as:

$$\frac{12.6v^*}{U_\infty} = \left(\frac{12.6\nu}{v^*\delta}\right)^{1/7} . \qquad (8.124)$$

Recasting this equation with respect to the friction velocity yields:

$$(v^*)^{8/7} = 0.114 U_\infty \left(\frac{\nu}{\delta}\right)^{1/7} \quad , \tag{8.125}$$

and, in turn:

$$\left(\frac{\tau_w}{\rho}\right)^{4/7} = 0.114 U_\infty \left(\frac{\nu}{\delta}\right)^{1/7} \quad . \tag{8.126}$$

Recasting this equation with respect to the wall shear stress yields:

$$\tau_w \simeq 0.022 \rho U_\infty^2 Re_\delta^{-0.25} \quad , \tag{8.127}$$

with $Re_\delta = U_\infty \delta(x)/\nu$. Combining Eqs. (8.114) and (8.127), the expression sought for $\delta(x)$ can be obtained:

$$\delta(x) \simeq 0.37 \left(\frac{\nu}{U_\infty}\right)^{1/5} x^{4/5} \quad . \tag{8.128}$$

This relation shows that the thickness of the turbulent boundary layer increases proportionally to $x^{4/5}$, instead of $x^{1/2}$ as in the laminar case.

For the wall shear stress, it is found that:

$$\tau_w \simeq 0.0287 \rho U_\infty^2 \left(\frac{U_\infty x}{\nu}\right)^{-1/5} = 0.0287 \rho U_\infty^2 Re_x^{-1/5} \quad , \tag{8.129}$$

which shows that τ_w decreases along the plate as $x^{-1/5}$, instead of $x^{-1/2}$ as in the laminar case. The local friction coefficient, measured at a given position x along the plate, is:

$$C_{f,x}^t = \frac{\tau_w}{\frac{1}{2}\rho U_\infty^2} = 0.0574 Re_x^{-1/5} \quad , \tag{8.130}$$

while the average coefficient, given by the integral of C_f, x' between $x = 0$ (leading edge of the plate) and $x = L$ (endpoint of the plate, with L length of the plate), is:

$$C_f^t = 0.072 Re_L^{-1/5} \quad , \tag{8.131}$$

with $Re_L = U_\infty L/\nu$. The integral of C_f, x' is calculated assuming that the entire boundary layer is turbulent and that the laminar portion of the boundary layer occupies a small portion of the plate length, L. Figure 8.7 shows the behavior of the local friction coefficient for the turbulent case, Eq. (8.130), comparing it with that for the

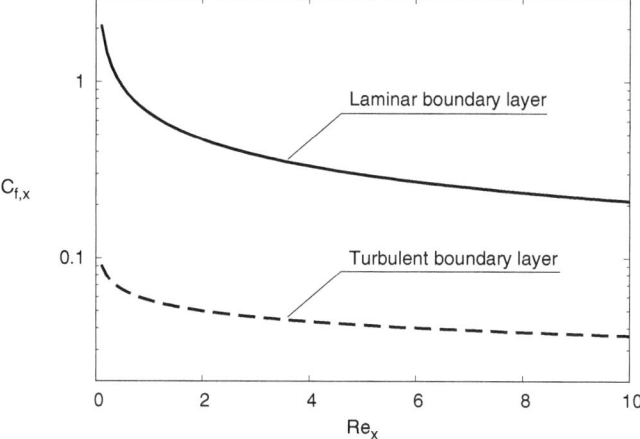

Fig. 8.7 Behavior of the local friction coefficient, $C_{f,x}$, as a function of the dimensionless position along the plate, Re_x, for turbulent boundary layer (dashed line) and for laminar boundary layer (solid line). The boundary layer is always initially laminar and, depending on the flow conditions, may become turbulent. The dashed curve refers to the situation in which the laminar portion of the boundary layer is negligible

laminar case, Eq. (7.50). It turns out that the local friction coefficient for the turbulent case is significantly smaller for the same Re_x, that is for a given position along the plate.

Problems

$\boxed{a}$ Multiply by v'_y the x component of the *N-S* equations and average over time the resulting equation.

$\boxed{b}$ Water at 20 °C flows with a mean velocity of 15 m/s in a pipe with diameter $D = 0.15$ m.

1. Determine the thickness of the viscous sublayer.
2. Assuming that, from the edge of the viscous sublayer (located at distance $y_{0,+}$ from the wall) to the center of the pipe, the following law can be used to express the velocity profile:

$$v_+ = 2.8 \ln\left(\frac{y_+}{y_{0,+}}\right) + y_{0,+} \quad y_{0,+} \leq y_+ \leq \frac{r_0}{\nu}\sqrt{\frac{\tau_0}{\rho}} \quad , \tag{8.132}$$

derive a relation between the friction factor $f = 2\tau_0/(\rho\langle v_z\rangle^2)$ and the Reynolds number. Neglect the presence of the viscous sublayer and simplify the result for large values of $r_0/\nu \sqrt{\tau_0/\rho}$.

$\boxed{c}$ Calculate the power required to pump water ($\nu = 10^{-6}$ m^2/s, $\rho = 10^3$ kg/m^3) at $Re = 1.8 \cdot 10^6$ in a pipe of diameter $D = 1.5$ m and length $L = 10$ km. The dimensionless velocity profile is:

$$v_{z,+}(y_+) = 2.5 \ \ln y_+ + 5.5 \ . \tag{8.133}$$

$\boxed{d}$ Derive an expression for the mixing length l_+ as a function of y_+ for the center region of a turbulent pipe flow, considering the following dimensionless velocity profile:

$$v_{z,+}(y_+) = 2.5 \ \ln y_+ + 5.5 \ . \tag{8.134}$$

$\boxed{e}$ Derive the time-averaged continuity equation for a compressible fluid. Perform the derivation considering that density can be a function of both space and time.

Part III
Design of One-Dimensional Flow Systems

Chapter 9
Macroscopic Balance Equations

The objective of this chapter is to derive the conservation equations in macroscopic form. Examples of applications for these equations will be presented in Chaps. 10 and 11. The conservation law, as formulated in Chap. 3, can be applied to *one-dimensional flow systems*, characterized by a single inlet section through which the fluid enters the system and a single outlet section from which the fluid exits the system. In this case, the control volume is the volume of fluid comprised between the two sections (see Fig. 9.1). Let Γ be the quantity that is conserved (e.g., energy). Denote with γ the value of Γ per unit mass, and with $\rho\gamma$ the value of Γ per unit volume. Consider an elemental area $d\mathbf{A}$ on the inlet or outlet section of the control volume ($d\mathbf{A}$ being a vector normal to the surface of the section, of magnitude dA and oriented *according to the flow*). Let $\mathbf{v}$ be the velocity of the fluid when it crosses $d\mathbf{A}$ and v its magnitude. Denote with v_n the component of $\mathbf{v}$ normal to the surface, so that the flowrate of the incoming fluid through $d\mathbf{A}$ is:

$$dw = \rho\mathbf{v} \cdot d\mathbf{A} = \rho v_n dA \quad , \tag{9.1}$$

and the flowrate of Γ is $dw_\gamma = \gamma\rho v_n dA$.

The total flowrate through an arbitrary surface S are, respectively:

$$w = \int_S \rho v_n dA, \tag{9.2}$$

$$w_\gamma = \int_S \gamma\rho v_n dA \quad . \tag{9.3}$$

In general, v_n is different from the velocity magnitude v. In the following, however, we will assume that v_n coincides with v for simplicity.

© The Author(s), under exclusive license to Springer Nature Switzerland AG 2024
A. Soldati and C. Marchioli, *Fluid Mechanics for Mechanical Engineers*,
https://doi.org/10.1007/978-3-031-53950-3_9

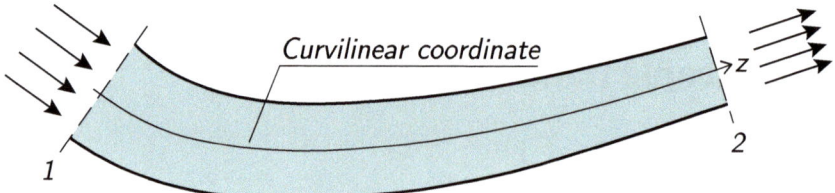

Fig. 9.1 Control volume for the derivation of the macroscopic balance equations

The mean value of any quantity defined on the surface S is given by:

$$\langle \psi \rangle = \frac{1}{A} \int_S \psi \, \mathrm{d}A \quad , \tag{9.4}$$

and, therefore:

$$w_\gamma = \langle \gamma \rho v \rangle A \quad . \tag{9.5}$$

The total quantity q_γ of γ contained in the control volume is equal to:

$$q_\gamma = \int_{z_1}^{z_2} \int_S \rho \gamma \mathrm{d}A \mathrm{d}z = \int_{z_1}^{z_2} \langle \rho \gamma \rangle A \mathrm{d}z \quad , \tag{9.6}$$

where the coordinates z_1 and z_2 refer to the inlet and outlet sections of the control volume.

9.1 Mass Conservation

If Γ represents the mass of fluid in the control volume, then $\gamma = 1$ and mass conservation can be expressed in differential form as follows:

$$\frac{\partial(\langle \rho \rangle A)}{\partial t} = -\frac{\partial(\langle \rho v \rangle A)}{\partial z} \quad . \tag{9.7}$$

Integration of this equation between sections 1 and 2 yields:

$$\frac{\mathrm{d}}{\mathrm{d}t} \int_{z_1}^{z_2} \langle \rho \rangle A \mathrm{d}z = \langle \rho v \rangle_1 A_1 - \langle \rho v \rangle_2 A_2 = w_1 - w_2 \quad , \tag{9.8}$$

where w_1 and w_2 are the mass flowrates that enter and leave the control volume, respectively.

9.2 Energy Conservation

The *energy* conservation law, stated in Sect. 3.5, imposes that the sum of the internal energy, kinetic energy and potential energy is conserved. It follows that, in this case, γ is given by:

$$\gamma = e + \frac{1}{2}v^2 + gh \quad . \tag{9.9}$$

In analogy with mass conservation, the conservation of energy in differential form becomes:

$$\frac{\partial}{\partial t}\left[\left\langle \rho\left(e + \frac{1}{2}v^2 + gh\right)\right\rangle A\right] = -\frac{\partial}{\partial z}\left[\left\langle \rho\left(e + \frac{1}{2}v^2 + gh\right)v\right\rangle A\right] +$$

$$-\frac{\partial}{\partial z}\left(\langle pv\rangle A\right) + \frac{\partial \dot{q}}{\partial z} + \frac{\partial \dot{w}_s}{\partial z} \quad . \tag{9.10}$$

Integration between sections 1 and 2 yields:

$$\frac{d}{dt}\int_{z_1}^{z_2}\left\langle \rho\left(e + \frac{1}{2}v^2 + gh\right)\right\rangle A dz = \left\langle \rho\left(e + \frac{1}{2}v^2 + gh\right)v\right\rangle_1 A_1 +$$

$$-\left\langle \rho\left(e + \frac{1}{2}v^2 + gh\right)v\right\rangle_2 A_2 + \langle pv\rangle_1 A_1 - \langle pv\rangle_2 A_2 + \dot{q} + \dot{w}_S \quad . \tag{9.11}$$

In Eqs. (9.10) and (9.11), the term $\dot{q}$ represents the heat exchanged with the environment. The term representing the work done on the system can be divided into two contributions:

1. Work per unit time done to move the fluid in and out of the control volume:

$$\langle pv\rangle_1 A_1 - \langle pv\rangle_2 A_2 \quad . \tag{9.12}$$

2. Work done on the fluid within the control volume, $\dot{w}_S$.

The conservation of energy can be greatly simplified in the case of:

1. steady flow;
2. uniform density and uniform temperature over the section;
3. uniform pressure over the section.

With these approximations, considering that the hypothesis of steady flow implies $\rho_1\langle v_1\rangle A_1 = \rho_2\langle v_2\rangle A_2$, and setting:

$$\alpha = \frac{\langle v^3\rangle}{\langle v\rangle^3} \quad , \tag{9.13}$$

Equation (9.11) becomes:

$$e_2 + \frac{1}{2}\alpha_2 v_2^2 + gh_2 + \frac{p_2}{\rho_2} - e_1 - \frac{1}{2}\alpha_1 v_1^2 - gh_1 - \frac{p_1}{\rho_1} = q + w_s \quad , \qquad (9.14)$$

where $q = \dot{q}/w$ and $w_s = \dot{w}_S/w$ represent the heat flux and the work done per unit mass of fluid flowing through the control volume, respectively. In Eq. (9.14), the angular brackets, which represent mean values, are omitted for ease of notation.

9.2.1   Bernoulli Equation

The total energy balance does not allow to distinguish between mechanical energy and thermal energy, since part of the mechanical energy is converted over time into thermal energy and only the sum of the two contributions is conserved.

Equation (9.14) can be written in the following differential form if the inlet section (subscript 1) and the outlet section (subscript 2) of the control volume are very close to each other:

$$d(e) + \frac{1}{2}d(\alpha v^2) + gdh + d\left(\frac{p}{\rho}\right) = dq + dw_s \quad . \qquad (9.15)$$

In terms of state functions of a thermodynamic system, $d(e)$ is given by:

$$d(e) = Tds - pd\left(\frac{1}{\rho}\right) \quad , \qquad (9.16)$$

where s is the entropy of the system.

The second law of thermodynamics can be expressed as:

$$dl_v = Tds - dq \geq 0 \quad , \qquad (9.17)$$

where dl_v represents the amount of mechanical energy that is irreversibly dissipated into heat. Therefore, the energy balance becomes:

$$\frac{1}{2}d(\alpha v^2) + gdh + \frac{dp}{\rho} = dw_s - dl_v \quad . \qquad (9.18)$$

Equation (9.18) is one of the forms in which the Bernoulli equation can be expressed. The equation was named after the Swiss mathematician and physicist Daniel Bernoulli (1700–1782).

9.3 Conservation of Momentum

The conservation of momentum can be greatly simplified for one-dimensional flows characterized by plane inlet and outlet sections and uniform physical properties as well as uniform pressure over the entire section. In this case, γ is given by the momentum per unit mass, equal to the vector $\mathbf{v}$, and the momentum conservation equation in differential form is:

$$\frac{\partial}{\partial t}(\rho \langle \mathbf{v} \rangle A) = -\frac{\partial}{\partial z}(\rho \langle v v \rangle A) - \frac{\partial}{\partial z}(p\mathbf{A}) - \frac{\partial}{\partial z}\mathbf{F} + \rho \mathbf{g} A \quad . \qquad (9.19)$$

In integral form:

$$\frac{\mathrm{d}}{\mathrm{d}t}\int_{z_1}^{z_2}\rho \langle \mathbf{v} \rangle A \,\mathrm{d}z = \rho_1 \langle v v \rangle_1 A_1 \ +$$

$$- \ \rho_2 \langle v v \rangle_2 A_2 + p_1 \mathbf{A}_1 - p_2 \mathbf{A}_2 - \mathbf{F} + \left(\int_{z_1}^{z_2} \rho A \mathrm{d}z \right) \mathbf{g}, \qquad (9.20)$$

where $\mathbf{F}$ is the force exerted by the fluid on the surface of the control volume. If the velocity vector is oriented in the direction normal to the inlet and outlet section and the coefficient:

$$\beta \equiv \frac{\langle v^2 \rangle}{\langle v \rangle^2} \quad , \qquad (9.21)$$

is introduced, then Eq. (9.19) at steady state becomes:

$$0 = w(\beta_1 \mathbf{v}_1 - \beta_2 \mathbf{v}_2) + p_1 \mathbf{A}_1 - p_2 \mathbf{A}_2 - \mathbf{F} + \left(\int_{z_1}^{z_2} \rho A \mathrm{d}z \right) \mathbf{g} \quad , \qquad (9.22)$$

where $w = \rho v A$. In this integral form of the momentum conservation equation, angular brackets identifying mean values are omitted for ease of notation.

Chapter 10
Analysis and Design of One-Dimensional Flow Systems

10.1 Velocity Measurement

Fluid flow measurements are necessary in a very wide range of applications that require the determination of the flow velocity, the mass flowrate or the volumetric flowrate. As far as the measurement of the velocity of a fluid is concerned, many devices have been developed. The most common is the Pitot tube, named after its inventor, the French engineer Henri Pitot (1695–1771).

10.1.1 Pitot Tube

The Pitot tube is used to measure the local velocity of a fluid. As shown schematically in Fig. 10.1, the device consists of a small-diameter center tube pointing directly into the incoming fluid flow, corresponding to section 2 in Fig. 10.1, and an outer tube that has two static pressure taps, corresponding to section 3 in Fig. 10.1. The center tube (high pressure branch of the device) and the pressure taps (belonging to the low pressure branch of the device) are connected to a differential manometer.

To analyze the working principle of a Pitot tube, it suffices to consider the Bernoulli equation. By neglecting the losses of mechanical energy, the Bernoulli equation written along the streamline that begins far upstream of the tube (section 1) and comes to rest in the mouth of the center tube (section 2) reads as:

$$\frac{1}{2}v_1^2 + \frac{p_1 - p_2}{\rho} = 0 \quad .$$

(10.1)

Consider now a streamline that is close to the one previously considered, begins far upstream of the tube (section 1) but ends in section 3 at some distance from the tube. The Bernoulli equation becomes:

© The Author(s), under exclusive license to Springer Nature Switzerland AG 2024   273
A. Soldati and C. Marchioli, *Fluid Mechanics for Mechanical Engineers*,
https://doi.org/10.1007/978-3-031-53950-3_10

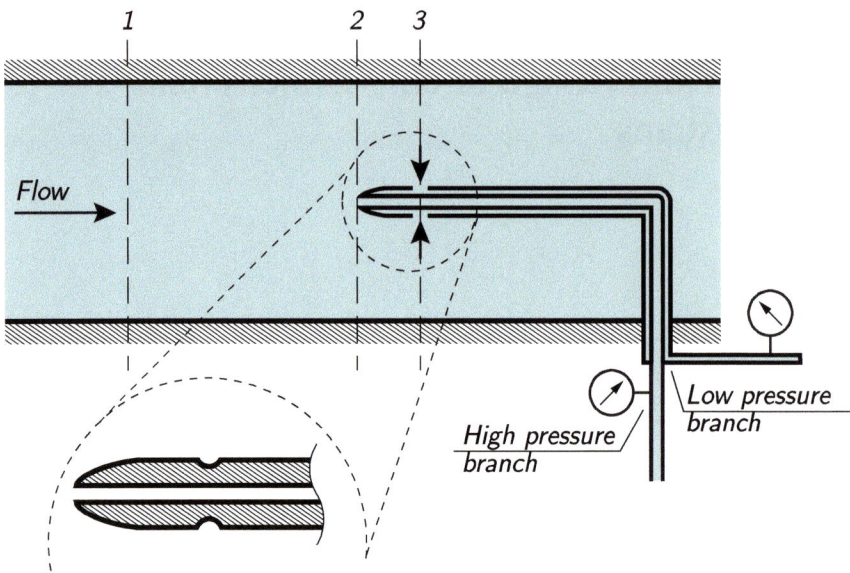

Fig. 10.1 The Pitot tube evaluates the velocity of a fluid by measuring the difference between the pressure sensed by the tube and the pressure of the surrounding flow

$$p_1 = p_3 \quad , \tag{10.2}$$

under the (reasonable) assumption of negligible losses of mechanical energy between sections 1 and 3. Pressures p_1, p_2 and p_3 are, by definition, uniform on the sections they are applied to. Therefore, by combining Eqs. (10.1) and (10.2), the velocity v_1 can be obtained as:

$$v_1 = \sqrt{\frac{2(p_2 - p_3)}{\rho}} \quad . \tag{10.3}$$

This equation allows to measure the local fluid velocity with good accuracy.

10.1.2   Examples

$\boxed{1}$ A Pitot tube is immersed in a fluid that moves with velocity v. If the fluid is air and the liquid in the differential manometer (to which the center tube and the pressure taps are connected) is water, determine the velocity of the fluid when the height difference of the manometric liquid is $\Delta h = 0.65$ cm.

Solution The velocity can be obtained from Eq. (10.4) once the pressure difference $p_2 - p_3$ is known. This difference is related to Δh by the law of fluid statics, which imposes:

$$p_2 - p_3 = \rho_{H_2O} g \Delta h = 638 \text{ Pa} \quad . \tag{10.4}$$

Assuming a density of $\rho = 1.3 \text{ kg/m}^3$ for air, Eq. (10.4) yields $v = 31.3$ m/s.

10.2 Flowrate Measurement

In addition to the devices developed for measuring the velocity of a fluid, there are devices (called flowmeters) that can be used to determine the flowrate of a fluid. Similar to the Pitot tube, the working principle of flowmeters is based on the Bernoulli equation and exploits pressure measurements to determine the flowrate.

10.2.1 Calibrated Orifice

The calibrated orifice consist of a thin plate with a hole in it, which is used to measure the flowrate in ducts and pipes by forcing the fluid to converge to pass through the hole. The operating principle of this measurement device is shown in Fig. 10.2. Note that, while section 1 is located upstream of the orifice plate, section 2 delimiting the control volume is located immediately downstream of the orifice, at a position in which the fluid velocity is close to that in the hole. This type of flowmeter generates a non-negligible pressure drop, generated by the flow through the hole and associated with swirling motions occurring downstream of the section.

An expression for the pressure drop can be obtained using the Bernoulli equation, written in integral form:

$$\frac{1}{2}v_2^2 = \frac{1}{2}v_1^2 + \frac{p_1 - p_2}{\rho} - l_v \quad . \tag{10.5}$$

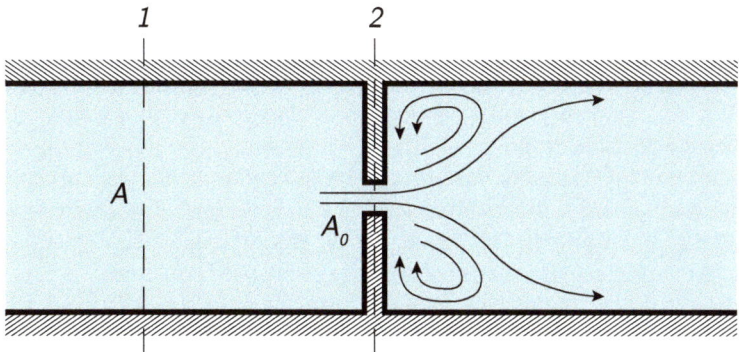

Fig. 10.2 Flow through a calibrated orifice

Assuming that the fluid is incompressible and the flow is turbulent, the continuity equation becomes:

$$Av_1 = A_0v_2 \quad , \tag{10.6}$$

and, combined with Eq. (10.5), yields:

$$\frac{p_1 - p_2}{\rho} = \frac{1}{2}v_2^2\left[1 - \left(\frac{A_0}{A}\right)^2\right] + l_v \quad . \tag{10.7}$$

Experimental observations indicate that l_v is not negligible. By assuming l_v proportional to the kinetic energy per unit mass of the fluid:

$$l_v = \frac{1}{2}v_2^2k \quad , \tag{10.8}$$

for high Reynolds numbers, the coefficient k can be expressed as:

$$k = 1.6\left[1 - \left(\frac{A_0}{A}\right)^2\right] \quad . \tag{10.9}$$

This expression, combined with Eq. (10.7), allows to express the flowrate Q as a function of the orifice geometry and the pressure drop, which can be easily measured. The expression is:

$$Q = 0.62A_0\sqrt{\frac{2(p_1 - p_2)}{[1 - (A_0/A)^2]\rho}} \quad . \tag{10.10}$$

10.2.2   Venturi Tube (Venturimeter)

The Venturi tube, or *Venturimeter*, named after the Italian physicist Giovanni Battista Venturi (1746–1822), allows for the measurement of the flowrate in ducts and pipes by means of a smooth gradual contraction of the cross section, which allows to minimize the pressure drop.

As shown in Fig. 10.3, the Venturimeter consists of a smooth gradual contraction from the main pipe size to the throat section, which prevents flow detachment from the pipe wall and is followed by a much longer gradual enlargement from the throat section to the original pipe diameter, which prevents the formation of eddies within the decelerated flow. In this way, pressure losses due to changes in the fluid velocity can be assumed negligible. Therefore, the Bernoulli equation applied between sections 1 and 2 reads as:

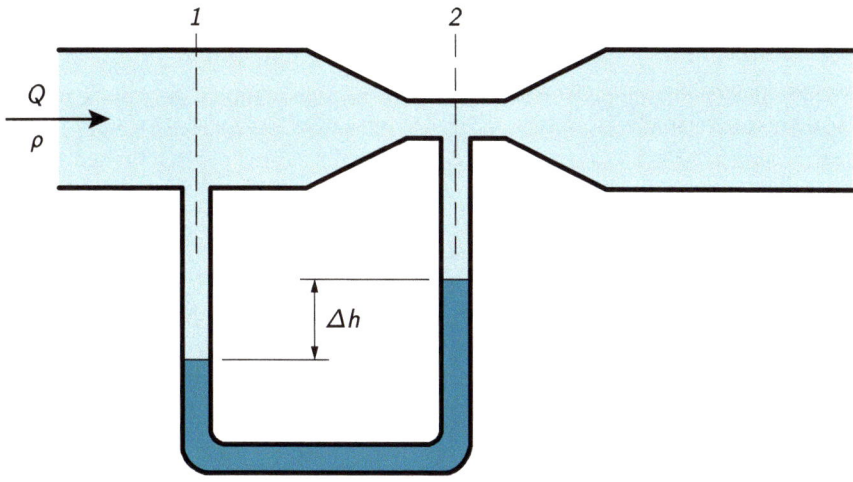

Fig. 10.3 Schematic representation of a Venturi tube

$$\frac{1}{2}v_2^2 = \frac{1}{2}v_1^2 + \frac{p_1 - p_2}{\rho} \quad . \tag{10.11}$$

Assuming, as for the calibrated orifice, that the fluid is incompressible and the flow is turbulent, from the continuity equation it follows that:

$$A_1 v_1 = A_2 v_2 \quad , \tag{10.12}$$

and, in combination with Eq. (10.11), the flowrate can be expressed as:

$$Q = \sqrt{\frac{2(p_1 - p_2)}{[1 - (A_2/A_1)^2]\rho}} \quad . \tag{10.13}$$

The pressure difference $p_1 - p_2$ can be easily measured, for example, by using a U-shaped manometer (or differential manometer). The manometer consists of a graduated U-shaped tube placed perpendicular to the flow direction and filled with a manometric liquid (e.g. mercury). The pressure difference $p_1 - p_2$ is proportional to the change in height ΔH that the manometric liquid reaches in each branch of the tube:

$$\Delta H = \frac{|p_1 - p_2|}{\rho_m g} \quad , \tag{10.14}$$

with ρ_m density of the manometric fluid.

10.2.3 *Rotameter*

Another device commonly used to make flow measurements is the rotameter. The
rotameter consists of a slightly conical, tapered tube with a shaped weight (called
float) inside. The axial position of the float within the tube is determined by the
flowrate. With reference to the control volume shown in Fig. 10.4, it can be noticed
how the cross-sectional area of tube at sections 1 and 2 is practically the same. The
average fluid velocities in these sections can be expressed as:

$$\langle v \rangle_1 = \langle v \rangle_2 = \frac{Q}{A_T} \quad . \tag{10.15}$$

For section 1, it is possible to set $\alpha_1 = \beta_1 = 1$, while for section 2, the two coefficients
read as:

$$\beta_2 = \frac{\langle v^2 \rangle_2}{\langle v \rangle_2^2} = \frac{\left(\frac{Q}{A_T - A_G}\right)^2 \frac{A_T - A_G}{A_T}}{\left(\frac{Q}{A_T}\right)^2} = \frac{A_T}{A_T - A_G} \quad , \tag{10.16}$$

Fig. 10.4 Working
principle of a rotameter.
The tube is tapered, with a
slightly conical shape

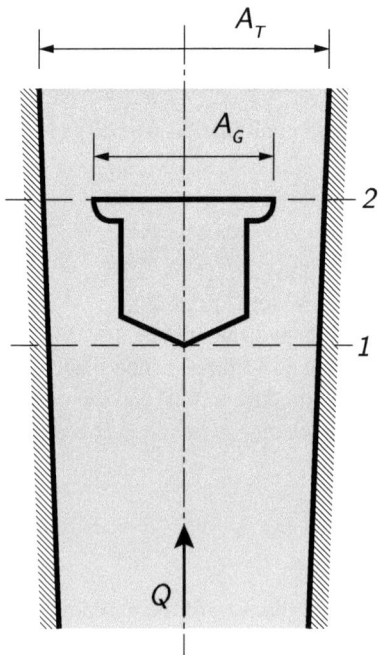

$$\alpha_2 = \frac{\langle v^3 \rangle_2}{\langle v \rangle_2^3} = \left(\frac{A_T}{A_T - A_G} \right)^2 . \tag{10.17}$$

The Bernoulli equation becomes:

$$\frac{1}{2} \left[\alpha_1 \langle v \rangle_1^2 - \alpha_2 \langle v \rangle_2^2 \right] + \frac{p_1 - p_2}{\rho} + g(h_1 - h_2) - l_v = 0 . \tag{10.18}$$

In Eq. (10.18), the energy losses can be expressed as:

$$l_v = \frac{1}{2} k_R \left(\frac{Q}{A_T - A_G} \right)^2 . \tag{10.19}$$

The momentum conservation equation, neglecting the effect of the tangential forces acting on the pipe wall and on the float, can be written as follows:

$$\rho Q \Big[\beta_1 \langle v \rangle_1 - \beta_2 \langle v \rangle_2 \Big] + (p_1 - p_2) A_T +$$

$$- \rho g [(h_2 - h_1) A_T - V_G] - \rho_G g V_G = 0 , \tag{10.20}$$

where V_G is the volume and ρ_G the density of the float. Replacing the expressions for α_1 and α_2 in Eq. (10.18), the expressions for β_1 and β_2 in Eq. (10.20) and eliminating from these two equations the term $(p_1 - p_2)$, the flowrate can be obtained as:

$$Q = (A_T - A_G) \sqrt{\frac{A_T/A_G}{1 + k_R (A_T/A_G)^2}} \sqrt{\frac{2g V_G}{A_G} \left(\frac{\rho_G}{\rho} - 1 \right)} . \tag{10.21}$$

In this equation, the term k_R must be determined experimentally. It is found that k_R is a function of the Reynolds number calculated with reference to the flow section between the float and the pipe.

10.2.4 Examples

1 It is necessary to measure the volumetric flowrate of an oil of density $\rho = 750$ kg/m^3 in a pipe of diameter $D = 0.05$ m, using a horizontally-mounted orifice, as shown in Fig. 10.2. The contraction ratio of the orifice is 6.25. Determine the value of the volumetric flowrate if a differential manometer placed between sections 1 and 2 measures a pressure difference $\Delta p = 9700$ Pa.

Fig. 10.5 Schematic
diagram of a
vertically-mounted
Venturimeter

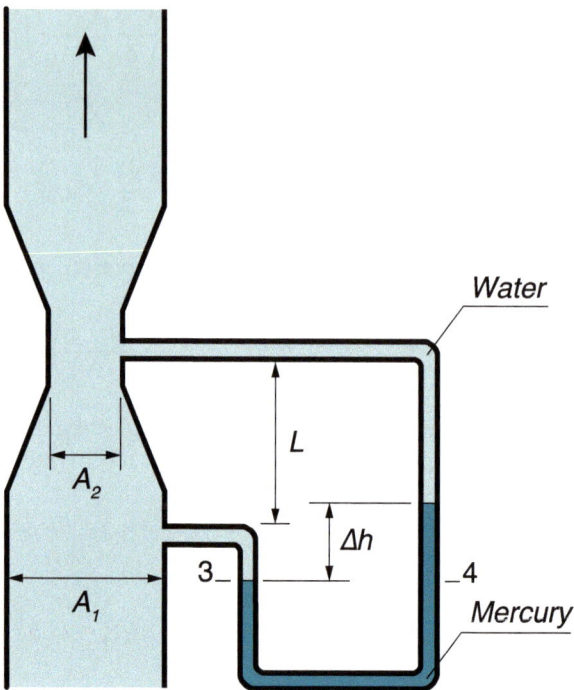

Solution The flowrate can be calculated directly from Eq. (10.10):

$$Q = 0.62A_0\sqrt{\frac{2(p_1 - p_2)}{[1 - (A_0/A)^2]\rho}} \quad . \tag{10.22}$$

Being $A_0 = A/6.25 = 3.14 \cdot 10^{-4}$ m^2, the flowrate is:

$$Q = 0.62 \cdot 3.14 \cdot 10^{-4}\sqrt{\frac{2 \cdot 9700}{[1 - 0.16^2] \cdot 750}} = 0.001 \text{ m}^3/\text{s} \quad . \tag{10.23}$$

$\boxed{2}$ Consider the flow of water ($\rho = 1000$ kg/m^3, $\mu = 10^{-3}$ Pa s) in the vertically-mounted Venturimeter shown in Fig. 10.5. The ratio A_1/A_2 is equal to 2 and the diameter D_1 is equal to 100 mm. The manometric liquid is mercury ($\rho_m = 1.3 \cdot 10^4$ kg/m^3) and a manometric height difference $\Delta h = 0.12$ m is measured. Determine the flowrate of water.

Solution The Bernoulli equation, written between sections 1 and 2, reads as:

$$p_1 + \rho g h_1 + \frac{1}{2}\rho v_1^2 = p_2 + \rho g h_2 + \frac{1}{2}\rho v_2^2 \quad . \tag{10.24}$$

The continuity equation, written between the same two sections, reads as:

$$A_1 v_1 = A_2 v_2 \quad . \tag{10.25}$$

From these two equations, it follows that:

$$p_1 - p_2 = \rho g(h_2 - h_1) + \frac{\rho}{2}(v_2^2 - v_1^2) =$$

$$= \rho g L + \frac{\rho}{2}v_2^2\left[1 - \left(\frac{A_2}{A_1}\right)^2\right] \quad . \tag{10.26}$$

Using subscript 3 to denote the section at the free surface of the mercury in the left branch of the manometer and subscript 4 to denote the same height in the right branch, it follows that:

$$p_3 = p_4 \quad , \tag{10.27}$$

where pressure p_3 is given by:

$$p_3 = p_1 + \rho g h_1 \quad , \tag{10.28}$$

with h_1 height of the water column above the mercury in the left branch, while pressure p_4 is given by:

$$p_4 = p_2 + \rho g h_2 + \rho_m g \Delta h \quad , \tag{10.29}$$

with h_2 height of the water column above the mercury in the right branch. Considering that:

$$\Delta h + h_2 = L + h_1 \quad , \tag{10.30}$$

comparing the expressions written for p_3 and p_4 and solving for the difference between p_1 and p_2, the following expression can be obtained:

$$p_1 - p_2 = \rho g(L - \Delta h) + \rho_m g \Delta h \quad . \tag{10.31}$$

Substituting this expression into Eq. (10.26) yields:

$$\rho g L + \frac{\rho}{2} v_2^2 \left[1 - \left(\frac{A_2}{A_1} \right)^2 \right] = \rho g (L - \Delta h) + \rho_m g \Delta h \quad , \qquad (10.32)$$

from which it is possible to express v_2 in the form:

$$v_2 = \sqrt{\frac{2g\Delta h(\rho_m - \rho)}{\rho} \left[1 - \left(\frac{A_2}{A_1} \right)^2 \right]^{-1}} \quad , \qquad (10.33)$$

and, in turn, calculate the water flowrate through the Venturimeter:

$$Q = A_2 v_2 = A_2 \sqrt{\frac{2g\Delta h(\rho_m - \rho)}{\rho} \left[1 - \left(\frac{A_2}{A_1} \right)^2 \right]^{-1}} =$$

$$= \frac{1}{2} \frac{\pi \cdot 0.1^2}{4} \sqrt{2 \cdot 9.81 \cdot 0.12 \frac{(13 - 1)}{1.}(1 - 0.5^2)^{-1}} = 2.4 \cdot 10^{-2} \ \mathrm{m^3/s} \quad . \quad (10.34)$$

10.3   Examples of One-Dimensional Flow Systems

There are three possible unknowns in the conservation equations: the pressure drop, the viscous losses and the force exerted by the fluid on the wall of a duct. In general, the fluid dynamics problems dealing with one-dimensional flow systems are classified according to the unknowns that have to be determined. The following sections discuss a number of examples of practical interest, which are characterized by one or more of the unknowns mentioned above.

10.3.1   Pressure Loss Due to a Sudden Section Enlargement

The problem is illustrated in Fig. 10.6. It is assumed that the fluid is incompressible and the flow is turbulent ($\alpha = \beta = 1$). The continuity equation and the momentum conservation equation give:

$$A_1 v_1 = A_2 v_2 \quad , \qquad (10.35)$$

$$\rho A_1 v_1^2 - \rho A_2 v_2^2 + p_1 A_1 - p_2 A_2 - F = 0 \quad . \qquad (10.36)$$

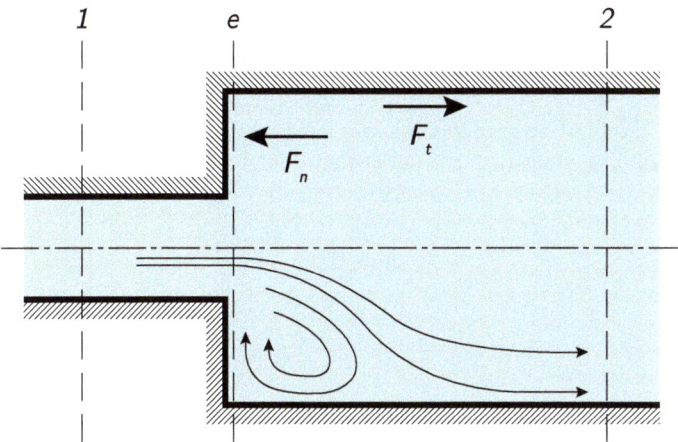

Fig. 10.6 Pressure loss in a sudden section enlargement

The force F consists of two contributions. One contribution, given by the component F_n normal to the surface $A_2 - A_1$, is significant. The other, given by the tangential component F_t, is instead negligible for small axial distances. The normal component, F_n, can be expressed as:

$$F_n = -p_e(A_2 - A_1) \ .$$
(10.37)

It is possible to assume $p_e \cong p_1$, since the cross section of the flow is essentially the same. Thus, from Eq. (10.36), it follows that:

$$\frac{p_1 - p_2}{\rho} = v_2^2 \left(1 - \frac{A_2}{A_1}\right) \ .$$
(10.38)

The Bernoulli equation allows for the calculation of the losses of mechanical energy, l_v, which in this case are localized (or concentrated) being produced by the flow through the section enlargement:

$$l_v = \frac{p_1 - p_2}{\rho} + \frac{1}{2}\left(v_1^2 - v_2^2\right) \ .$$
(10.39)

The energy loss can be obtained by substituting Eq. (10.38) into Eq. (10.39):

$$l_v = \frac{v_2^2}{2}\left(\frac{A_2}{A_1} - 1\right)^2 \ .$$
(10.40)

10.3.2   Force on a Pipe Bend

With reference to Fig. 10.7, consider the turbulent flow ($\alpha = \beta = 1$) in a horizontal pipe bend. The flow enters the bend with a velocity v_1 and leaves it at velocity v_2, the corresponding areas of the cross section being A_1 and A_2, respectively. The equations of mass and energy conservation yield:

$$A_1 v_1 = A_2 v_2 \quad , \tag{10.41}$$

$$\frac{1}{2} v_2^2 + \frac{p_2}{\rho} = \frac{1}{2} v_1^2 + \frac{p_1}{\rho} - l_v \quad . \tag{10.42}$$

The x and y components of the momentum conservation equations are:

x component : $\qquad 0 = \rho A_1 v_1^2 + p_1 A_1 - F_x \quad , \tag{10.43}$

y component : $\qquad 0 = -\rho A_2 v_2^2 - p_2 A_2 - F_y \quad , \tag{10.44}$

where F_x and F_y are the components of the force $\mathbf{F}$ exerted by the fluid on the wall. Experimentally, it is found that:

$$l_v = \frac{1}{2} k v_2^2 \quad , \tag{10.45}$$

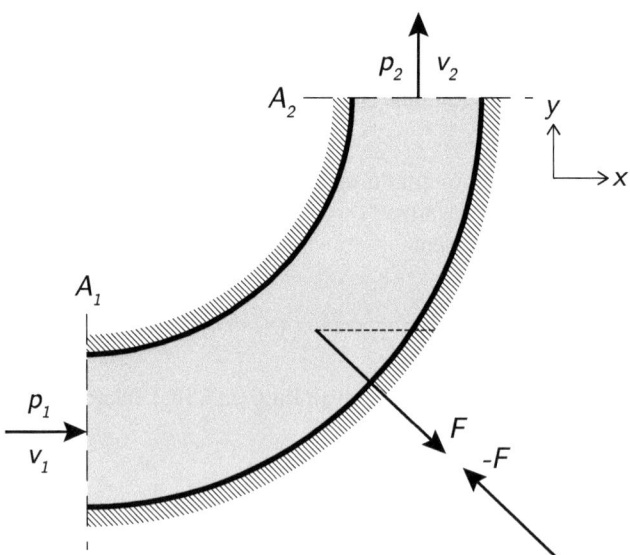

Fig. 10.7 Control volume for the calculation of the force acting on a pipe bend

with $k \approx 3/4$. With this value, the force components F_x and F_y can be expressed as:

$$F_x = A_1(p_1 + \rho v_1^2) \ , \tag{10.46}$$

$$F_y = -A_2 \left\{ p_1 + \frac{1}{2}\rho v_1^2 \left[1 + (1-k)\left(\frac{A_1}{A_2}\right)^2 \right] \right\} \ . \tag{10.47}$$

10.3.3 Jet Pump

A jet pump differs from other types of pumps because it has no moving parts. In a jet pump, shown schematically in Fig. 10.8, a high-velocity fluid jet, which has velocity v_j, is used to increase the pressure of the fluid to be pumped, which has velocity v_s. The average fluid velocity in section 1 is equal to:

$$\langle v \rangle_1 = \lambda^2 v_j + (1 - \lambda^2)v_s = \langle v \rangle_2 \ , \tag{10.48}$$

with λD the inner pipe diameter and D the outer pipe diameter. The viscous losses downstream of section 1 are likely to be large because of the mixing between the two streams, which can be considered as complete only downstream of section 2 in the schematic diagram of Fig. 10.8. As a consequence, in this case, the conservation of energy cannot be used to compute the pressure drop Δp between the two sections. On the other hand, the friction at the wall appears to be negligible. This allows to use the momentum conservation equation:

$$(p_2 - p_1)A = \rho A \beta_1 \langle v \rangle_1^2 - \rho A \beta_2 \langle v \rangle_2^2 \ , \tag{10.49}$$

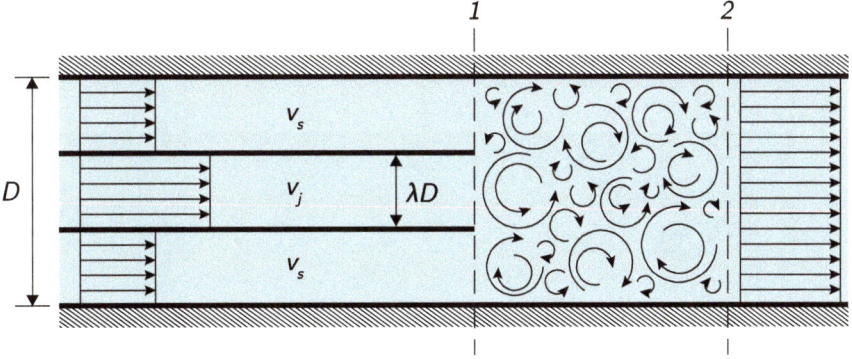

Fig. 10.8 Schematic diagram of a jet pump

where β_1 is different from unity since the velocity profile at section 1 is not uniform, while $\beta_2 = 1$ since the profile is uniform at section 2. The coefficient β_1 can be calculated as:

$$\beta_1 = \frac{\langle v^2 \rangle_1}{\langle v \rangle_1^2} = \frac{\lambda^2 v_j^2 + (1 - \lambda^2) v_s^2}{\langle v \rangle_1^2} \quad . \tag{10.50}$$

By substituting Eq. (10.50) into Eq. (10.49), the following expression is obtained for the increase of pressure provided by the pump:

$$p_2 - p_1 = \lambda^2 (1 - \lambda^2) \rho (v_j - v_s)^2 \quad , \tag{10.51}$$

where v_j is the velocity of the inner fluid and v_s the velocity of the outer fluid.

10.3.4 Flow Distribution in Manifolds

Manifolds are systems used to collect (or distribute) a fluid stream through a series of lateral inlets (or outlets). The design of a distribution manifold may, for example, require that the flowrates at the outlets are all equal. The simplest configuration consists of a T-shaped distribution manifold with a side outlet. With reference to the schematic shown in Fig. 10.9, assuming that the flow at the outlet is perpendicular to the main flow direction, the momentum conservation equation yields:

$$p_2 - p_1 = \rho (v_1^2 - v_2^2) \quad , \tag{10.52}$$

where friction at the wall has been neglected. Note that, in Eq. (10.52), $v_1 > v_2$ and therefore $p_2 > p_1$, which implies that the pressure increases in the portion of the pipe located downstream of the side outlet.

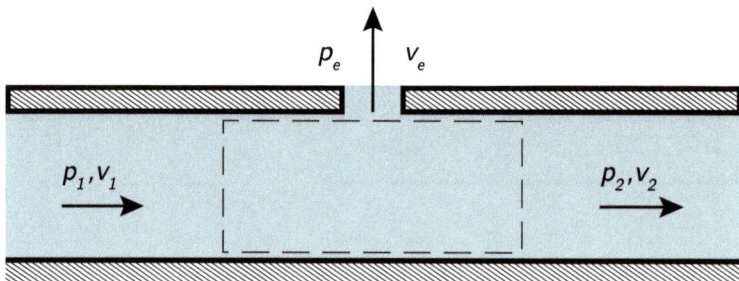

Fig. 10.9 Flow distribution in manifolds: Control volume for the application of the momentum conservation equation

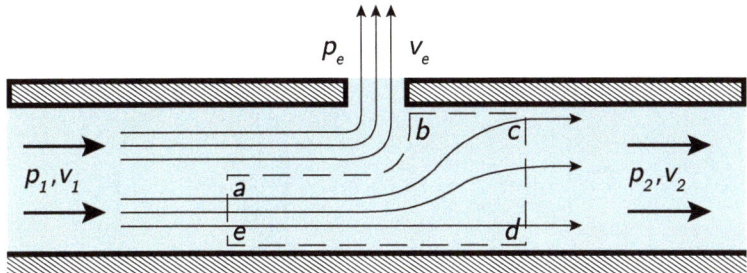

Fig. 10.10 Flow distribution in manifolds: Control volume for the application of the Bernoulli equation

The Bernoulli equation can be applied to the system. Using the control volume (*a b c d e a*) shown in Fig. 10.10 and neglecting all losses, the equation reads as:

$$p_2 - p_1 = \frac{\rho}{2}(v_1^2 - v_2^2) \quad , \tag{10.53}$$

and predicts a pressure recovery equal to half of the pressure recovery predicted by momentum conservation. Experimentally, the correct value is the one given by the Bernoulli equation.

To estimate the outgoing flowrate, the conservation of mass can be considered:

$$D^2 v_1 = D^2 v_2 + d^2 v_e \quad . \tag{10.54}$$

Assuming that the side outlet behaves like an orifice, the exit velocity can be derived from Eq. (10.10):

$$v_e = 0.62 \sqrt{\frac{2}{\rho}\left(\frac{p_1 + p_2}{2} - p_e\right)} \quad , \tag{10.55}$$

where it is assumed that $A_0/A \ll 1$ and that the pressure inside the manifold at the outlet is equal to the average between the pressure before the outlet and the pressure after the outlet.

The system of Eqs. (10.53), (10.54) and (10.55) makes it possible to determine the unknown quantities p_2, v_2 and v_e.

10.3.5 Examples

1 A siphon tube of diameter $D = 0.3$ m is used to remove water from a container as shown in Fig. 10.11 Estimate the pressure that must be exerted at the outlet section of the tube to start the siphon. In addition, calculate the pressure in B and the

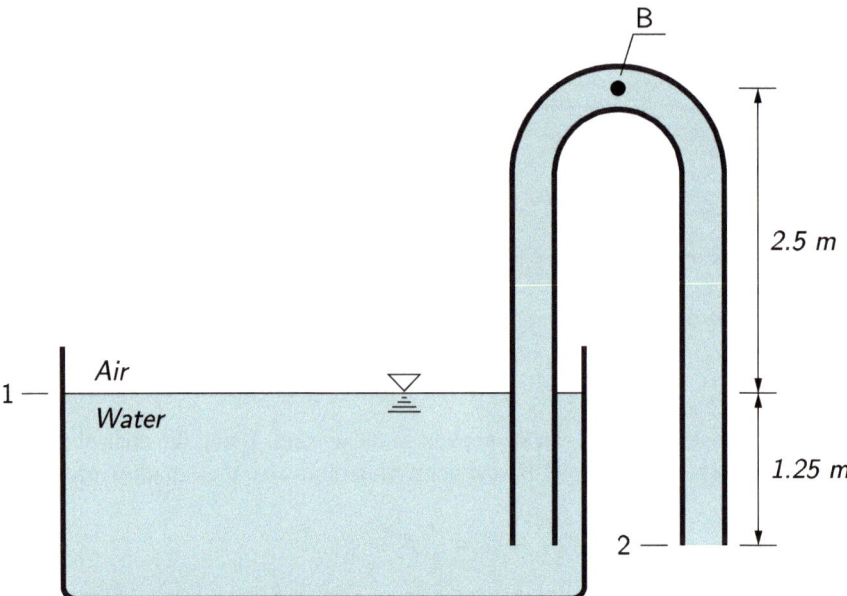

B

2.5 m

Air

1 —

Water

2.5 m

1.25 m

2 —

Fig. 10.11 Schematic of a siphon tube

corresponding flowrate when a steady flow is established in the siphon tube (neglect viscous losses).

Solution To start the siphon tube, the liquid must rise to the highest section of the pipe, indicated by section B in Fig. 10.11. When the liquid is at this position, the Bernoulli equation yields:

$$p_1 + \rho g h_1 = p_B + \rho g h_B \quad , \tag{10.56}$$

from which the pressure difference $p_1 - p_B$ can be obtained as:

$$p_1 - p_B = \rho g (h_B - h_1) = 10^3 \cdot 9.8 \cdot 2.5 = 2.45 \cdot 10^4 \text{ Pa} \quad . \tag{10.57}$$

The pressure p_1 is equal to the atmospheric pressure and so, in the present problem, a depression equal to about one quarter of the atmospheric pressure is required to start the siphon tube. Once the flow is established, the Bernoulli equation between sections 1 and 2 in the absence of significant losses can be written as follows:

$$\frac{p_1}{\rho} + g h_1 = \frac{p_2}{\rho} + \frac{1}{2} v^2 + g h_2 \quad . \tag{10.58}$$

Since $p_1 = p_2 = p_{atm}$, the exit velocity and mass flowrate at section 2 are, respectively:

$$v_2 = \sqrt{2g(h_1 - h_2)} = 4.95 \text{ m/s} \quad , \tag{10.59}$$

$$w = \frac{\pi D^2}{4} \rho v_2 = 349.8 \text{ kg/s} \quad . \tag{10.60}$$

The fluid velocity does not change from section B to section 2 and the Bernoulli equation reads as:

$$\frac{p_B}{\rho} + gh_B = \frac{p_2}{\rho} + gh_2 \quad , \tag{10.61}$$

which yields:

$$p_B - p_2 = \rho g(h_2 - h_B) = -3.7 \cdot 10^4 \text{ Pa} \quad . \tag{10.62}$$

This result shows that, once flow conditions are established in the siphon tube, the pressure at section B decreases appreciably, although no cavitation phenomena is produced by the pressure decrease.

2 During a fire, a firefighter holds a fire hydrant tilted by 45° upward. The hydrant is fed with a flowrate $Q = 2 \text{ m}^3/\text{min}$ and produces a jet that reaches a height of 20 m. Determine the force exerted by the hydrant on the firefighter (which acts to push the firefighter to the ground).

Solution In a Cartesian frame of reference, in which the x axis is horizontal and the y axis is vertical and directed upward, the force F_y that pushes the firefighter toward the ground is due to the momentum of the jet in the y direction:

$$F_y = wv_y \quad , \tag{10.63}$$

where w is the mass flowrate of the jet and v_y is the vertical component of the velocity at the hose exit. The mass flowrate is:

$$w = Q\rho = \frac{2}{60} 10^3 = 33.3 \text{ kg/s} \quad . \tag{10.64}$$

To calculate v_y, it is convenient to apply the Bernoulli equation between the hose outlet and the point of highest elevation reached by the jet:

$$\frac{1}{2} v_o^2 = g(h_1 - h_0) + \frac{1}{2} v_1^2 \quad . \tag{10.65}$$

In this case, $v_o^2 = v_{o,x}^2 + v_{o,y}^2$, and $v_1^2 = v_{1,x}^2$ because $v_{1,y} = 0$ at the position of maximum jet elevation. Assuming that v_x does not change appreciably because of the resistance of the air surrounding the jet, it is possible to set $v_{o,x} = v_{1,x}$ and rewrite Eq. (10.65) as:

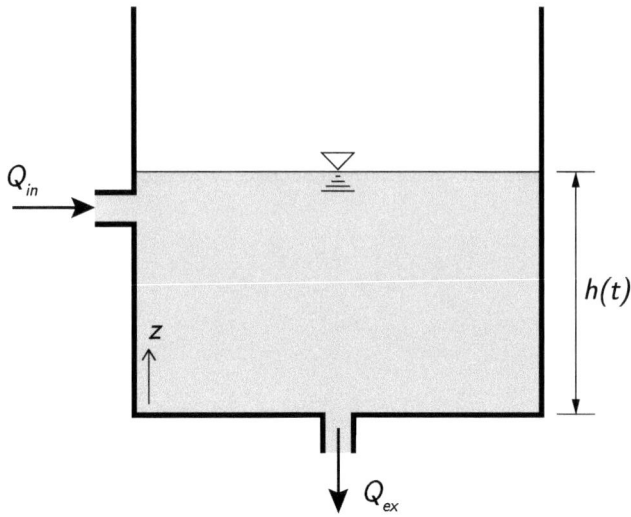

Fig. 10.12 Emptying of a tank through a drainage hole at the bottom

$$\frac{1}{2}v_{yo}^2 = g(h_1 - h_0) \quad . \tag{10.66}$$

For the given value $h_1 - h_0 = 20\ m$, the vertical velocity of the jet at the hose exit is:

$$v_y = \sqrt{2g\Delta h} = 19.81 \text{ m/s} \quad , \tag{10.67}$$

and $F_y = 660$ N.

3 The open cylindrical tank, shown in Fig. 10.12, has volume $V = 50\ m^3$, diameter $D = 4$ m, and initially contains 6 m^3 of water. The tank is fed with a volumetric flowrate $Q_{in} = 0.05\ m^3/s$. The water in the tank exits from a drainage hole of diameter $d = 0.20$ m located at the bottom of the tank. Calculate the level of the water in the tank at equilibrium and the time required for the water to reach a level equal to the equilibrium value $\pm 10\%$.

Solution The problem can be solved by imposing the following mass balance:

$$\frac{dV(t)}{dt} = Q_{in} - Q_{ex}(t) \quad . \tag{10.68}$$

The volume of water contained in the tank at a given time t is equal to $V(t) = Ah(t)$, with $A = \pi D^2/4$ base area of the tank and $h(t)$ water level in the tank. On the other hand, the flowrate discharged through the drainage hole is equal to $Q_{ex}(t) = v_{ex}(t)A_{ex}$, with v_{ex} the exit velocity of the water through the drainage hole. The Bernoulli equation allows to calculate v_{ex} as:

$$\frac{1}{2}v_{ex}^2(t) = gh(t) \quad \longrightarrow \quad v_{ex}(t) = \sqrt{2gh(t)} \quad , \tag{10.69}$$

and $Q_{ex}(t) = k\sqrt{h(t)}$ with $k = 0.14$. By calculating the initial level of water, $h_{in} = h(t = 0) = 0.48$ m, it is apparent that the initial flowrate discharged through the drainage hole is larger than the flowrate received by the tank. As a consequence, the water level drops. The equilibrium level (which is reached when the two flowrates become equal) can be calculated as:

$$Q_{in} = 0.05 = 0.14\sqrt{h_{eq}} = Q_{ex,eq} \quad \longrightarrow \quad h_{eq} = \left(\frac{0.05}{0.14}\right)^2 = 0.13 \text{ m} \quad . \tag{10.70}$$

It is now possible compute the time required for the water to reach a level equal to the equilibrium value $\pm 10\%$, which in the present case is equal to $\hat{h} = 0.13 + 0.13 \cdot 0.1 = 0.143$ m (Note that, in the present problem, the water cannot reach a level equal to -10% of the equilibrium value). The time required for the water to reach the level $\hat{h}$ can be calculated by integrating the balance equation:

$$\frac{dV(t)}{dt} = Q_{in} - Q_{ex}(t) \quad \longrightarrow \quad \frac{dh(t)}{dt} = \frac{Q_{in}}{A} - \frac{k}{A}\sqrt{h(t)} \quad . \tag{10.71}$$

By setting $\mathcal{A} = Q_{in}/A$ and $\mathcal{B} = k/A$, the equation can be solved by separation of variables:

$$\frac{dh}{\mathcal{A} - \mathcal{B}\sqrt{h}} = dt \quad \longrightarrow \quad \int_{h_{in}}^{\hat{h}} \frac{dh}{\mathcal{A} - \mathcal{B}\sqrt{h}} = \int_0^{\hat{t}} dt \quad , \tag{10.72}$$

with $\hat{t}$ time required to reach the level $\hat{h}$. Integration of the equation yields:

$$2\frac{A}{k}\left[\left(\sqrt{h_{in}} - \sqrt{\hat{h}}\right) + \frac{Q_{in}}{k}\ln\frac{k\sqrt{h_{in}} - Q_{in}}{k\sqrt{\hat{h}} - Q_{in}}\right] = \hat{t} = 234 \text{ s} \quad . \tag{10.73}$$

4 A flow distribution manifold must be designed so that the N outlets discharge the same flowrate. To this end, the diameter of the manifold must change after each outlet. If D is the initial diameter of the main pipe of the distribution manifold and d the (constant) diameter of the outlets, determine the law of variation of the diameter of the main pipe. It can be assumed that:

1. The pressure at each distribution outlet is the same ($p_{1e} = p_{2e} = \cdots = p_{Ne}$).
2. The distribution manifold is mounted horizontally.
3. The losses are negligible.

Solution Let the subscript n denote the quantities referred to the portion of the distribution manifold comprised between the nth outlet and the $(n + 1)$th outlet.

Assume that the diameter d of the outlet is much smaller than the diameter D_n of the main pipe of the distribution manifold, whatever n is, so that $d^4/D_n^4 \simeq 0$. Furthermore, at the nth outlet, let the external pressure be constant (p_e) and the internal pressure be equal to the average between the pressures p_{n-1} and p_n. Let the diameter of the main pipe be constant in between two successive outlets and let D_n be just slightly smaller than D_{n-1}. If the exit velocity of the fluid is purely axial with respect to the outlet, then Eq. (10.55) applies and can be recast as:

$$v_e = 0.62 \sqrt{\frac{2}{\rho} \left[\frac{p_n + p_{n-1}}{2} - p_e \right]} . \tag{10.74}$$

Since it must be $v_e = $ constant, it follows that $(p_n + p_{n-1}) = constant$ whatever the value of n is and, in turn, that the pressure along the main pipe can be assumed constant. The momentum balance equation (neglecting losses) yields:

$$p_n - p_{n-1} = \rho(v_{n-1}^2 - v_n^2) , \tag{10.75}$$

which implies that the velocity along the main pipe must remain constant. Based on these considerations and on the mass conservation equation:

$$v_{n-1} D_{n-1}^2 = v_e d^2 + v_n D_n^2 , \tag{10.76}$$

it can be concluded that the diameter of the main pipe changes as:

$$D_n = \sqrt{D_{n-1}^2 - \frac{v_e}{v} d^2} . \tag{10.77}$$

5 A plane jet of a Newtonian fluid of density ρ and viscosity μ impinges on a flat plate and then separates into two fluid streams. Let the plate be inclined at an angle θ with respect to the direction of the jet (as shown in Fig. 10.13). Immediately before of impact, the flowrate of the jet is Q_0, its thickness h_0 and its width W. Determine the force imparted by the jet on the plate and the flowrate of the two streams.

Solution The following assumptions can be made:

1. The flow system is on the horizontal plane.
2. The friction losses are negligible.
3. The flow is steady.

From the continuity equation:

$$A_0 v_0 = A_1 v_1 + A_2 v_2 , \tag{10.78}$$

since $A_i = h_i W$, it follows that:

$$h_0 v_0 = h_1 v_1 + h_2 v_2 . \tag{10.79}$$

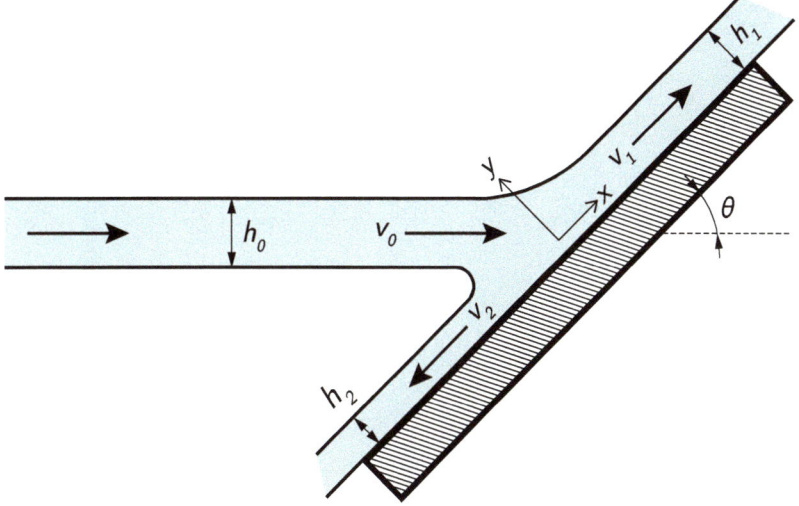

Fig. 10.13 Plane jet impinging on an inclined flat plate

From the Bernoulli equation, neglecting the friction losses and considering that $p_0 = p_1 = p_2 = p$, it follows that:

$$v_0 = v_1 = v_2 \quad , \tag{10.80}$$

and, from Eq. (10.79):

$$h_0 = h_1 + h_2 \quad . \tag{10.81}$$

Consider now a frame of reference in which the x axis is parallel to the plate and the y axis is perpendicular to it. In this frame of reference, the two components v_x and v_y of the velocity v_0 are:

$$v_y = -v_0 \sin \theta \quad , \tag{10.82}$$

$$v_x = v_0 \cos \theta \quad . \tag{10.83}$$

The momentum balance equation along y gives:

$$F_y = \rho Q_0 v_y = -\frac{\rho Q_0^2}{W h_0} \sin \theta \quad . \tag{10.84}$$

Since the component F_x is equal to zero (due to the assumption of negligible friction forces), the momentum balance equation along x yields:

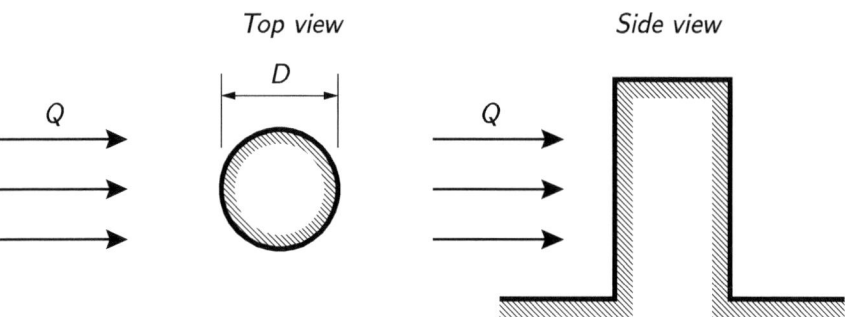

Fig. 10.14 Two-dimensional jet used for stone cutting

$$F_x = 0 \quad \longrightarrow \quad Q_0 v_0 \cos \theta = Q_1 v_1 - Q_2 v_2 \quad, \tag{10.85}$$

and hence, according to Eq. (10.80):

$$h_0 \cos \theta = h_1 - h_2 \quad . \tag{10.86}$$

The values of h_1 and h_2 can be obtained from Eqs. (10.81) and (10.86):

$$h_1 = \frac{h_0}{2}(1 + \cos \theta) ; \qquad h_2 = \frac{h_0}{2}(1 - \cos \theta) \quad . \tag{10.87}$$

The force exerted by the fluid on the plate is equal to F_y.

$\boxed{6}$ A high-velocity water jet is used to cut a cylindrical stone of diameter D, as shown in Fig. 10.14. Denote with T_f the specific force (given by the force normal to the cylinder axis divided by the cross-sectional area of the cylinder). Determine the water flowrate required to cut the stone if the jet diameter is d.

Solution Assuming that after the impact on the cylinder, the jet of water has zero momentum along x (the x axis being parallel to the direction of the velocity of the water), the momentum balance equation along x yields:

$$F = \rho v^2 DL \quad , \tag{10.88}$$

where L is the height of the portion of cylinder impacted by the jet. If A is the cross-sectional area of the cylinder, then the specific force is:

$$T_f = \frac{F}{A} = \frac{4F}{\pi D^2} = 4\rho v^2 \frac{L}{\pi D} \quad, \tag{10.89}$$

and, therefore:

$$v = \left(\frac{T_f \pi D}{4 \rho L} \right)^{1/2} \quad . \tag{10.90}$$

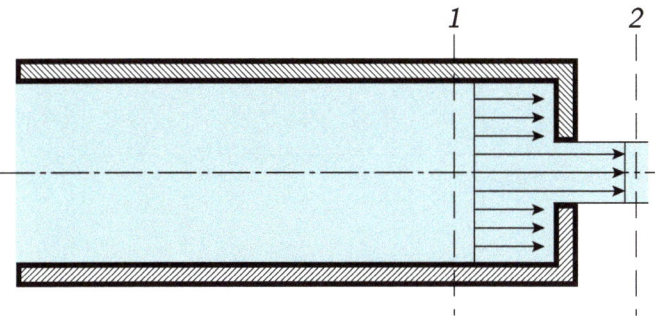

Fig. 10.15 Jet issuing from a pipe collar

Finally, the water flowrate is:

$$Q = \frac{\pi d^2}{4} \left(\frac{T_f \pi D}{4 \rho L} \right)^{1/2} . \tag{10.91}$$

$\boxed{7}$ In a pipe of diameter $D_1 = 0.03$ m, water flows with an average velocity $v = 0.1$ m/s, as shown in Fig. 10.15. At the end of the pipe, a pipe collar is placed, which produces a contraction of the cross section (characterized by diameter D_2) with a $5 \div 1$ ratio between the areas (the collar is geometrically equivalent to a plug drilled in the center and thus the contraction is sudden, not gradual). Determine the force acting on the collar if the pressure outside of the pipe is equal to the atmospheric pressure.

Solution Since $Re = \rho D_1 v_1 / \mu > 2100$, the flow is turbulent ($\alpha = 1$). If the outlet section of the control volume is placed immediately downstream of the collar (namely at a position in which the fluid velocity is close to the velocity in the contraction and pressure is $p = p_{atm}$), the mass conservation equation gives:

$$v_1 D_1^2 = v_2 D_2^2 , \tag{10.92}$$

from which $v_2 = 5v_1 = 0.5$ m/s. The Bernoulli equation gives:

$$\frac{1}{2} \rho v_1^2 + p_1 - \frac{1}{2} \rho v_2^2 - p_2 - \rho l_v = 0 , \tag{10.93}$$

with:

$$l_v = \frac{1}{2} v_2^2 k_a , \tag{10.94}$$

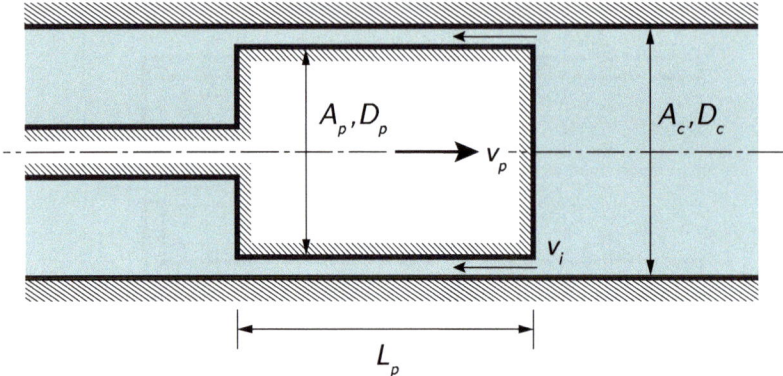

Fig. 10.16 Piston moving in a pipe with clearance

where k_a is the pressure loss coefficient due to the sudden contraction. This coefficient can be calculated as in the case of flow through an orifice, Eq. (10.9), which yields $k_a = 1.54$ and, in turn, $l_v = 0.1928$ m²/s².

From the Bernoulli equation, it follows that $p_1 - p_2 = p_1 - p_{atm} = 312$. Pa. Finally, the momentum conservation equation yields:

$$F = \rho A_1 v_1 (v_1 - v_2) + (p_1 - p_{atm}) A_1 = 0.192 \text{ N} \quad . \tag{10.95}$$

⑧ A cylindrical piston of cross-section A_p moves (with clearance) into a cylindrical pipe of cross section A_c, as schematically illustrated in Fig. 10.16. The clearance (gap) between the piston and the pipe is filled with oil. Let v_p be the velocity of the piston. Determine the force exerted on the piston as a function of v_p and of the geometry. Assume that the flow of oil is turbulent in the clearance between the piston and the pipe and that the oil velocity is $v_i \gg v_p$.

Solution The force exerted on the piston, which moves with uniform rectilinear motion at velocity v_p, is given by:

$$F = F_p + F_t = \Delta p \, A_p + \tau_w \, \pi D_p \, L_p \quad , \tag{10.96}$$

where F_p is the pressure force, F_t is the shear stress force, Δp is the pressure drop due to the frictional losses between two sections located immediately upstream and downstream of the piston, and τ_w is the shear stress at the wall of the piston. Considering that $v_i \gg v_p$, the pressure drop is equal to:

$$\Delta p = \rho l_v = 2\rho \frac{L_p}{D_i} v_i^2 f \quad , \tag{10.97}$$

where D_i is the hydraulic diameter of the gap between the cylinder and the piston. The hydraulic diameter can be calculated from the following relation (see Sect. 2.2.5):

$$D_i = \frac{4A_i}{P_i} \quad , \tag{10.98}$$

where $A_i = (A_c - A_p)$ is the cross-sectional area of the gap and P_i is the wetted perimeter associated to A_i. Denoting by D_p and D_c the diameters of the piston and of the cylinder, respectively, the wetted perimeter can be expressed as:

$$P_i = \pi(D_c + D_p) \quad . \tag{10.99}$$

Since:

$$A_i = \frac{\pi}{4}(D_c^2 - D_p^2) = \frac{\pi}{4}(D_c + D_p)(D_c - D_p) \quad , \tag{10.100}$$

it follows that $D_i = D_c - D_p$. From the mass conservation equation, the velocity of the oil in the gap, v_i, is given by:

$$v_i = v_p \frac{A_p}{A_i} = v_p \frac{A_p}{A_c - A_p} = v_p \frac{D_p^2}{D_c^2 - D_p^2} \quad . \tag{10.101}$$

Substituting in Eq. (10.97) gives:

$$\Delta p = \frac{1}{2}\rho \frac{L_p}{D_c - D_p} \left(v_p \frac{D_p^2}{D_c^2 - D_p^2} \right)^2 f \quad . \tag{10.102}$$

The friction factor can be expressed as $f = 0.079 Re^{-1/4}$ with $Re = \rho v_i (D_c - D_p)/\mu$. A force balance on the volume of fluid in the gap yields:

$$\Delta p A_i = P_i L_p \tau_w \quad \longrightarrow \quad \tau_w = \Delta p \frac{A_i}{P_i L_p} \quad . \tag{10.103}$$

Substituting in Eq. (10.96) gives:

$$F = F_p + F_t = \pi D_p \Delta p \left(\frac{D_p}{4} + \frac{D_i}{4} \right) \quad . \tag{10.104}$$

9 In the **T**-pipe shown in Fig. 10.17, water flows with velocity $v_1 = 6$ m/s and pressure $p_1 = 1.5 \cdot 10^5$ Pa upstream of the branch. Calculate the velocities v_2 and v_3 on the basis of proper assumptions. The main branch of the **T**-pipe has diameter $D = 100$ mm, the side branch has diameter $D = 50$ mm. Assume that the side branch 3 discharges the flow directly at atmospheric pressure.

Solution The system is equivalent to a distribution manifold. It is assumed that the frictional losses at the wall are negligible. The unknowns are v_2, v_3 and p_2. Therefore, three equations are necessary:

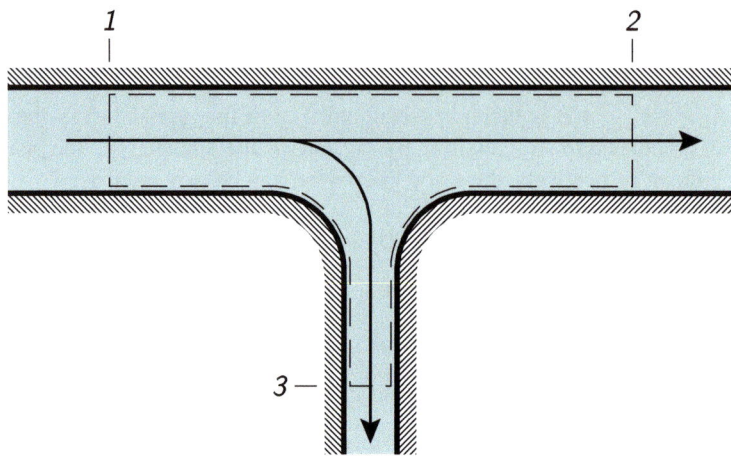

Fig. 10.17 Flow in a plane T-pipe

1. Conservation of mass:

$$A_1 v_1 = A_2 v_2 + A_3 v_3 \quad , \tag{10.105}$$

from which it follows that $v_1 = v_2 + v_3(A_3/A_1)$, being $A_1 = A_2$.

2. Bernoulli equation, written with respect to the horizontal line in Fig. 10.17 (experiments show that the Bernoulli equation and not the momentum conservation equation leads to correct results):

$$p_2 - p_1 = \frac{\rho}{2}(v_1^2 - v_2^2) \quad . \tag{10.106}$$

3. Assuming that the side branch behaves as an orifice:

$$v_3 = 0.62 \sqrt{\frac{2}{\rho}\left(\frac{p_1 + p_2}{2} - p_3\right)} \quad , \tag{10.107}$$

where $p_3 = p_{atm}$. It is assumed that the pressure in the portion of the pipe where branching occurs is equal to the average between pressure p_1 upstream of the branching and pressure p_2 downstream of the branching, as shown in Fig. 10.17.

To solve this system of equations, it is convenient to calculate first p_2 from Eq. (10.106) and then v_3 from Eq. (10.105). Substitution of the values so obtained into Eq. (10.107) yields:

$$(v_1 - v_2)\frac{A_1}{A_3} = 0.62 \sqrt{\frac{2}{\rho}\left[\frac{2p_1 + \rho(v_1^2 - v_2^2)/2}{2} - p_3\right]} \quad . \tag{10.108}$$

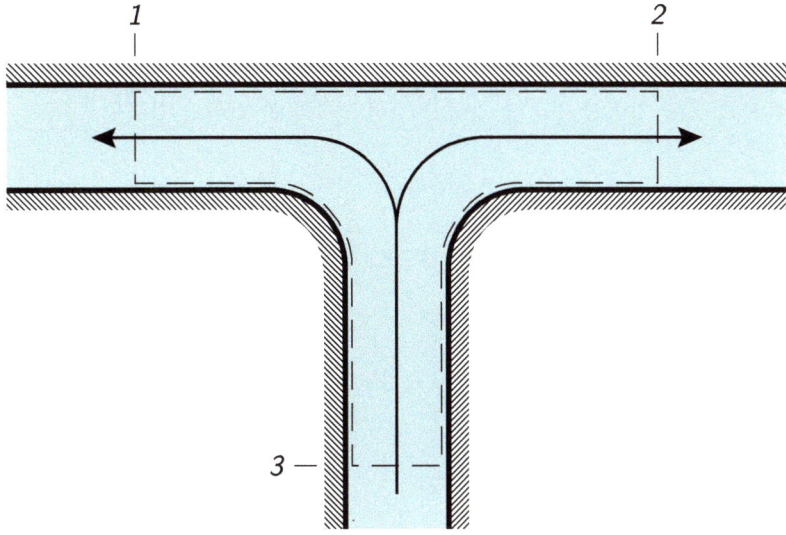

Fig. 10.18 Flow splitting in a **T**-pipe connection

In this equation, the only unknown is v_2. By taking the square of Eq. (10.108) and solving the resulting quadratic equation in v_2, the value $v_2 = 4.3$ m/s is obtained. It follows that $v_3 = 6.47$ m/s and $p_2 = 1.59 \cdot 10^5$ Pa.

$\boxed{10}$ Consider the flow of a fluid of density $\rho = 10^3$ kg/m^3 in a **T**-pipe connection, as shown in Fig. 10.18. The **T**-pipe connection has a diameter of 200 mm and splits the flowrate equally into two streams. For an inlet fluid velocity of 6 m/s and an absolute pressure at position 3 equal to $p_3 = 5 \cdot 10^5$ Pa, determine the pressure at the exit of the **T**-pipe connection and the force exerted on it.

Solution The pressure at the exit of the **T**-pipe connection can be determined by assuming that the flowrate entering the **T**-pipe connection is divided exactly in half and that the resulting streams are each confined in a separate streamtube, so that the Bernoulli equation can be applied. Application of the Bernoulli equation to the streamtube that has inlet cross section 3 and outlet cross section 2 yields:

$$\frac{1}{2}v_2^2 + \frac{p_2}{\rho} = \frac{1}{2}v_3^2 + \frac{p_3}{\rho} - l_v = \frac{1}{2}v_3^2 + \frac{p_3}{\rho} - 0.75\frac{1}{2}v_2^2 \ , \qquad (10.109)$$

where the concentrated losses can be calculated from Table 11.1 as losses in a 90° bend, whereas the distributed losses are neglected given the small distance between sections 2 and 3.

Taking into account that mass conservation imposes $v_3 A_3 = 2v_2 A_2$:

$$p_2 = p_3 + \frac{1}{2}\rho v_3^2 - \frac{1}{2}\rho\left(v_2^2 + 0.75v_2^2\right) = p_3 + \frac{1}{2}\rho\left(v_3^2 - 1.75v_2^2\right) + $$

$$= p_3 + \frac{1}{2}\rho\, 2.25v_2^2 = 5.045 \cdot 10^5 \text{ Pa} \quad . \tag{10.110}$$

The force exerted on the **T** is calculated by applying the momentum balance equation:

$$w(\mathbf{v}_3 - \mathbf{v}_2) + p_3\mathbf{A}_3 - p_2\mathbf{A}_2 - \mathbf{F} = 0 \quad . \tag{10.111}$$

The force in the x direction (direction of the incoming stream) exerted by the fluid on the inner wall of the pipe is:

$$F_x = \rho v_3 A_3 v_3 + p_3 A_3 = 16210.7 \text{ N} \quad . \tag{10.112}$$

The net force acting on the **T**-pipe is:

$$F_{x,net} = F_x - p_{atm} A_3 = 13069 \text{ N} \quad . \tag{10.113}$$

being $p_{atm} A_3$ the force produced by the atmospheric pressure on the portion of area A_3 of the outer pipe wall. Such force acts in the direction opposite to F_x and is the only unbalanced force between the inside and the outside of the pipe in the x direction.

11 Water ($\mu = 1 \cdot 10^{-3}$ Pa s, $\rho = 10^3$ kg/m^3) flows in a U-shaped bend with velocity equal to 10 m/s, as shown in Fig. 10.19. The pipe diameter is 150 mm, the inlet pressure is $2.5 \cdot 10^5$ Pa and the distributed losses are negligible. Calculate the force that the fluid exerts on the walls of the bend.

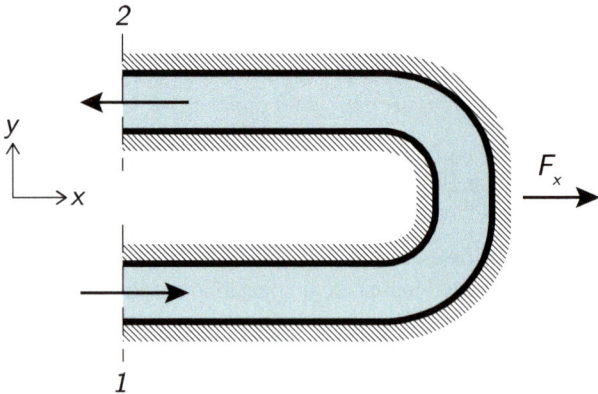

Fig. 10.19 Fluid flowing in a U-shaped bend

Solution The momentum conservation for the present problem reads as:

$$w(\mathbf{v}_1 - \mathbf{v}_2) + p_1\mathbf{A}_1 - p_2\mathbf{A}_2 - \mathbf{F} = 0 \quad . \tag{10.114}$$

Considering that $\mathbf{v}_2 = -\mathbf{v}_1$ and $\mathbf{A}_2 = -\mathbf{A}_1$, this equation gives the following expression for the force F_x:

$$F_x = w(v_1 + v_2) + A(p_1 + p_2) \quad , \tag{10.115}$$

while, in the y direction (normal to x and tangential to the curve): $F_y = 0$. Mass conservation yields $v_1 = v_2 = v$ and the concentrated losses are equal to $\Delta p = 0.75\rho v^2$, according to Table 11.1. The force exerted by the fluid on the walls of the bend is directed along the symmetry axis (x) and is equal to:

$$F_x = \rho\pi\frac{D^2}{4}v(v+v) + \pi\frac{D^2}{4}(2p_1 - \Delta p) +$$

$$= 10^3\pi\frac{D^2}{4}\left[200 + 2\frac{250 \cdot 10^3}{10^3} - 0.75 \cdot 10^2\right] = 11045\,\text{N} \quad . \tag{10.116}$$

⎡12⎤ A containment wall must be designed to collect a flammable liquid ($\rho = 800\,\text{kg/m}^3$) that can spill out in case of leakage from a cylindrical tank. Given that the wall must be far enough from the tank to avoid overflows and high enough to contain all the liquid contained in the tank, determine the minimum radial distance at which the wall should be placed and the minimum height that it must have with reference to the dimensions given in Fig. 10.20.

Solution The distance at which the wall should be built can be determined by the product of the jet exit velocity, v_e, and the flight time, which is the time taken by the jet to reach the ground as a result of a leakage occurred at height z. The exit velocity (in

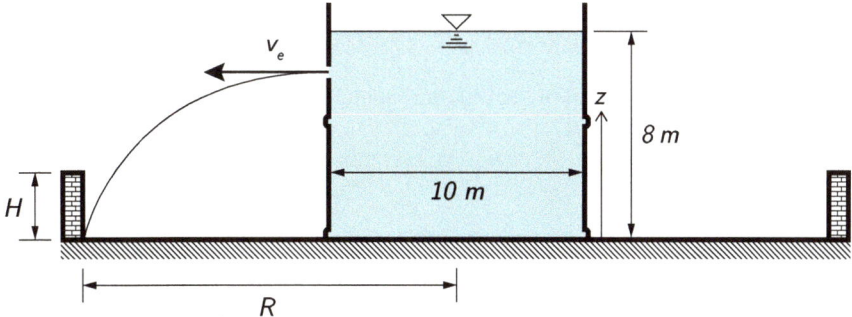

Fig. 10.20 Design of a containment wall for a tank filled with flammable liquid

the horizontal direction) is determined using the Bernoulli equation. This equation, for a hole at height h in the tank, gives for v_e:

$$\frac{1}{2}v_e^2 = g(h - z) \rightarrow v_e = \sqrt{2g(h - z)} \quad . \tag{10.117}$$

The flight time, t_c, is derived from Galileo's law of free-fall:

$$z = g\frac{t_c^2}{2} \rightarrow t_c = \sqrt{\frac{2z}{g}} \quad . \tag{10.118}$$

It follows that the horizontal distance s traveled by the jet is a function of the liquid height in the tank and the height z at which the leakage occurs:

$$s = v_e \, t_c = \sqrt{\frac{2z}{g}2g(h - z)} = 2\sqrt{hz - z^2} \quad . \tag{10.119}$$

The maximum value for this distance is obtained by setting the derivative with respect to z equal to zero:

$$\frac{ds}{dz} = \frac{1}{\sqrt{hz - z^2}}(h - 2z) = 0 \rightarrow z_m = \frac{h}{2} \quad . \tag{10.120}$$

Substitution into Eq. (10.119) yields $s_m = 8$ m. Therefore, the wall should be built at a distance of 8 m from the tank, namely at a radial distance $R = 13$ m.

In case of leakage, the wall must be able to contain the entire volume of liquid in the tank, which is equal to $V_s = 628.3$ m^3. Therefore, the minimum height H must be:

$$V_s = \pi R^2 H \rightarrow H = \frac{V_s}{\pi R^2} = 1.18 \text{ m} \quad . \tag{10.121}$$

13 A fire hose like the one shown in Fig. 10.21 discharges a flowrate of water equal to 10 kg/s. The flowrate is supplied by pipe of diameter equal to 50 mm while the outlet diameter of the fire hose is 25 mm. Calculate:

1. the horizontal component of the force F_e that must be exerted to hold the fire hose in place.
2. The force F_i exerted in the coupling between the hose and the lance (assume inviscid flow).

Solution To calculate the force required to hold the fire hose in place, it is convenient to apply the momentum conservation equation between section 1 (before the elbow) and section 3 (immediately after the jet outlet) in the x direction, which is parallel to the jet:

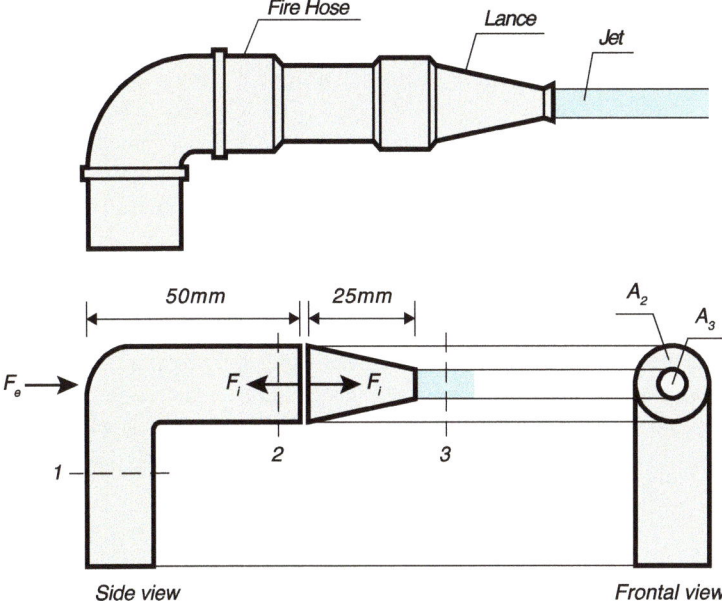

Fig. 10.21 Schematic diagram of a fire hose

$$w(v_{1,x} - v_{3,x}) + p_1 A_{1,x} - p_3 A_{3,x} + F_x = 0 \quad, \tag{10.122}$$

where F_x is the force exerted by the fluid on the inner wall of the fire hose and $w = 10$ kg/s is the mass flowrate. In the above equation, $A_{1,x} = 0$, $A_{3,x} = A_3$, $p_3 = p_{atm}$ and:

$$v_{1,x} = 0 \quad, \quad v_{3,x} = v_3 = \frac{4 w}{\rho \pi D_3^2} = 20.4 \text{ m/s} \quad. \tag{10.123}$$

Substitution of the values in Eq. (10.122) yields:

$$F_x = -w v_3 - p_{atm} A_3 = -253.6 \text{ N} \quad. \tag{10.124}$$

This force acts in the direction opposite to x and tends to push the hose backwards. This force is only partially balanced by the force $F_{atm,ext}$ produced in the x direction by the atmospheric pressure on the outer wall of the hose. Therefore, it is necessary to exert an additional external force F_e to balance F_x. The resulting force balance is:

$$F_x + F_e + F_{atm,ext} = 0 \quad, \tag{10.125}$$

with $F_{atm,ext} = p_{atm} A_3$: This is the only contribution due to the atmospheric pressure that is unbalanced between the inside and outside of the hose along the jet direction. The outlet section of the lance, of area A_3, is open and therefore the atmospheric pressure cannot generate any force, as it can do on the same area at the elbow of the hose. Substituting the values, the force F_e is obtained:

$$F_e = -F_x - p_{atm} A_3 = 203.7 \text{ N} \quad . \tag{10.126}$$

The velocity $v_3 = 20.4$ m/s has been already calculated. Similarly:

$$v_1 = v_2 = \frac{4\,w}{\rho \pi D_1^2} = 5.09 \text{ m/s} \quad . \tag{10.127}$$

Taking the atmospheric pressure as a reference and applying the Bernoulli equation between section 2 (immediately upstream of the lance) and section 3 (outlet section of the jet, immediately downstream of the lance), it is possible to write:

$$p_2 + \frac{1}{2}\rho v_2^2 = p_3 + \frac{1}{2}\rho v_3^2 \quad . \tag{10.128}$$

which gives $p_2 = 2.9 \cdot 10^5$ Pa. From the conservation of momentum in the x direction:

$$w(v_2 - v_3) + p_2 A_2 - p_3 A_3 - F_x^{2-3} = 0 \quad , \tag{10.129}$$

it is possible to derive:

$$F_x^{2-3} = w(v_2 - v_3) + p_2 A_2 - p_3 A_3 \quad , \tag{10.130}$$

with F_x^{2-3} force exerted by water on the inner wall at the coupling between the hose and the lance. As for the previous point, this force is only partially balanced by the external force, denoted as $F_{atm,ext}^i$, produced in the x direction by atmospheric pressure: The force F_i will therefore be given by the difference between F_x^{2-3} and $F_{atm,ext}^i$:

$$F_1 = F_x^{2-3} - F_{atm,ext}^i = F_x^{2-3} - p_{atm}(A_2 - A_3) = 220 \text{ N} \quad . \tag{10.131}$$

14 A paddle wheel of radius $R = 1$ m must be set into rotation at angular velocity $\Omega = 5$ s^{-1}. For this purpose, a pump draws water from a storage tank and feeds it to a pipe of length $L = 2$ m and diameter $d = 0.1$ m. The jet of water issuing from the pipe hits the paddles of the wheel. Knowing that the efficiency of the wheel, given by the ratio of the power available at the paddle of the wheel to the power supplied by the water jet, is equal to 0.5, calculate:

1. the velocity of the jet,
2. the discharge pressure required for the paddle wheel to rotate at angular velocity Ω.

Solution The velocity can be calculated directly for the definition of efficiency:

$$\eta = \frac{\text{Available power}}{\text{Supplied power}} = \frac{P_a}{P_s} \quad . \tag{10.132}$$

The available power can be calculated determining the momentum actually transferred from the jet to the paddles. Denoting as $v = \Omega R = 5$ m/s the velocity of the paddles and as u the velocity of the jet, the power is given by:

$$P_a = F \cdot v = \frac{d(mu)}{dt} \cdot v = \left(u \frac{dm}{dt} \right) \cdot v = (uw)v =$$

$$= u \cdot v \left[\rho(u - v) \frac{\pi d^2}{4} \right] = \rho\pi \frac{d^2}{4} u(u - v)v \quad . \tag{10.133}$$

In this case, it has been considered that the *actual* flowrate is reduced if each paddle moves with a constant velocity instead of being stationary. In such a case, the jet has to travel a longer distance, thus wasting a fraction of the flowrate (which does not contribute to the transfer of momentum to the paddles). The power supplied by the water jet, on the other hand, can be calculated by determining its kinetic energy:

$$P_s = \frac{d}{dt}\left(\frac{1}{2}mu^2 \right) = \frac{1}{2}u^2 \frac{dm}{dt} = \frac{1}{2}u^2 \left(\rho u \frac{\pi d^2}{4} \right) = \frac{1}{8}\rho\pi d^2 u^3 \quad . \tag{10.134}$$

Substituting these expression in the efficiency yields:

$$\eta = \frac{\rho(\pi d^2/4)[u(u - v)v]}{\rho\pi d^2 u^3/8} = \frac{v(u - v)}{u^2/2} = 0.5 \quad , \tag{10.135}$$

from which it follows that $u = 2v = 10$ m/s.

The discharge pressure provided by the pump must be high enough to overcome the pressure drop along the pipe and generate the desired jet velocity at its outlet. Applying the Bernoulli equation between the outlet section of the pump (denoted as section 1) and the outlet section of the jet (denoted as section 2), the following relation is obtained:

$$p_1 = p_2 + \rho l_v \quad , \tag{10.136}$$

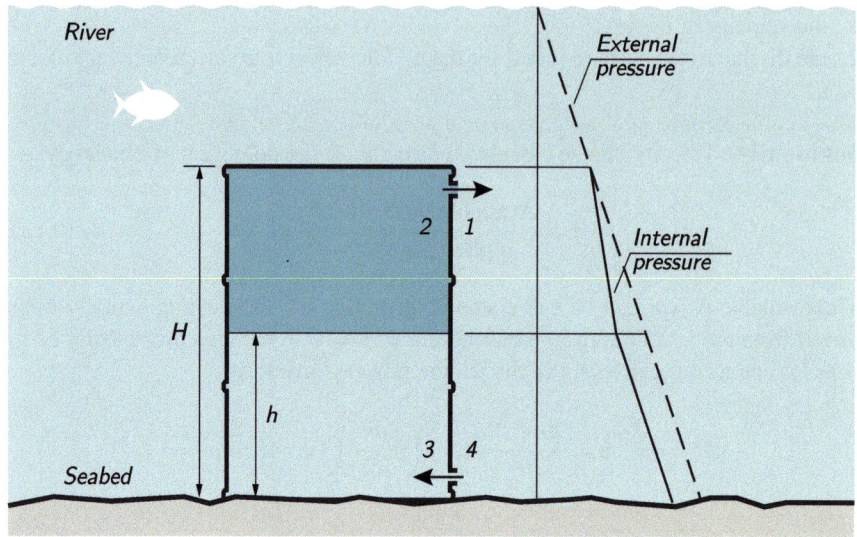

Fig. 10.22 Possible scenario for calculating the emptying time of a waste barrel filled with toxic liquid and fallen at the bottom of a river

where the loss term l_v can be obtained as:

$$l_v = 2f\frac{L}{d}u^2 \quad ,$$

(10.137)

with f given by the Blasius equation. Since the Reynolds number is $Re = ud/v = 10^6$, the friction factor is $f = 0.0316$, the loss term is $l_v = 126.4$ m^2/s^2 and, finally, the discharge pressure is $p_1 = 2.26 \cdot 10^5$ Pa, being $p_2 = p_{atm}$.

15 A toxic waste barrel filled with toxic liquid of density $\rho_t < \rho_w$ with ρ_w density of water, falls into a river (Fig. 10.22). Two holes of equal area located at the two ends of the barrel, at distance H, allow the liquid to flow out of the barrel and the water to flow into it. Knowing the dimensions of the barrel and the areas of the holes, determine the law with which the level h of water in the barrel changes over time (assume negligible viscous forces).

Solution As the toxic liquid has a lower density, it will occupy the upper part of the barrel. Therefore, the toxic liquid will outflow from the upper hole while water will flow into the barrel from the lower hole. Let us denote as 1 and 3 the sections where the toxic liquid outflows and the water enters the barrel and as 2 and 4 two points, one inside and one outside the barrel, respectively placed at the same height as 1 and 3, as shown in Fig. 10.22. Also shown in the figure is the increase of hydrostatic pressure inside (solid line) and outside the barrel. It can be seen that the pressure on the outside is always higher than the pressure on the inside, and also that the pressure difference is larger at the bottom hole. This favors the entry of water into the lower part of the barrel and the outflow of toxic liquid from the upper hole.

Let V_{in} be the volume of water that enters into the barrel in a given time interval Δt and V_{out} the volume of toxic liquid that exits from the barrel in the same time interval. It must be $V_{in} = V_{out}$ and the volumetric flowrates $Q_{in} = V_{in}/\Delta t$ and $Q_{out} = V_{out}/\Delta t$ must be equal. Expressing these flowrates as a function of the water inlet velocity, $v_{in} \equiv v_1$, and of the toxic liquid outflow velocity, $v_{out} \equiv v_3$, it is easy to find that, independent of the interval Δt considered:

$$Q_{out} = A v_1 = A v_3 = Q_{in} \quad \rightarrow \quad v_1 = v_3 = v \ , \tag{10.138}$$

being the areas of the two holes equal, $A_{in} = A_{out} = A$. This result implies that the total mass of water and liquid instantaneously contained in the barrel does not remain constant during the emptying, since the mass flowrate of water that enters, $m_{in} = \rho_w v_{in} A_{in}$, is certainly larger than the mass flowrate of toxic liquid that exits, $m_{out} = \rho_t v_{out} A_{out}$, being $\rho_w > \rho_t$.

Assuming that the velocity is everywhere zero inside the barrel, the Bernoulli equation between points 1 and 2 (outflow of toxic liquid from the barrel) gives:

$$p_2 = p_1 + \rho_t \frac{v_1^2}{2} = p_1 + \rho_t \frac{v^2}{2} \ . \tag{10.139}$$

Assuming also that the velocity of water in the river is zero, the Bernoulli equation between points 4 and 3 (inflow of water from the river into the barrel) gives:

$$p_4 = p_3 + \rho_w \frac{v_3^2}{2} = p_3 + \rho_w \frac{v^2}{2} \ . \tag{10.140}$$

The pressure difference between points 1 and 4, outside the barrel, can be expressed as:

$$p_4 = p_1 + \rho g H \ . \tag{10.141}$$

A similar expression applies to the pressure difference between points 2 and 3 inside the barrel. Denoting as h the level of water with respect to the entry point:

$$p_3 = p_2 + \rho_t g (H - h) + \rho g h \ . \tag{10.142}$$

By combining the expressions derived so far, it is possible to obtain an expression for the velocity. Adding Eqs. (10.139) and (10.140) and taking into account Eq. (10.138):

$$p_2 + p_4 = p_1 + p_3 + (\rho + \rho_t) \frac{v^2}{2} \ . \tag{10.143}$$

Substituting Eqs. (10.141) and (10.142), after some simplifications, the velocity is obtained as:

$$v = \sqrt{2g\frac{\rho - \rho_t}{\rho + \rho_t}(H - h)} \ .$$
(10.144)

This latter expression allows to determine the change in water level h inside the barrel. It is sufficient to use it in combination with the conservation of mass for the inflow of water, which reads as:

$$\frac{dV(t)}{dt} = Q_{in}(t) - \underbrace{Q_{out}(t)}_{=0} \ ,$$
(10.145)

where $V(t) = S\,h(t)$ is the volume of water in the barrel at a given time t, with S the base area of the barrel, and $Q_{in}(t)$ is the volumetric flowrate of water that enters the barrel, already expressed as $Q_{in} = Av(t)$. Equation (10.145) can then be written in the following form:

$$\frac{dh}{dt} = \frac{A}{S}v = \frac{A}{S}\sqrt{2g\frac{\rho - \rho_t}{\rho + \rho_t}(H - h)} = K\sqrt{H - h} \ .$$
(10.146)

By integrating the equation with the boundary condition $h(t = 0) = 0$, the law with which the level of water in the barrel changes with time can be obtained as:

$$h(t) = K\sqrt{H t} - \frac{K^2}{4}t^2 \ .$$
(10.147)

This equation is applicable as long as $h(t) < H$, that is for $t \le 2\sqrt{H}/K$, since there is no outflow from the barrel for $t > 2\sqrt{H}/K$.

16   An underwater oil well (A) produces light oil ($\rho = 750$ kg/m^3, $\mu = 3 \cdot 10^{-3}$ kg/ms) at pressure $p_A = 4.5 \cdot 10^5$ Pa. A jet pump fed with this oil is used to increase the oil pressure produced from a second well (B), available at pressure $p_B = 3 \cdot 10^5$ Pa, and to carry the oil from the two wells to a platform that is located at a distance of 2 km and at an height of 32 m above the two wells. To transport the oil, a pipeline with a diameter of 0.4 m and negligible roughness is used. With reference to the geometry shown in Fig. 10.23, determine:

1. the pressure at the inlet of the transport pipeline (position 3) for $w_A = 80$ kg/s and the flowrate from the well B if the losses in the supply lines of the pump are neglected,
2. the pressure losses in the transport pipeline downstream of the pump for $w_A = w_B = 80$ kg/s.

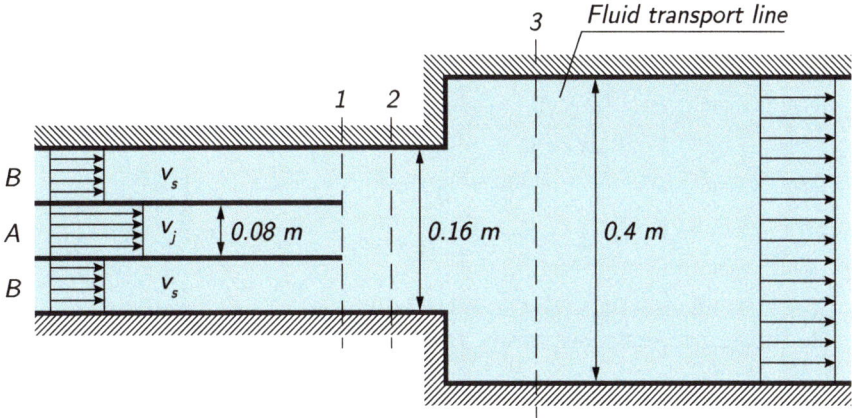

Fig. 10.23 Jet pump used to transport oil from an underwater well

Solution

1. For the given geometry and flowrate w_A, it is easy to derive the ratio λ between the diameter of the inner pipe that generates the high-speed fluid jet and the diameter of the outer suction pipe inside the pump:

$$\lambda = \frac{D_A}{D_B} = 0.5 \quad , \tag{10.148}$$

and the velocity of the high-speed fluid:

$$v_A = v_j = 4 \frac{w_A}{\rho \pi D_A^2} = 21.2 \text{ m/s} \quad . \tag{10.149}$$

By writing the Bernoulli equation for the high-speed fluid coming from the well at pressure $p_A = 4.5 \cdot 10^5$ Pa, and neglecting losses, the pressure at point 1 can be obtained:

$$p_1 = p_A - \frac{1}{2}\rho v_j^2 = 2.815 \cdot 10^5 \text{ Pa} \quad . \tag{10.150}$$

The fluid from well B is fed into the pump with inlet velocity v_s. Neglecting losses, this velocity can be derived from the Bernoulli equation:

$$\frac{1}{2}\rho v_s^2 = p_B - p_1 \quad \longrightarrow \quad v_s = \sqrt{\frac{2}{\rho}(p_B - p_1)} = 7.02 \text{ m/s} \quad . \tag{10.151}$$

Once v_j and v_s are known, it is easy to calculate the velocity in sections 2 and 3:

$$v_2 = \lambda^2 v_j + (1 - \lambda^2)v_s = 10.6 \text{ m/s} \quad , \tag{10.152}$$

$$v_3 = \frac{A_2}{A_3}v_2 = 1.7 \text{ m/s} \quad , \tag{10.153}$$

as well as the pressure p_2 at the pump outlet from Eq. (10.51):

$$p_2 = p_1 + \rho\lambda^2(1 - \lambda^2)(v_j - v_s)^2 = 3.1 \cdot 10^5 \text{ Pa} \quad . \tag{10.154}$$

The pressure in section 3 is obtained by taking into account the pressure recovery due to the pipe section expansion. Using Eq. (10.38):

$$\frac{p_3 - p_2}{\rho} = v_3^2\left(\frac{A_3}{A_2} - 1\right) \quad \longrightarrow \quad p_3 = 3.25 \cdot 10^5 \text{ Pa} \quad . \tag{10.155}$$

Finally, the flowrate of fluid B that can be transported using fluid A is:

$$w_B = \frac{\pi}{4}\rho(D_B^2 - D_A^2)v_s = 80 \text{ kg/s} \quad . \tag{10.156}$$

2. The Bernoulli equation, written between section 3 and the outlet section, allows to obtain the pressure difference Δp between the two sections as:

$$\Delta p = \rho gh + 2f\rho\frac{L}{D}v_3^2 \quad , \tag{10.157}$$

where the second term on the right-hand side represents the pressure loss. This term can be calculated from the available data ($L = 2000$ m, $D = 0.4$ m) taking into account that the friction factor f must be calculated using the most appropriate relation for the flow regime that is established in the pipe. In the present case, since:

$$Re = \frac{\rho v D}{\mu} = 1.7 \cdot 10^5 \quad , \tag{10.158}$$

the flow is turbulent and f is given by:

$$f = 0.079 Re^{-0.25} = 3.9 \cdot 10^{-3} \quad , \tag{10.159}$$

since roughness is negligible. The pressure losses are:

$$2f\rho\frac{L}{D}v_3^2 = 0.84 \cdot 10^5 \text{ Pa} \quad , \tag{10.160}$$

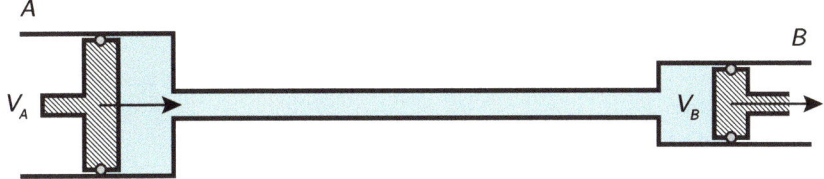

Fig. 10.24 Flow between two reservoirs connected by a pipe

and the pressure difference is:

$$\Delta p = 2.35 \cdot 10^5 + 0.84 \cdot 10^5 = 3.19 \cdot 10^5 \text{ Pa} \quad . \tag{10.161}$$

The pressure available at section 3 ($3.25 \cdot 10^5$ Pa) is therefore sufficient to transport the oil.

$\boxed{17}$ As shown in Fig. 10.24, the two cylindrical reservoirs A and B, having diameter $D_A = 200$ mm and $D_B = 100$ mm respectively, are filled with a fluid of density $\rho = 800 \text{ kg/m}^3$ and viscosity $\mu = 1.2 \cdot 10^{-1}$ Pa s. The two reservoirs are connected by a pipe of length $L = 1$ m and inner diameter $d_T = 5$ mm.

1. Determine the velocity of piston B and the flowrate in the connecting pipe if the velocity of piston A is 20 mm/min.
2. Neglecting the motion of the fluid inside the reservoirs as well as inlet and outlet effects in the connecting pipe, calculate the power to be applied to piston A.

Solution

1. To determine the velocity of piston B, it suffices to apply the continuity equation between sections A and B of the reservoirs:

$$v_A A_A = v_B A_B \quad . \tag{10.162}$$

The velocity in A ($v_A = 20$ mm/min = $3.33 \cdot 10^{-4}$ m/s) and the cross-sectional areas are given. Therefore:

$$v_B = v_A \frac{A_A}{A_B} = v_A \frac{D_A^2}{D_B^2} = 1.33 \cdot 10^{-3} \text{ m/s} \quad . \tag{10.163}$$

To determine the flowrate in the connecting pipe, it is again possible to use the continuity equation, written between section A and a generic section T in the connecting pipe:

$$Q_T = v_T A_T = v_A A_A = v_A \frac{\pi D_A^2}{4} = 1.05 \cdot 10^{-5} \text{ m}^3/\text{s} \quad , \tag{10.164}$$

where the volumetric flowrate Q_T corresponds to a mass flowrate:

$$w_T = \rho Q_T = 8.4 \cdot 10^{-3} \text{ kg/s} \quad . \tag{10.165}$$

2. The velocity of the fluid in the connecting pipe is:

$$v_T = \frac{Q_T}{A_T} = 0.53 \text{ m/s} \quad . \tag{10.166}$$

The pressure drop in the connecting pipe depends on this velocity. Such pressure drop can be expressed using the Hagen-Poiseuille equation (the flow in the pipe is laminar since $Re_T = \rho v_T d_T / \mu = 17.8$) and neglecting inlet and outlet effects:

$$\Delta p = 8 \mu L \frac{v_T}{R^2} = 0.814 \cdot 10^5 \text{ Pa} \quad . \tag{10.167}$$

The power to be applied to piston A to realize the transport of fluid is:

$$P = F_A v_A \quad , \tag{10.168}$$

where F_A is the force that must be applied to the piston to overcome the pressure drop in the connecting pipe:

$$F_A = A_A \Delta p \quad . \tag{10.169}$$

From Eqs. (10.168) and (10.169), it follows that:

$$P = 0.85 \text{ W} \quad . \tag{10.170}$$

---

18   The damper shown in Fig. 10.25 consists of a cylinder with a diameter of 200 mm and six channels of diameter 5 mm that connect the chambers A and B. The damper is filled with oil of viscosity 10^{-2} Pa s and density 800 kg/m³.

1. Determine the pressure losses of the oil in the channels for a constant piston velocity equal to 10^{-2} m/s. Neglect inlet and outlet effects and assume fully-developed flow.
2. Determine the force F required to move the piston, neglecting the friction at the cylinder wall.

Solution The Bernoulli equation, applied between two sections at the left and at the right of the piston, allows to derive the pressure difference between chambers A and B:

$$\Delta p = 2 \rho v_c^2 \frac{L}{d} f \quad , \tag{10.171}$$

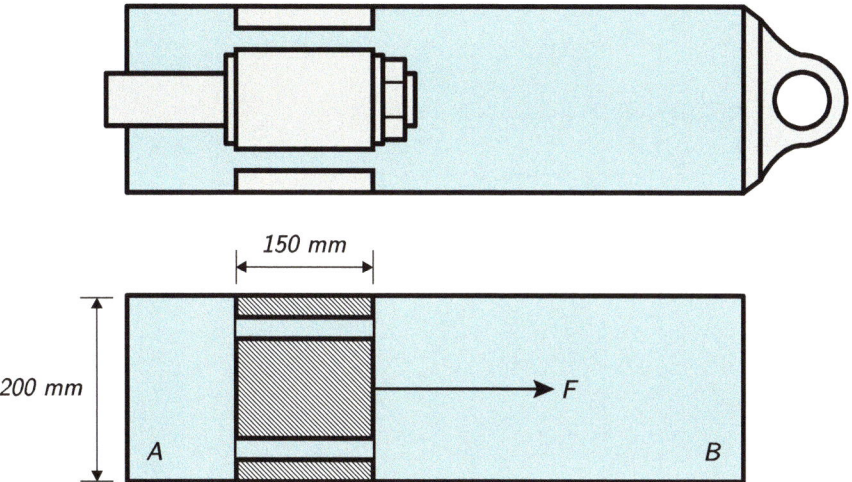

Fig. 10.25 Schematic diagram of a damper

where v_c is the velocity of the fluid in the six channels. The force required to move the piston, F_P, is equal to:

$$F_P = \Delta p A_C \quad, \tag{10.172}$$

with A_C cross section of the cylinder. For a displacement Δx of the piston, the volume of fluid that must flow through the connecting channels is equal to $A_C \Delta x$. If the piston moves with a constant velocity, v_P, then it follows that $\Delta x = v_P \Delta t$, and the flowrate through the channels is given by:

$$w_c = v_P \rho A_C = 0.25 \text{ kg/s} \quad. \tag{10.173}$$

The fluid velocity in the channels is:

$$v_c = \frac{w_c}{6 A_c \rho} = 2.66 \text{ m/s} \quad. \tag{10.174}$$

The Reynolds number in the channels is $Re = \rho v_c d / \mu = 1066$. The flow is laminar and the friction factor can be calculated as $f = 16 Re^{-1} = 1.5 \cdot 10^{-2}$. The pressure drop in the channels is therefore:

$$\Delta p_c = 2 v_c^2 \frac{L}{d} f \rho = 5.12 \cdot 10^3 \text{ Pa} \quad, \tag{10.175}$$

and the force required to move the piston is given by:

$$F_P = \Delta p_c A_C = 1.61 \cdot 10^2 \text{ N} \quad. \tag{10.176}$$

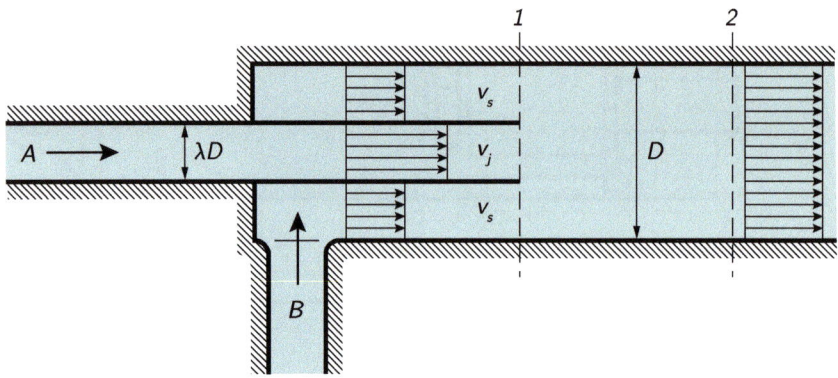

Fig. 10.26 Jet pump

Note that, in the calculation of this force, the contribution due to the shear stress exerted by the fluid flowing in the channels does not appear explicitly. In fact, this contribution is included in the term $\Delta p_c A_C$, since the pressure difference Δp_c is applied to the entire section of the piston.

19 A jet pump is fed with liquid A ($\rho = 10^3$ kg/m^3, $\mu = 10^{-3}$ Pa s), available at pressure $5 \cdot 10^5$ Pa, and is used to pump a liquid B with the same physical properties, available at atmospheric pressure ($p = 1 \cdot 10^5$ Pa), as shown in Fig. 10.26.

Calculate the total liquid flowrate that the pump can lift at an elevation of 12 m neglecting the losses of mechanical energy outside of the pump. Consider a feed flowrate of liquid A equal to 10 kg/s and a ratio of the diameters, D_A and D_B, of the pump equal to 0.5.

Solution Neglecting the losses outside of the pump, it is possible to determine the velocity in the suction section (which is the maximum theoretical velocity that can be achieved in the pump) using the Bernoulli equation:

$$\frac{1}{2}\rho v_j^2 = p_A - p_1 \longrightarrow v_j = \sqrt{2\frac{p_A - p_1}{\rho}} =$$

$$= \sqrt{2\frac{5 \cdot 10^5 - 1 \cdot 10^5}{1000}} = 28.3 \text{ m/s} \quad , \tag{10.177}$$

where it is assumed that the pressure in section 1 is very close to the atmospheric pressure:

$$(p_A - p_1) \simeq (p_A - p_B) \quad \longrightarrow \quad p_1 \simeq p_B \quad . \tag{10.178}$$

For the given flowrate of fluid A, the diameter D_A is:

$$D_A = \sqrt{\frac{4}{\pi} \frac{w_A}{\rho v_j}} = 0.021 \text{ m} \quad . \tag{10.179}$$

For the given value of the diameter ratio λ, it follows that:

$$D_B = \frac{D_A}{\lambda} = 0.042 \text{ m} \quad . \tag{10.180}$$

Assuming again that $p_1 \simeq p_{atm}$ and denoting by p_1, p_2 and p_3 the pressure values in the pump throat, in section 2 and at the target elevation of 12 m, the following relations can be written (neglecting pressure losses):

$$p_2 = p_3 + \rho g h \;, \qquad p_1 = p_3 = p_{atm} \quad . \tag{10.181}$$

Equation (10.51) becomes:

$$p_2 - p_1 = \lambda^2(1 - \lambda^2)\rho(v_j - v_s)^2 \quad \longrightarrow \quad \rho g h = \Lambda\rho(v_j - v_s)^2 \;, \tag{10.182}$$

with $\Lambda = 0.5^2(1 - 0.5^2) = 0.1875$. From Eq. (10.182), the velocity v_s is:

$$v_s = v_j - \sqrt{\frac{gh}{\Lambda}} = 3.23 \text{ m/s} \quad . \tag{10.183}$$

It is now possible to verify the hypothesis $p_1 \simeq p_B$. Application of the Bernoulli equation between sections B and 1 yields (neglecting losses):

$$p_B = p_1 + \frac{1}{2}\rho v_s^2 \rightarrow p_B - p_1 = 5.2 \cdot 10^3 \text{ Pa} \quad . \tag{10.184}$$

This pressure difference is actually negligible compared to the pressure jump $p_A - p_1 = 4 \cdot 10^5$ Pa, which is almost 80 times larger. The total flowrate (of fluid A and fluid B) is:

$$w_{A+B} = \frac{\pi D^2}{4}\rho v_2 = \frac{\pi D^2}{4}\rho\left[\lambda^2 v_j + (1 - \lambda^2)v_s\right] = 13.4 \text{ kg/s} \quad . \tag{10.185}$$

20 A conical funnel, whose geometry is shown in Fig. 10.27, is initially filled with a fluid. At time $t = 0$, the fluid starts to flow out of the funnel through the opening located at height $z = z_2$, under the action of the gravity force. Determine the emptying time of the funnel assuming that the fluid has a low viscosity and that the losses of mechanical energy can be neglected.

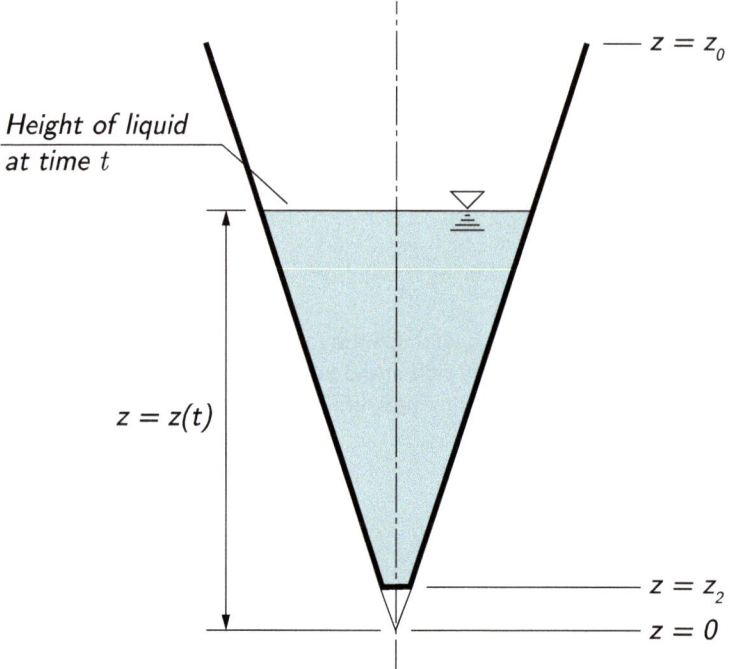

Fig. 10.27 Emptying of a funnel

Solution From the given geometric of the system, it follows that:

$$\frac{r_0}{z_0} = \frac{r}{z} = \frac{r_2}{z_2} \quad . \tag{10.186}$$

Therefore, the mass balance:

$$\rho \frac{dV}{dt} = -w \quad , \tag{10.187}$$

with:

$$V = \frac{1}{3}\pi(r^2 z - r_2^2 z_2) \quad , \tag{10.188}$$

and:

$$w = \rho \pi r_2^2 v_2 \quad , \tag{10.189}$$

can be expressed in the form:

$$r^2\frac{dz}{dt} = -3r_2^2 v_2 \quad .$$
(10.190)

Neglecting friction losses, the Bernoulli equation, applied between the free surface of the liquid in the funnel and the outlet section of the opening, becomes:

$$\frac{1}{2}\rho v^2 + p + \rho g z = \frac{1}{2}\rho v_2^2 + p_2 + \rho g z_2 \quad ,$$
(10.191)

where $p = p_2 = p_{atm}$. Furthermore, $v \simeq 0$ for $z \gg z_2$. Therefore:

$$v_2 = \sqrt{2g(z - z_2)} \simeq \sqrt{2gz} \quad .$$
(10.192)

Equations (10.190) and (10.192) yield:

$$r^2\frac{dz}{dt} = -3r_2^2\sqrt{2gz} \quad .$$
(10.193)

For Eq. (10.186), $r/r_2 = z/z_2$ and, substituting:

$$\frac{z^2}{z_2^2}\frac{dz}{dt} = -3\sqrt{2gz} \quad .$$
(10.194)

Integration of this equation yields the law with which the height of liquid in the funnel changes over time:

$$z_0^{5/2} - z^{5/2} = \frac{15}{2}z_2^2\sqrt{2g}\, t \quad .$$
(10.195)

Problems

[a] An open cylindrical tank of diameter $D = 30$ m, is filled with oil up to a height of $H = 5$ m. A pipe of diameter $d = 0.10$ m is used to transport the oil from the tank. The pipe is broken at a distance $L = 30$ m from the tank. Calculate the time it takes for the oil (viscosity $\mu = 0.01$Pa s and density $\rho = 800$ kg/m^3) to completely flow out of the tank.

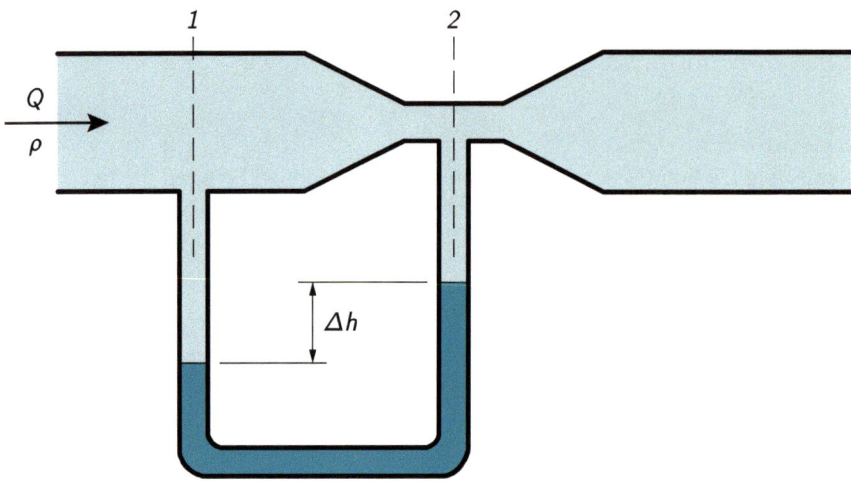

Fig. 10.28 Schematic diagram of a flowmeter based on the Venturi tube

b The flowmeter is a device used to measure the flowrate of a fluid in a pipe. Determine for the Venturimeter schematically shown in Fig. 10.28 the relation between the flowrate Q and Δh. Let ρ be the density of the fluid in the pipe and ρ_m the density of the fluid in the differential manometer. Neglect the viscous losses and assume known cross-sectional areas A_1 and A_2.

c The garden waterer shown in Fig. 10.29 discharges a flowrate of water equal to $2 \, \text{m}^3/\text{min}$. Water is discharged tangentially through nozzles of diameter $d = 40$ mm. Calculate the angular velocity if the rotating unit exerts a resistance torque $T = 53 \, \Omega^2$ Nm, with Ω the angular velocity in rad/s.

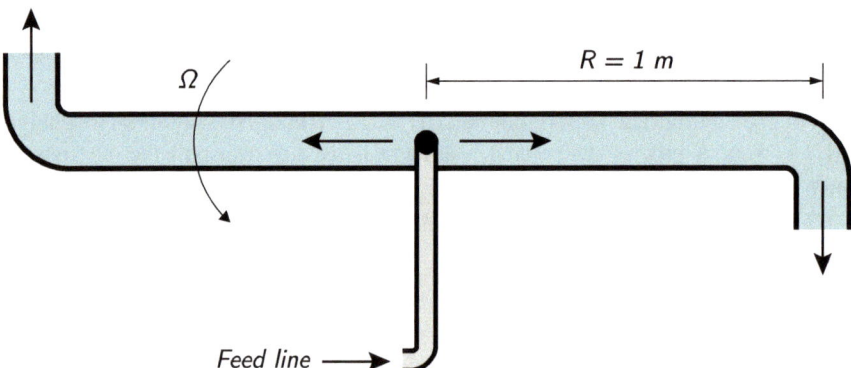

Fig. 10.29 Schematic illustration of the operating principle of a garden waterer

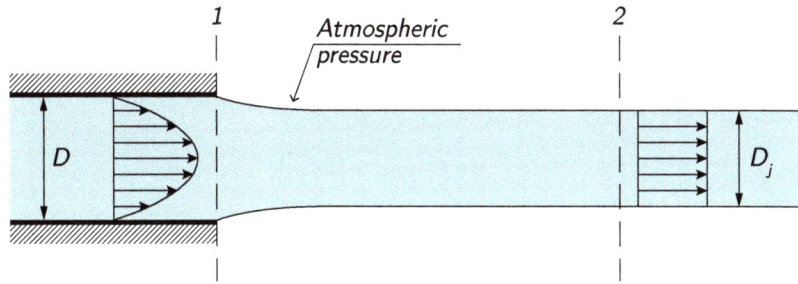

Fig. 10.30 Schematic diagram of a free jet. The control volume to be used for the application of the macroscopic balance equations is also shown

$\boxed{d}$ Consider the laminar flow of an incompressible Newtonian fluid in a pipe of diameter D. At the outlet pipe section, the fluid forms a horizontal free jet that is issued at atmospheric pressure, as shown in Fig. 10.30. After a short distance, the jet velocity redistributes and the velocity profile becomes uniform. Using a force balance on the jet, determine the diameter of the downstream part of the jet.

$\boxed{e}$ Evaluate the losses generated by a flow of gas inside a pipe that is characterized by a gradual increase in cross section, like the one illustrated in Fig. 10.31. Assume that the pressure change along the pipe is linear (note that the pressure difference is unknown). In addition, evaluate the losses associated to the gradual increase of cross section by assuming that, for each axial position along the pipe, the flow behaves as a turbulent flow in a pipe of the same diameter (calculate the losses for an infinitesimal portion of the pipe from the relations that apply to smooth pipes, and then integrate with respect to the length of the expanding portion of the pipe).

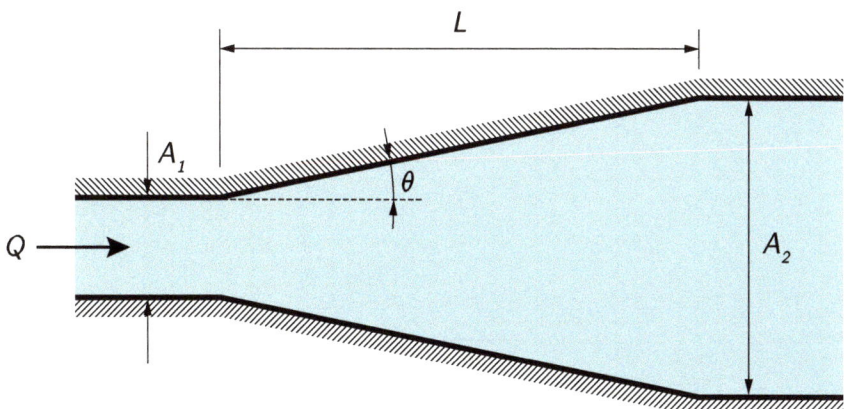

Fig. 10.31 Schematic representation of the gradual expansion of a gas

Chapter 11
Fluid Transport in Piping Systems

Many hydraulic systems consist of pipeline networks in which pipes are connected together to enable the transport of a fluid. Examples of such systems are the supply pipelines for the distribution of drinking water and the pipelines for the transport of oil and natural gas. The simplest piping systems are those consisting of pipelines in series or in parallel, which are also referred to as *hydraulic circuits* as they share several similarities with electrical circuits. Consider, for example, the relation between pressure drop Δp and flowrate Q in a pipe of diameter D and length L, given by Eq. (2.10). This relation can be recast as $\Delta p = \tilde{R} Q^2$ where $\tilde{R} = 32 \rho f / \pi^2 D^2$ depends on the friction factor f and provides a measure of the flow resistance. This relation is analogous to Ohm's law, which correlates the electric current through a conductor between two points with the voltage across the two points via the electric resistance of the conductor. The difference is that Ohm's law is linear (the current intensity doubles if the voltage doubles) whereas the relation $\Delta p = \tilde{R} Q^2$ is not always linear (the flowrate doubles if the pressure drop doubles only in the laminar flow regime).

In this chapter, the main criteria for the design and sizing of pipeline networks are presented. Typically, the main input parameters for the sizing are the flowrate of the fluid to be transported (in addition to the geometrical characteristics of the pipeline such as length of the pipes, bends and/or valves) and the main objective is to determine the pressure drop and the velocity of the fluid along the pipeline, based on economic design criteria. Large flowrates, that is high velocities, require high pressure drops ($\Delta p \propto Q^2$), which result in high consumption of electric power, typically associated with the use of hydraulic machines (pumps) to transport the fluid. It is therefore necessary to evaluate the optimal economic design conditions on a case-by-case basis. The complexity of such evaluation depends on the complex the network, for example in terms of the number of inter-connected pipes in branching pipelines, the number of pipe junction nodes (connecting two or more pipes), the number of flowrate inlet points and the number of flowrate delivery points.

In general, design and sizing are based on the use of the continuity equation, which can be applied to each node of the piping system by imposing that the sum of all the incoming flowrates and all the outgoing flowrates is zero, and the Bernoulli equation, which can be applied to each branch of the network. In the case of n nodes and m branches, a system of $n + m$ equations can be written, in which the unknowns are, typically, the pressures at the nodes and the velocities of the fluid in each branch (the velocities may vary if, for example, the length and/or the diameter of the pipe varies).

11.1   Distributed and Localised Losses

The total pressure drop of a pipeline network consists of two contributions. A first contribution is a distributed one, due to dissipation along the pipeline associated with the presence of a shear stress at the pipe wall. This contribution is precisely the one already analyzed in Chap. 2 by applying the Bernoulli equation between two sections of a horizontal pipe with a constant diameter into which an incompressible fluid is flowing. In this case, the Bernoulli equation yields:

$$\frac{p_1 - p_2}{\rho} = l_v \quad , \tag{11.1}$$

and the term l_v corresponds to the distributed head loss generated between the sections. If the pressure difference Δp is expressed as a function of the friction factor f via Eq. (2.8), the term l_v reads as:

$$l_v = 2v^2 f \frac{L}{D} \quad . \tag{11.2}$$

In addition to the distributed head loss, there may also be a localized contribution if the pipeline network is characterized by the presence of bends, fittings of various types, or changes in the pipe cross section, which all represent localised geometrical discontinuities along the pipeline. Localised losses can be expressed as:

$$l_v = \frac{1}{2}v^2 k_f \quad , \tag{11.3}$$

where the quantity $v^2/2$ is usually referred to as *kinetic head* and k_f is the number of kinetic heads lost. The values of k_f associated with the different type of geometrical discontinuities are given in Table 11.1. In summary, the total head losses in a pipeline network can be expressed as:

$$l_v = \sum_{i=1}^{m} \frac{2v_i^2 L_i f_i}{D_i} + \sum_{j=1}^{p} \frac{1}{2}v_j^2 k_{f,j} \quad , \tag{11.4}$$

Table 11.1 Localised head losses. The number of kinetic heads k_f is expressed in terms of the fluid velocity downstream of the geometrical discontinuity that causes the localised loss

Type of loss		Number of kinetic heads, k_f
45° Bend		0.3
90° Bend	Circular pipe section	$0.6 \div 0.8$
	Squared pipe section	1.2
Gate valve	Fully open	0.15
	Open at 75%	1
	Open at 50%	4
	Open at 25%	16
Section enlargement		$\left(\dfrac{A_2}{A_1} - 1\right)^2$
Section contraction		$\left(\dfrac{2}{m} - \dfrac{A_2}{A_1} - 1\right)^2$
	With m root of	$\left(\dfrac{m}{1.2}\right)^2 - \dfrac{1 - m(A_2/A_1)}{1 - (A_2/A_1)^2} = 0$

where the first summation refers to the m branches of the network and the second summation refers to the p localised losses of the network.

Therefore, the sizing requires to apply the following approximate formulation of the Bernoulli equation, expressed as in Eq. (9.18):

$$\frac{1}{2}\Delta v^2 + g\Delta h + \frac{\Delta p}{\rho} = w_s - \left(\sum_{i=1}^{m} \frac{2v_i^2 L_i f_i}{D_i} + \sum_{j=1}^{p} \frac{1}{2}v_j^2 k_{f,j}\right) \quad , \quad (11.5)$$

with Δv, Δh, and Δp differences in velocity, vertical height, and pressure between the initial and final sections of the pipeline network (typically corresponding to the flowrate inlet points and the flowrate delivery points, respectively). The term w_s, on the other hand, represents the net external work done per unit mass of fluid transported within the pipeline.

The following solved problems provide examples of the application of the equations presented above to complex piping systems and pipeline networks.

11.1.1 Examples

1 In Fig. 11.1, a pipeline network is shown. Calculate the pumping power required to convey 36 t/hr of fluid with density $\rho = 950 \, \text{kg/m}^3$ and viscosity $\mu = 9 \cdot 10^{-3} \text{Pa} \cdot \text{s}$. The pipes are made of steel, have diameter $D = 100$ mm and roughness $k = 5 \cdot 10^{-2}$mm.

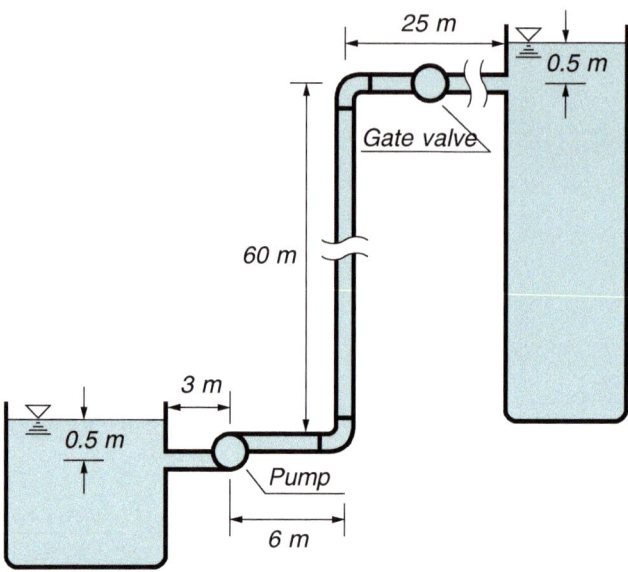

Fig. 11.1 Schematic for calculating the pumping power in a pipeline network

Solution Denoting by w_s the power that must be supplied by the pump and by w the flowrate, it follows that:

$$w_s = w(l_v + g\Delta h) \quad , \tag{11.6}$$

with $w = 36\,\mathrm{t/hr} = 10.\,\mathrm{kg/s}$ and $\Delta h = 60$ m. The head losses can be expressed as:

$$l_v = 2v^2 f \frac{L}{D} + \frac{1}{2}\sum_{i=1}^{n} k_i v_i^2 \quad , \tag{11.7}$$

where the summation refers the localised head losses. To calculate f, the Colebrook equation (2.23) must be used. For $Re = \rho v D/\mu = 1.41 \cdot 10^4$ and a first attempt value of f equal to $6 \cdot 10^{-3}$, the equation yields $f = 7.4 \cdot 10^{-3}$. The distributed head losses are therefore:

$$2v^2 f \frac{L}{D} = 27.6\,\mathrm{m^2/s^2} \quad . \tag{11.8}$$

The localised losses can be evaluated using Table 11.1:

1. Localised losses at the inlet: These losses can be computed by considering that, at the pipe inlet: $A_2/A_1 = 0 \rightarrow m = 1.2 \rightarrow k_1 = 0.44 \rightarrow \Delta p = 0.44\,\mathrm{m^2/s^2}$.
2. Localised losses in the bends: $k_2 = 0.75 \rightarrow \Delta p = 0.75\,\mathrm{m^2/s^2}$.
3. Localised losses in the valve gate: $k_3 = 0.15 \rightarrow \Delta p = 0.15\,\mathrm{m^2/s^2}$.

Fig. 11.2 Emptying of a demijohn

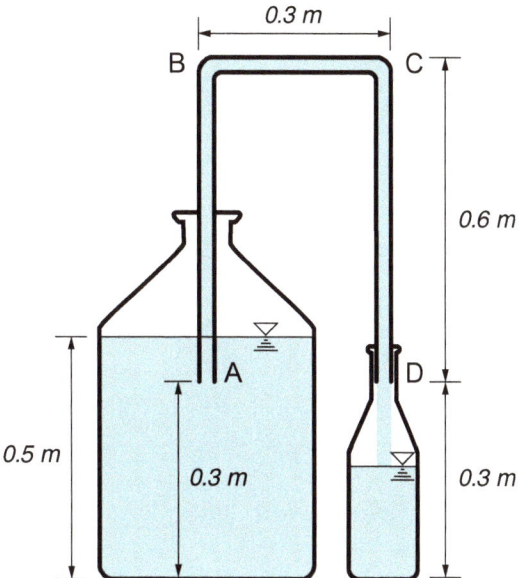

4. Localised losses at the outlet: These losses can be computed by considering that, at the pipe outlet: $A_1/A_2 = 0 \rightarrow k_4 = 1 \rightarrow \Delta p = 0.5 \text{ m}^2/\text{s}^2$.

Summing up all these losses gives $l_v = 29.4 \text{ m}^2/\text{s}^2$. Since the gravitational term is $g\Delta h = 588.6 \text{ m}^2/\text{s}^2$, it follows that the required pumping power is $w_s = 6.2 \text{ kW}$. Note that, in this example, the distributed head losses along the pipe are negligible with respect to the gravitational term.

2 The piping system shown in Fig. 11.2 is used to empty a demijohn. The total length of the pipe is 1.5 m and the diameter is 10 mm. The fluid (solution of ethyl alcohol in water) has density $\rho = 950 \text{ kg/m}^3$ and viscosity $\mu = 10^{-3}\text{Pa} \cdot \text{s}$. Determine:

1. the time required to fill the demijohn (having volume of 1 lt) assuming that the level of fluid in the demijohn does not change,
2. the minimum value of pressure in the pipe.

Solution

1. Taking section D as reference, The Bernoulli equation, applied between the free surface of the liquid in the demijohn and section D, reads as:

$$\rho g z_0 = \frac{1}{2}\rho v_D^2 + 2f\frac{L}{D}\rho v_D^2 + \frac{1}{2}\rho k_1 v_D^2 + 2\frac{1}{2}\rho k_C v_D^2 \quad , \tag{11.9}$$

where L is the total length of the pipe, z_0 is the liquid height measured with respect to the inlet section, $k_1 = 0.44$ (see Table 11.1) is the head loss coefficient at the inlet and $k_C = 0.8$ is the head loss coefficient due to a bend (two bends are shown in Fig. 11.2). Using the Blasius equation to calculate the friction factor, Eq. (11.9) turns out to be a function of v_D only:

$$\frac{1}{2}v_D^2 + \frac{1}{2}k_I v_D^2 + 2\frac{1}{2}k_C v_D^2 + 2 \cdot 0.079\frac{L}{D}\left(\frac{\rho D}{\mu}\right)^{-0.25}v_D^{1.75} = gz_0 \quad .(11.10)$$

Substituting the numerical values yields:

$$(0.22 + 0.5 + 0.8)v_D^2 + 2.4v_D^{1.75} = 1.96 \quad , \tag{11.11}$$

which, proceeding by trial and error, gives $v_D = 0.69$ m/s. It follows that the flowrate through the pipe is:

$$Q = Av_D = \frac{\pi D^2}{4}v_D = 5.42 \cdot 10^{-5}\text{m}^3/\text{s} \quad , \tag{11.12}$$

and the time required to fill a volume of 1 lt is:

$$t = \frac{V}{Q} = \frac{10^{-3}}{4.87 \cdot 10^{-5}} = 18.5\,\text{s} \quad , \tag{11.13}$$

since the supplied flowrate can be considered as constant if the level of fluid in the demijohn does not vary (at least in the time range required to fill the demijohn).

2. It is reasonable to assume that the point of minimum pressure is C. This is because point C is, among the points at highest elevation in the system, the one with the largest distributed head losses. The Bernoulli equation, applied between C and D, taking into account that $v_C = v_D$, gives:

$$p_C + \rho g h_C = 2f\frac{L'}{D}\rho v_D^2 + p_{atm} \quad . \tag{11.14}$$

Since $Re = \rho v_D D/\mu = 6655$, $f = 8.7 \cdot 10^{-3}$, $L' = 0.6$ m and $h_C = 0.6$ m, substitution of these numerical values yields:

$$p_C = 0.95 \cdot 10^5\,\text{Pa} \quad . \tag{11.15}$$

$\boxed{3}$ In the piping system shown in Fig. 11.3, the two tanks A and B feed by gravity tank C with a liquid of density 800 kg/m^3 and viscosity $1.2 \cdot 10^{-1}$Pa $\cdot$ s. Determine the flowrates in the different branches of the network using the dimensions given in Fig. 11.3 (pipe diameter 50 mm).

Solution Given the high value of fluid viscosity, it is reasonable to assume that the flow is laminar and, since the vertical portions of the pipeline are much shorter than the horizontal ones, that their contribution to the calculation of the distributed losses is negligible. The Bernoulli equation, written between sections A and D neglecting

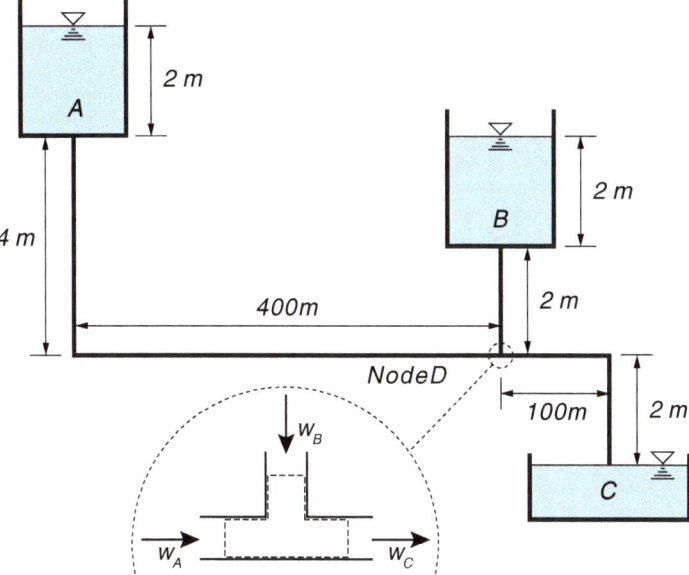

Fig. 11.3 Tanks feeding a third tank by gravity

the kinetic energy terms, gives:

$$\rho g h_A = \rho g h_D + p_D + \Delta p_{A-D} \quad , \tag{11.16}$$

from which:

$$p_D = \rho g (h_A - h_D) - \Delta p_{A-D} \quad . \tag{11.17}$$

Considering sections B and D, the Bernoulli equation gives:

$$\rho g h_B = \rho g h_D + p_D \quad , \tag{11.18}$$

from which:

$$p_D = \rho g (h_B - h_D) \quad . \tag{11.19}$$

Comparing the two expressions obtained for the pressure at node D and expressing the head loss Δp_{A-D} as a function of the mass flowrate w_A in the pipe as (laminar flow):

$$\Delta p_{A-D} = \frac{128 \mu L_{A-D} w_A}{\rho \pi D^4} \quad , \tag{11.20}$$

the following relation is obtained:

$$\frac{128\mu L_{A-D} w_A}{\rho\pi D^4} = \rho g(h_A - h_B) \quad , \tag{11.21}$$

from which the flowrate w_A can be calculated as:

$$w_A = \rho g(h_A - h_B)\frac{\rho\pi D^4}{128\mu L_{A-D}} = 3.95 \cdot 10^{-2}\text{kg/s} \quad . \tag{11.22}$$

Similarly, the Bernoulli equation written between D and C gives:

$$p_D + \rho g h_D = \rho g h_C + \Delta p_{D-C} \quad . \tag{11.23}$$

Using for p_D one of the expressions previously derived and expressing the head loss Δp_{D-C} as a function of the flowrate w_C, the following relation is obtained:

$$\frac{128\mu L_{D-C} w_C}{\rho\pi D^4} = \rho g(h_B - h_C) \quad , \tag{11.24}$$

from which:

$$w_C = \rho g(h_B - h_C)\frac{\rho\pi D^4}{128\mu L_{D-C}} = 0.482 \text{ kg/s} \quad . \tag{11.25}$$

The flowrate in the branch B-D can be obtained from the conservation of mass at node D:

$$w_A + w_B = w_C \quad , \tag{11.26}$$

which gives $w_B = 0.442$ kg/s. It is now possible to verify the assumption of laminar flow, made initially to express the head losses as a function of the flowrate, is indeed correct. The Reynolds number, calculated in the branch D-C where the flowrate is the highest, is equal to:

$$Re = \frac{\rho v D}{\mu} = \frac{4 w_C}{\pi\mu D} = 102 \quad . \tag{11.27}$$

This value is well below the critical value of 2100, so it can be concluded that the assumption is correct and so is the solution of the problem.

4 In the pipeline network shown in Fig. 11.4, the pump P is used to transport a flowrate of 35 m³/hr of fluid ($\rho = 1000$ kg/m³, $\mu = 10^{-3}$Pa · s) from tank A to tank B. All branches have diameter $D = 92$ mm and are smooth. For the dimensions given in Fig. 11.4:

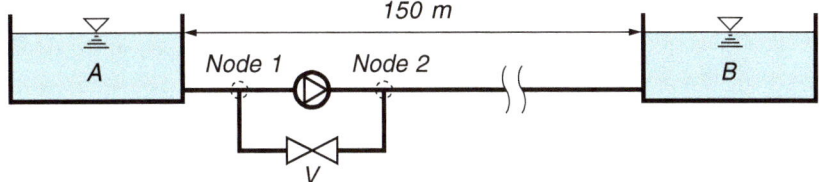

Fig. 11.4 Schematic diagram for the control of the flowrate in a pipeline network

1. determine the required pumping power, assuming efficiency equal to one and valve V closed;
2. determine the flowrate transferred to B for the same pumping power calculated at point 1. but now assuming the valve V is fully open. Assume for the branch 1-V-2 an equivalent length[2] of 20 m.

Solution

1. The required pumping power when the valve V is closed can be easily calculated by applying the Bernoulli equation between the two tanks:

$$\Delta p_P = \Delta p_{A-B} \quad , \tag{11.28}$$

where Δp_P is the head of the pump and Δp_{A-B} is the head loss, which can be expressed as:

$$\Delta p_{A-B} = 2f\frac{L_{A-B}}{D}\rho v^2 \quad . \tag{11.29}$$

The friction factor f can be calculated from the Blasius equation:

$$f = 0.079 Re^{-0.25} \quad . \tag{11.30}$$

Substitution of the numerical values gives $w = 9.72$ kg/s, $L_{A-B} = 150$ m, $Re = 1.35 \cdot 10^5$ (which implies that the flow is turbulent), $f = 4.12 \cdot 10^{-3}$, $v = 1.47$ m/s and, for the pump head:

$$\Delta p_P = 0.29 \cdot 10^5 \text{ Pa} \quad . \tag{11.31}$$

The corresponding pumping power is:

$$P = Q\Delta p_P = 283 \text{ W} \quad . \tag{11.32}$$

---

[2] The equivalent length is defined as the length that a given pipeline section should have in order to have $l_{v,eq} = l_{v,distributed} + l_{v,localised}$, namely to be able to express both the distributed losses and the localised losses as a single equivalent term of distributed losses

2. The pressure at node 2 downstream of the pump is higher than the pressure at node 1 upstream of the pump. When the valve V is open, this pressure difference causes a flow from 2 to 1, thereby decreasing the flowrate that can be transported to tank B. From mass conservation at node 2, considering that all the branches have the same diameter, the fluid velocity v downstream of the node is:

$$v = v_1 - v_2 \quad , \tag{11.33}$$

where, v_1 is the fluid velocity in the branch of the pump and v_2 the fluid velocity in the branch of the valve. The Bernoulli equation gives:

$$p_2 + \frac{1}{2}\rho v^2 = 2f\frac{L_{2-B}}{D}\rho v^2 \quad , \tag{11.34}$$

for the portion of the pipe between Sect. 2 and tank B, as well as:

$$0 = \frac{1}{2}\rho v^2 + p_1 + 2f\frac{L_{A-1}}{D}\rho v^2 \quad , \tag{11.35}$$

for the portion of the pipe between tank A and Sect. 1 and finally:

$$p_2 = p_1 + 2f_V\frac{L_{2-V-1}}{D}\rho v_2^2 \quad , \tag{11.36}$$

for the branch 1-V-2 that contains the valve. These three equations, combined with the continuity equation applied at node 2, form the system of equations that must be solved to derive the values of the unknown variables p_1, p_2, v and v_2, being $v_1 = 1.47$ m/s (the flowrate through the pump is the same as in the previous point by assumption). Summing up the first two equations and solving for the pressure difference $p_2 - p_1$ gives:

$$p_2 - p_1 = 2f_V\frac{L_{2-V-1}}{D}\rho v_2^2 \quad , \tag{11.37}$$

and:

$$p_2 - p_1 = 2f\frac{L_{A-1} + L_{2-B}}{D}\rho v^2 \quad . \tag{11.38}$$

If Eq. (11.30) is used to calculate the friction factors and the continuity equation is taken into account, the following expression can be derived by equating the right-hand sides of Eqs. (11.37) and (11.38):

$$0.158\frac{\rho}{D}\left(\frac{\rho D}{\mu}\right)^{-0.25}\left[L_{2-V-1}(v_1 - v)^{1.75} - (L_{A-B})v^{1.75}\right] = 0 \quad , \tag{11.39}$$

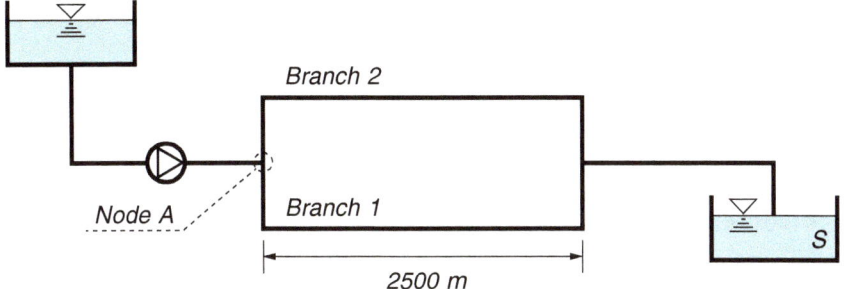

Fig. 11.5 Schematic diagram for calculating the maximum flowrate of the pipeline network

which can be rewritten as:

$$20(1.47 - v)^{1.75} - 150v^{1.75} = 0 \quad . \tag{11.40}$$

The solution of this equation is $v = 0.35$ m/s and, hence, the flowrate transferred from A to B is:

$$w = \rho v \frac{\pi D^2}{4} = 2.33 \text{ kg/s} \quad . \tag{11.41}$$

5 To supply water to the tank S shown in Fig. 11.5, two smooth pipes are used in parallel. The pipes have equal length $L = 2500$ m but different diameters, $D_1 = 200$ mm and $D_2 = 120$ mm respectively. For an available pumping power of 20 kW, determine the maximum theoretical flowrate that can be transferred.

Solution The following equations can be used to solve the problem:

$$\frac{\Delta p}{\rho} = l_{v1} = 2v_1^2 f_1 \frac{L}{D_1} \quad , \tag{11.42}$$

$$\frac{\Delta p}{\rho} = l_{v2} = 2v_2^2 f_2 \frac{L}{D_2} \quad , \tag{11.43}$$

$$w = w_1 + w_2 \quad . \tag{11.44}$$

The first two equations are obtained by applying the Bernoulli equation to the upper branch (labelled as branch 1) and to the lower branch (labelled as branch 2), respectively. The third equation comes from the conservation of mass at node A (located immediately downstream of the pump). Assuming turbulent flow in the pipeline, the Blasius equation can be used to compute the friction factor, $f = 0.079 Re^{-0.25}$. Therefore, the above equations give:

$$\frac{\Delta p}{\rho} = 2 \cdot 0.079 \left(\frac{\rho D_1 v_1}{\mu} \right)^{-0.25} \frac{L}{D_1} v_1^2 \ , \tag{11.45}$$

$$\frac{\Delta p}{\rho} = 2 \cdot 0.079 \left(\frac{\rho D_2 v_2}{\mu} \right)^{-0.25} \frac{L}{D_2} v_2^2 \ . \tag{11.46}$$

Equating the right-hand side of these two equations gives:

$$\frac{v_1}{v_2} = \left(\frac{D_1}{D_2} \right)^{\frac{1.25}{1.75}} \ , \tag{11.47}$$

or, equivalently:

$$\frac{w_1}{w_2} = \left(\frac{D_1}{D_2} \right)^{\frac{1.25}{1.75}+2} \ . \tag{11.48}$$

For the given values of D_1 and D_2, it follows that:

$$\frac{w_1}{w_2} = 4 \ , \tag{11.49}$$

and, in turn: $w = 5w_2$. The pumping power is given by the relation $P = w\Delta p/\rho$, and therefore:

$$P = 5\frac{w_2}{\rho} \left[2 \cdot 0.079 \left(\frac{\rho D_2 v_2}{\mu} \right)^{-0.25} \frac{L}{D_2} \rho v_2^2 \right] \ , \tag{11.50}$$

which gives $w_2 = 14.5$ kg/s and, finally, $w = 72.7$ kg/s, since $P = 20$ kW.

6 Consider the pipeline network shown in Fig. 11.6, which is used to transport water. At position A, the absolute pressure is constant and equal to $p_A = 2 \times 10^5$ Pa. At position C, the pressure is equal to the atmospheric pressure ($p_C = 10^5$ Pa). The pipes have inner diameter $D = 80$ mm and roughness equal 0.02 mm.

1. Determine the flowrate transferred to C when the pump P is turned off (no flow in the pipe branch A-P-B).
2. Determine the transferred flowrate as a function of the pumping power P.

Solution

1. To determine the transferred flowrate when the pump is turned off, the Bernoulli equation can be used. Application of this equation between points A and C yields:

$$\frac{p_A - p_C}{\rho} = l_v = \frac{2v^2 Lf}{D} \ . \tag{11.51}$$

Substituting the numerical values:

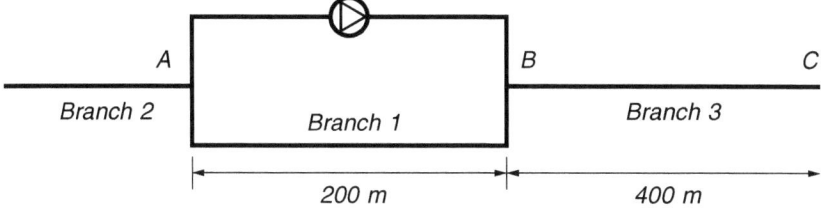

Fig. 11.6 Schematic diagram of a water supply pipeline network

$$\frac{2 \cdot 10^5 - 1 \cdot 10^5}{10^3} = 100 = \frac{2v^2 \, Lf}{D} \quad, \tag{11.52}$$

and, hence:

$$v^2 f = 100 \frac{D}{2\,L} = 6.7 \cdot 10^{-3} \quad. \tag{11.53}$$

The friction factor f can be expressed as a function of the Reynolds number by means of a relation that depends on the flow regime. Assuming that the flow is turbulent, and taking a first attempt value $f = 5 \cdot 10^{-3}$ for the friction factor, it follows that $v = 1.16$ m/s and, in turn, that $Re = 9.26 \cdot 10^4$. This value of Re confirms that the flow is turbulent. From the Colebrook equation (2.23), the given value of roughness yields:

$$\frac{1}{\sqrt{f}} = -4.0 \, \log_{10} \left(\frac{k}{D} + \frac{4.67}{Re\sqrt{f}} \right) + 2.28 \quad \rightarrow \quad f = 4.9 \cdot 10^{-3} \quad. \tag{11.54}$$

This value is sufficiently close to the first attempt value, and so it can be used to calculate the transferred mass flowrate:

$$w = 5.83 \, \text{kg/s} \quad. \tag{11.55}$$

2. When the pump is running, it is important to note that the pressure at position B may be higher than the pressure at position A, which would cause a recirculation of liquid from B to A in branch 1 of the network. To take this possibility into account, it is convenient to express the pressure difference in branch 1 as:

$$p_A - p_B = 2 f_1 \frac{L_1}{D} \rho v_1 |v_1| \quad, \tag{11.56}$$

where a positive value is assigned to the fluid velocity directed from A to B. Likewise, the friction factor will be defined as $f = f(Re)$ where $Re = (\rho |v| D)/\mu$. For the other branches, it is possible to write:

$$p_A - p_B + \Delta p_P = 2 f_2 \frac{L_2}{D} \rho v_2^2 \quad , \tag{11.57}$$

with $\Delta p_P = P/Q$ being the head provided by the pump, and:

$$p_B - p_C = 2 f_3 \frac{L_3}{D} \rho v_3^2 \quad . \tag{11.58}$$

From mass conservation, it also follows that:

$$v_3 = |v_1| + v_2 \quad . \tag{11.59}$$

The solution of the system of Eqs. (11.56)–(11.59) with unknowns v_1, v_2 and v_3 for a given pumping power P is not easy.

An easier procedure is to use an indirect method, which requires to solve the problem *step by step* by assigning different values to the flowrate in branch 1. Indeed, since the pressure in A is known, it is easy to calculate the pressure in B and, in turn, the flowrate in branch 3 since point C is at atmospheric pressure. Finally, having calculated by summation the flowrate in branch 2, it is possible to determine the pumping power. In the following, an example of this procedure is provided for three different values of the flowrate in branch 1: (**a**) zero flowrate; (**b**) positive flowrate; (**c**) negative flowrate.

(**a**) It is very simple to determine the pumping power for which the flowrate in branch 1 is zero. In this case, it must be:

$$p_B = p_A = 2 \cdot 10^5 \, \text{Pa}, \quad v_1 = 0 \quad . \tag{11.60}$$

In addition, Eq. (11.58), for $f = 5 \cdot 10^{-3}$, yields:

$$v_3 = \sqrt{\frac{p_B - p_C}{2 \rho f_3 L_3 / D}} = 1.41 \, \text{m/s} \quad , \tag{11.61}$$

and, in turn, $w_3 = 7.08$ kg/s.

Branch 2 is characterized by a velocity $v_2 = v_3 = 1.41$ m/s and the evaluation of the head losses allows to calculate the pumping power as:

$$P = \frac{w}{\rho} 2 f_2 \frac{L_2}{D} \rho v_2^2 = 354 \, \text{W} \quad . \tag{11.62}$$

Note that a constant value for f equal to $5 \cdot 10^{-3}$ has been used.

(**b**) Considering a flowrate of $w_1 = 3$ kg/s from A to B in the branch 1. Using Eq. (11.56), the head losses are equal to $p_A - p_B = 9 \cdot 10^3$ Pa (albeit the friction factor in this case is slightly higher than $5 \cdot 10^{-3}$, the same value is used for simplicity). Knowing p_B and using Eq. (11.61), it is possible to calculate $v_3 =$

1.35 m/s and, in turn, $v_2 = 0.75$ m/s. From Eq. (11.57), the work done per unit mass, Δp_P, can be calculated and, finally, the pumping power can be obtained as $P = 19.1$ W.

(c) Assuming instead $w_1 = -3$ kg/s gives $p_A - p_B = -9 \cdot 10^3$ Pa, from which $v_3 = 1.48$ m/s and, in turn, $v_2 = 2.08$ m/s. Using the Colebrook equation, the friction factor in the branch 2 can be obtained as $f = 3.3 \cdot 10^{-3}$. Finally, Eqs. (11.61) and (11.62) give $P = 840$ W.

11.2 Minimum Cost Pipe System Design

When considering the transport of a given flowrate in a pipe, a change in the pipe diameter corresponds to a change of the viscous forces acting on the pipe walls and, as a consequence, in the pumping power required to transfer the flowrate. On the other hand, the cost of the pipeline increases as the pipe diameter increases. It is therefore important to determine the *economic pipe diameter*, that is the diameter that minimizes the costs of the pipeline (per unit mass of transported fluid). There are several sophisticated procedures to determine the economic diameter. In the following, the simplest possible kind of economic analysis based on a cost balance is considered. This analysis is aimed at answering the question: which is the smallest pipe diameter that will carry the required flow with the available pressure drop, while minimizing the costs of the pipeline?

The simplest cost balance considers the so-called capital costs (initial pipe material, pump and installation costs) and the so-called operational costs (associated with energy consumption and maintenance of the pump).

The cost incurred to purchase the pipe material and build the pipeline network can be expressed as:

$$C_T = k_T DL \quad , \tag{11.63}$$

where $k_T D$ is the cost per meter of installed pipe. This cost is assumed to be proportional to the pipe diameter, D and to the pipe length, L. Note that the cost C_T includes the construction and installation costs as well as the costs of the different components of the pipeline.

In addition to C_T, the cost balance includes the cost of the pumping station, which is assumed proportional to the installed pumping power P:

$$C_P = k_P P \quad , \tag{11.64}$$

where k_P is the cost per Watt of the pump. If the pipeline network is expected to operate for N years, then the total capital cost per year of operation is:

$$C_C = \frac{C_T + C_P}{N} = \frac{k_T DL + k_P P}{N} \quad . \tag{11.65}$$

Finally, the cost balance includes the pumping cost per year of operation, C_E, equal to the electric power required by the pump for the hours of operation per year, H, times the hourly cost of the electric power, k_E:

$$C_E = k_E P H \quad , \tag{11.66}$$

where the pumping power to be installed is:

$$P = \frac{\Delta p Q}{\eta} \quad , \tag{11.67}$$

with Q the volumetric flowrate to be transferred and η the efficiency of the pump. For a turbulent flow in a smooth horizontal pipe (and, hence, for a pipeline in which all sections have the same height), the head loss is:

$$\Delta p = \frac{2 f \rho v^2 L}{D} \quad , \tag{11.68}$$

with:

$$f = 0.079 Re^{-0.25} \quad . \tag{11.69}$$

The total cost per year is given by:

$$C_{Tot} = C_I + C_E = \frac{k_T DL}{N} + P \left(\frac{k_P}{N} + k_E H \right) \quad . \tag{11.70}$$

By defining $\lambda_1 = k_T/N$ and $\lambda_2 = (K_P/N + K_E H)$ and using Eqs. (11.67), (11.68) and (11.69), the total C_{Tot} can be expressed as:

$$C_{Tot} = \lambda_1 DL + 0.24 \lambda_2 L \frac{\rho^{0.75} \mu^{0.25} Q^{2.75}}{D^{4.75}} \quad . \tag{11.71}$$

This function has a minimum that can be determined by differentiating the total cost with respect to diameter D and setting the derivative equal to zero. This yields the *economic diameter*:

$$D_{econ} = \left(\frac{1.14 \lambda_2}{\lambda_1} \rho^{0.75} \mu^{0.25} Q^{2.75} \right)^{\frac{1}{5.75}} \quad . \tag{11.72}$$

From this expression, substituting $Q = v_{opt} D_{opt}^2 \pi/4$, the *economic velocity* can be derived as:

$$v_{econ} = \frac{4}{\pi} \left(\frac{\lambda_1}{4.75\lambda_2} \right)^{1/2.75} \frac{D^{1/11}}{\rho^{3/11}\mu^{1/11}} \quad . \tag{11.73}$$

This relation shows that v_{econ} depends weakly on the diameter and on the physical properties of the fluid. It is found that, as the cost parameters λ_1 and λ_2 are changed, the economic velocity for low viscosity fluids is $1.0 \div 1.5$ m/s.

11.2.1 Examples

1 An incompressible Newtonian fluid must be pumped over a distance L with a flowrate Q_0 and the available pressure difference is $|\Delta p|$. Show that the total cost of invested capital is lower when a single large pipe is used rather than two small pipes, regardless of the flow regime (laminar or turbulent).

Solution The volumetric flow rate depends on the available pressure difference and on the pipe length according to the equation:

$$\Delta p = 2f \frac{L}{D} \rho \frac{16Q^2}{(\pi D^2)^2} \quad . \tag{11.74}$$

Taking into account the dependence of the friction factor on the Reynolds number, Eq. (11.74) yields $Q = CD^m$, with $m = 4$ for laminar flow and $m = 19/7$ for turbulent flow. The cost of Invested Capital (IC) is proportional to the diameter, so in the case of two pipes it must be:

$$IC = K(D_1 + D_2) \quad , \tag{11.75}$$

$$Q_0 = C(D_1^m + D_2^m) = \text{constant} \quad . \tag{11.76}$$

The aim is to minimize IC for a given flowrate Q_0. Figure 11.7 shows the curves corresponding to Eqs. (11.75) and (11.76) for a given value of IC (cost line). The functions are evaluated for laminar (solid line) and turbulent (dashed line) flow conditions. In the (D_1, D_2) plane, Eq. (11.76) provides the geometric locus of all pairs (D_1, D_2) that are compatible with the given value Q_0 of the flowrate. For each pair (D_1, D_2) that satisfies Eq. (11.76), there is a line defined by Eq. (11.75) to which corresponds a given value of IC, being IC/K the value of the intercept on both the x-axis and the y-axis. Among the infinite curves that can be obtained by varying IC, only the one that intersects the curve of Eq. (11.76) and is as close as possible to the origin (in order for IC to be minimum) is of interest. Such a curve is the one passing through points $(D_{10}, 0)$ and $(0, D_{20})$ where D_{10} and D_{20} are the values at the intersections of the curve corresponding to Eq. (11.76) with the axes. The figure shows clearly that IC is minimized when a single pipe is used.

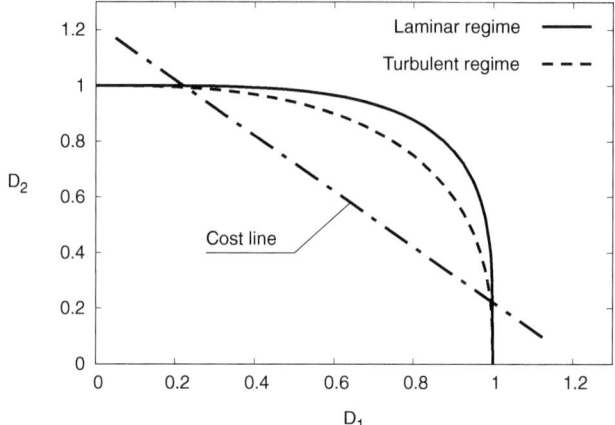

Fig. 11.7 Cost optimization of a pipeline network. The costs associated with the use of a single pipe or two parallel pipes for laminar flow, Eq. (11.76) with $m = 4$, and for turbulent flow, Eq. (11.76) with $m = 19/7$, are compared

2   The flowrate of a water supply pipeline network must be doubled from 400 to 800 m³/h ($\rho = 10^3$ kg/m³, $\mu = 10^{-3}$Pa · s). Three alternatives are possible: installing a new pump in the existing pipeline (alternative 1), building a new pipeline to replace the existing one (alternative 2), pairing the existing pipeline with a parallel pipeline (alternative 3). The diagram of the water supply network is shown in Fig. 11.8. The diameter of all pipes is $D = 0.35\,m$. Considering that the costs are:

- $K_T = 5 \cdot 10^2\ D$ €/m (D is the diameter of the pipe),
- $K_P = 3 \cdot 10^3$ €/kW,
- $K_E = 0.2$ €/kWh.

and the payback period is 10 years with 6000 h of operation per year:

1. determine the power of the installed pump,
2. determine the most cost-effective solution.

Solution

1. The power of the installed pump is given by:

$$P = Q\Delta p_{tot} \quad , \tag{11.77}$$

with Δp_{tot} energy per unit volume of fluid that must be supplied to transport the given flowrate Q. This energy is required to balance the distributed head losses and height difference associated with the vertical portion of the pipeline:

$$\Delta p_{tot} = \Delta p_L + \Delta p_h \quad , \tag{11.78}$$

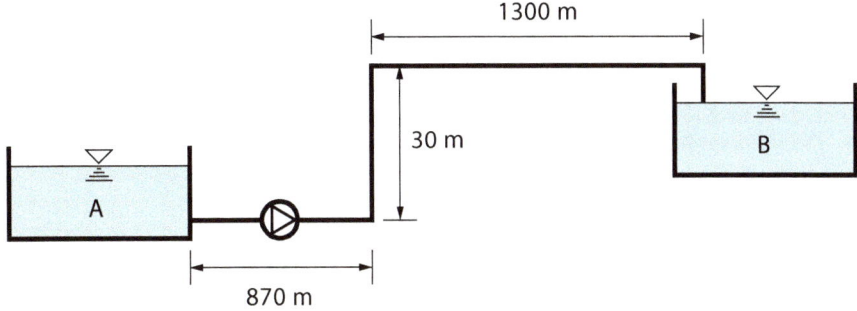

Fig. 11.8 Schematic diagram of a water supply pipeline network

with:

$$\Delta p_L = 2f\frac{L}{D}\rho v^2 \quad , \tag{11.79}$$

$$\Delta p_h = \rho g h \quad . \tag{11.80}$$

The fluid velocity v is:

$$v = \frac{4Q}{\pi D^2} = \frac{400\,(4/3600)}{\pi\,0.35^2} = 1.15 \text{ m/s} \quad . \tag{11.81}$$

The total length of the pipeline is:

$$L = 870 + 30 + 1300 = 2200 \text{ m} \quad , \tag{11.82}$$

and the friction coefficient f can be calculated (assuming turbulent flow and smooth pipe) as a function of the Reynolds number, $Re = (vD\rho)/\mu$, using the Blasius equation:

$$f = 0.079 Re^{-0.25} \quad , \tag{11.83}$$

which gives $f = 3.1 \cdot 10^{-3}$. Using these values, Δp_L and Δp_h are equal to:

$$\Delta p_L = 2 \cdot 3.1 \cdot 10^{-3}\frac{2.2 \cdot 10^3}{0.35} 10^3 \cdot 1.15^2 = 0.515 \cdot 10^5 \text{ Pa} \quad , \tag{11.84}$$

$$\Delta p_h = 10^3 \cdot 9.81 \cdot 30 = 2.94 \cdot 10^5 \text{ Pa} \quad , \tag{11.85}$$

and, in turn:

$$\Delta p_{tot} = \Delta p_L + \Delta p_h = 3.46 \cdot 10^5 \text{ Pa} \quad . \tag{11.86}$$

The corresponding pumping power is:

$$P = 3.46 \cdot 10^5 \left(\frac{400}{3600}\right) = 38.5 \text{ kW} \quad . \tag{11.87}$$

2. To determine the most cost-effective solution, it is necessary to calculate the total cost associated with each of the proposed alternative solutions.
   Option 1: Installation of a new pump in the existing pipeline. To calculate the pumping power required to transport the new (doubled) flowrate, it is possible to follow the same procedure used in the previous step, but now with:

$$v' = 2v = 2.31 \text{ m/s} \quad , \tag{11.88}$$

and:

$$Re' = 2Re \quad , \tag{11.89}$$

which gives:

$$f' = 0.079(Re')^{-0.25} = 2.6 \cdot 10^{-3} \quad . \tag{11.90}$$

The distributed head losses (the only losses that change as a result of a change in the flowrate) are equal to:

$$\Delta p_L = 1.764 \cdot 10^5 \text{ Pa} \quad . \tag{11.91}$$

The total head that the pump must provide is therefore equal to:

$$\Delta p_{tot} = 1.764 \cdot 10^5 + 2.94 \cdot 10^5 = 4.7 \cdot 10^5 \text{ Pa} \quad , \tag{11.92}$$

and the corresponding pumping power is:

$$P' = \Delta p_{tot} Q' = 104.5 \text{ kW} \quad . \tag{11.93}$$

Since the pump already installed in the piping system is only capable of supplying a power $P = 38.5$ kW, it can be concluded that the pump must be replaced.

The costs associated with this alternative solution are given by the sum of the capital cost of the new pump to be installed (C_P), which scales with the additional pumping power required, and the operational costs incurred over the payback period of 10 years (C_E), proportional to the total pumping power installed. These two costs are respectively equal to:

$$C_P = 3 \cdot 10^3 \cdot 104.5 = 313.5 \cdot 10^3 \ € \quad , \tag{11.94}$$

$$C_E = 0.2 \cdot 10 \cdot 6000 \cdot 104.5 = 1.255 \cdot 10^6 \ € \quad . \tag{11.95}$$

The total cost is, therefore, $C_{Tot} = 1.568 \cdot 10^6 \ €$.

Option 2: building a new pipeline to replace the existing one. Since the diameter of the new pipeline is not specified, it will be chosen from those available on the market in order to minimize the total costs of the pipeline. These costs can be attributed not only to the standard operational costs but also to the installation of the new pump and to the purchase of the pipe material required by the new pipeline. All of these costs can be derived as a function of the pipe diameter. Being:

$$v = \frac{4(800/3600)}{\pi D^2} = \frac{0.283}{D^2} \ \text{m/s} \quad , \tag{11.96}$$

$$Re = \frac{vD\rho}{\mu} = \frac{2.83 \cdot 10^5}{D} \quad , \tag{11.97}$$

$$f = 0.079 \left(\frac{2.83 \cdot 10^5}{D} \right)^{-0.25} = 3.43 \cdot 10^{-3} \ D^{0.25} \quad , \tag{11.98}$$

$$\Delta p_L = 2 \cdot 3.43 \cdot 10^{-3} D^{0.25} \frac{2.2 \cdot 10^3}{D} 10^3 \left(\frac{0.283}{D^2} \right)^2 =$$

$$= 1.208 \cdot 10^3 \ D^{-4.75} \quad , \tag{11.99}$$

the pumping power can be obtained as:

$$P = \frac{800}{3600} \left(2.94 \cdot 10^5 + 1.208 \cdot 10^3 D^{-4.75} \right) =$$

$$= 6.53 \cdot 10^4 + 268 D^{-4.75} \quad . \tag{11.100}$$

The capital cost of the pump, C_P, is then:

$$C_P = 3 \cdot 10^3 \frac{6.53 \cdot 10^4 + 268 D^{-4.75}}{10^3} \quad , \tag{11.101}$$

which, added to the capital cost incurred to install the new pipeline:

$$C_T = 5 \cdot 10^2 \cdot 2.2 \cdot 10^3 \ D \quad , \tag{11.102}$$

and to the operational cost, C_E, which is a function of the total pumping power installed:

$$C_E = 0.2 \frac{6.53 \cdot 10^4 + 268 D^{-4.75}}{10^3} 10 \cdot 6000 \quad , \qquad (11.103)$$

yields, for the total cost:

$$C_{Tot} = 1.1 \cdot 10^6 D + 15(6.53 \cdot 10^4 + 268 D^{-4.75}) \quad . \qquad (11.104)$$

Taking the derivative with respect to D and setting it equal to zero, the economic pipe diameter can be obtained and is equal to $D_{econ} = 0.494$ m. The nearest commercial pipe diameter available on the market is $D = 0.5$ m, which is selected to calculate the total cost as:

$$C_{Tot} = 1.64 \cdot 10^3 \text{ €} \quad . \qquad (11.105)$$

Option 3: pairing the existing pipeline with a parallel pipeline. Again, the diameter of the new pipeline is not specified but can be determined so as to minimize the total costs. Proceeding as in the previous case, it is found that:

$$v = \frac{4}{\pi D^2} \frac{(800 - 400)}{3600} = \frac{0.1415}{D^2} \text{ m/s} \quad , \qquad (11.106)$$

$$Re = \frac{v D \rho}{\mu} = \frac{1.415 \cdot 10^5}{D} \quad , \qquad (11.107)$$

$$f = 0.079 Re^{-0.25} = 4.07 \cdot 10^{-3} D^{0.25} \quad , \qquad (11.108)$$

$$\Delta p_L = 2 \cdot 4.07 \cdot 10^{-3} D^{0.25} \frac{2.2 \cdot 10^3}{D} 10^3 \left(\frac{0.1415}{D^2}\right)^2 =$$

$$= 359 D^{-4.75} \quad , \qquad (11.109)$$

and the pumping power is:

$$P = \frac{400}{3600} \left(2.94 \cdot 10^5 + 359 D^{-4.75}\right) =$$

$$= 3.267 \cdot 10^4 + 40 D^{-4.75} \quad . \qquad (11.110)$$

The capital cost of the pump, C_P, is then:

$$C_P = 3 \cdot 10^3 \frac{3.267 \cdot 10^4 + 40D^{-4.75}}{10^3} \quad , \qquad (11.111)$$

which, added to the capital cost, C_T, incurred to install the new pipeline:

$$C_T = 5 \cdot 10^2 \cdot 2.2 \cdot 10^3 \, D \quad , \qquad (11.112)$$

and the operational cost, C_E, which is a function of the total pumping power installed:

$$C_E = 0.2 \frac{3.85 \cdot 10^4 + 3.267 \cdot 10^4 + 40D^{-4.75}}{10^3} 10 \cdot 6000 \quad , \qquad (11.113)$$

yields, for the total cost:

$$C_{Tot} = 1.1 \cdot 10^6 D + 952 \cdot 10^3 + 600D^{-4.75} \quad . \qquad (11.114)$$

Taking the derivative with respect to D and setting it equal to zero, the economic pipe diameter can be obtained and is equal to $D_{econ} = 0.355$ m. The nearest commercial pipe diameter available on the market is $D = 0.35$ m, which is selected to calculate the total cost as:

$$C_{tot} = 1.425 \cdot 10^6 € \quad . \qquad (11.115)$$

By comparing the total costs of the three alternative solutions, it is apparent that the most cost-effective one is option 3.

⎯⎯⎯
$\boxed{3}$ It is necessary to design a water supply pipeline network having the dimensions shown in Fig. 11.9, where the vertical sections of the pipeline indicate height differences. Assume pipes with an absolute roughness equal to 0.4 mm and consider that:

- the capital cost of the installed pipe is $K_T = 1.8 \cdot 10^3 \, D$ €/m, where D is the pipe diameter,
- the capital cost of the pumping station is $K_P = 3.5 \cdot 10^3$ €/kW,
- the operational cost is $K_E = 0.15$ €/kWh.

with a payback period is 20 years and 5000 h of operation per year. Determine:

1. the economic diameter for a flowrate of 2 m^3/s,
2. the pumping power P using the economic diameter or the nearest commercial pipe diameter available.

Note: The minimum pressure in the pipeline must be higher than 10^5 Pa for sanitary reasons ($p_{min} = 1.2 \cdot 10^5$ Pa).

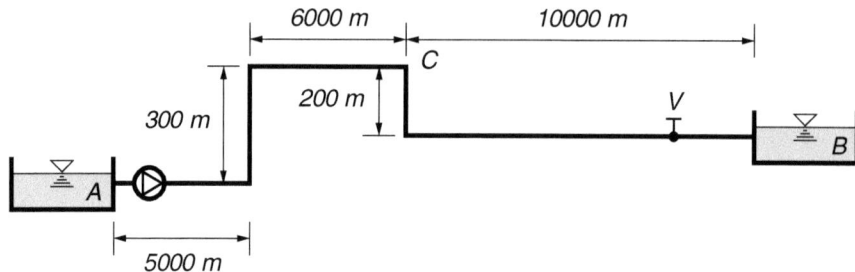

Fig. 11.9 Schematic diagram for the calculation of the economic diameter and the pumping power a water supply pipeline network

Solution

1. To calculate the economic diameter, it is necessary to express each cost as a function of the pipe diameter.

   (a) The capital cost, C_T, incurred to purchase and install the pipe material is:

   $$C_T = 1.8 \cdot 10^3 DL_{tot} = 3.87 \cdot 10^7 D \quad , \tag{11.116}$$

   with $L_{tot} = 2.15 \cdot 10^4$ m the total length of the pipeline.

   (b) The capital cost of the pumping station, C_P, is a function of the head of the pump, which must ensure that the minimum pressure along the pipeline is equal to $1.2 \cdot 10^6$ Pa. The most likely locations of minimum pressure are two: point C at the end of the 6000 m long section (where the maximum value of the distributed losses at the highest elevation is reached) and the point immediately downstream of the valve V (where the absolute maximum value of the distributed losses). Let us assume that C is the point of minimum pressure. The Bernoulli equation written between A and C yields:

   $$\rho g h_A + p_A + \Delta p_P = \rho g h_C + p_C + \Delta p_{A-C} \quad , \tag{11.117}$$

   where Δp_P is the head provided by the pump, Δp_{A-C} is the pressure drop due to the distributed losses between section A and section C, and the contribution of the kinetic terms has been neglected. The distributed losses are given by:

   $$\Delta p_{A-C} = 2f \frac{L_{A-C}}{D} \rho v^2 \quad , \tag{11.118}$$

   where the friction factor f for a turbulent flow in a commercial pipeline can be calculated as:

   $$f = 0.04 Re^{-0.16} \quad . \tag{11.119}$$

By replacing this expression into Eq. (11.118) and using:

$$v = \frac{4 Q}{\pi D^2} \quad , \tag{11.120}$$

for the velocity and the value:

$$L_{A-C} = 5000 + 300 + 6000 = 1.13 \cdot 10^4 \text{ m} \quad , \tag{11.121}$$

the following expression is obtained for Δp_{A-C}:

$$\Delta p_{A-C} = 2 \cdot 0.04 \left(\frac{4\rho Q}{\mu \pi D}\right)^{-0.16} \frac{L_{A-C}}{D} \rho \left(\frac{4Q}{\pi D^2}\right)^2 \quad , \tag{11.122}$$

$$\Delta p_{A-C} = \beta D^{-4.84} \quad , \tag{11.123}$$

with $\beta = 5.53 \cdot 10^5$ Pa $\cdot$ m$^{4.84}$. Solving the Bernoulli equation with respect to the head of the pump, and taking into account that the height difference between points C and A is 300 m and that the pressures in A and C are 10^5 and $1.2 \cdot 10^5$ Pa respectively, gives:

$$\Delta p_P = \alpha + \beta D^{-4.84} \quad , \tag{11.124}$$

with $\alpha = 9.8 \cdot 300 \cdot 10^3 + 0.2 \cdot 10^5 = 2.96 \cdot 10^6$ Pa. The resulting pumping power is:

$$P = Q \Delta p_P = 2(\alpha + \beta D^{-4.84}) \quad , \tag{11.125}$$

and the cost of the pumping station can thus be expressed as:

$$C_P = 3.5 \cdot 10^3 \cdot 2(\alpha + \beta D^{-4.84}) 10^{-3} =$$

$$= 3.5(2\alpha + 2\beta D^{-4.84}) \quad . \tag{11.126}$$

(c) The operational cost is proportional to the installed pumping power and the hours of operation hours of the pipeline (5000 h/year for 20 years):

$$C_E = 0.15(2\alpha + 2\beta D^{-4.84}) \cdot 10^{-3} \cdot 5000 \cdot 20 =$$

$$= 15(2\alpha + 2\beta D^{-4.84}) \quad . \tag{11.127}$$

Therefore, the total cost is:

$$C_{Tot} = C_T + C_P + C_E \quad , \tag{11.128}$$

$$C_{Tot} = 3.87 \cdot 10^7 D + 18.5(2\alpha + 2\beta D^{-4.84}) \quad . \tag{11.129}$$

Taking the derivative with respect to D and setting it equal to zero, the economic pipe diameter can be obtained and is equal to $D_{econ} = 1.17$ m. It is now necessary to verify that the assumptions made to obtain the solution are correct. For the calculation of f, the assumption of turbulent flow was made. Since the Reynolds number based on D_{econ} is equal to $2.2 \cdot 10^6$, it follows that the assumption was correct. It was also assumed that C was the point of minimum pressure along the pipeline. Application of the Bernoulli between point C and the point downstream of the valve V yields:

$$p_V = p_C + \rho g(h_C - h_V) - 2f\frac{L_{C-V}}{D}\rho v^2 = 2.04 \cdot 10^6 \text{ Pa} \quad , \tag{11.130}$$

and since $p_V > p_C$, it follows that this assumption was also correct.
2. The calculated economic diameter D_{econ} is not available on the market and therefore the nearest commercial pipe diameter must be selected. Suppose this diameter is $D = 1.20$ m. Using this value in Eq. (11.125) allows to calculate the theoretical pumping power, which is equal to $P = 6.38$ MW.

☐4 It is necessary to design the piping system shown in Fig. 11.10, such that the total flowrate to be transferred to the tanks B and C is equal to 120 kg/s and the flowrate to be delivered to B must be double than the flowrate delivered to C. The available commercial pipe diameters are 275, 316, 348, 400 mm and their absolute roughness is 0.4 mm. The entire pipeline is at the same height and the concentrated head losses are negligible.

1. Determine the pumping power P for $D = 348$ mm.
2. Choose, among the available pipe diameters, the one that minimizes the total cost for the following data:

   - $K_T = 8 \cdot 10^2 \cdot D$ €/m (with D diameter of the pipe),
   - $K_P = 3 \cdot 10^3$ €/kW,
   - $K_E = 0.2$ €/kWh.
   - payback period of 10 years with 6000 h of operation per year.

3. How does the flowrate split and what is the required pumping power when both valves are fully open?

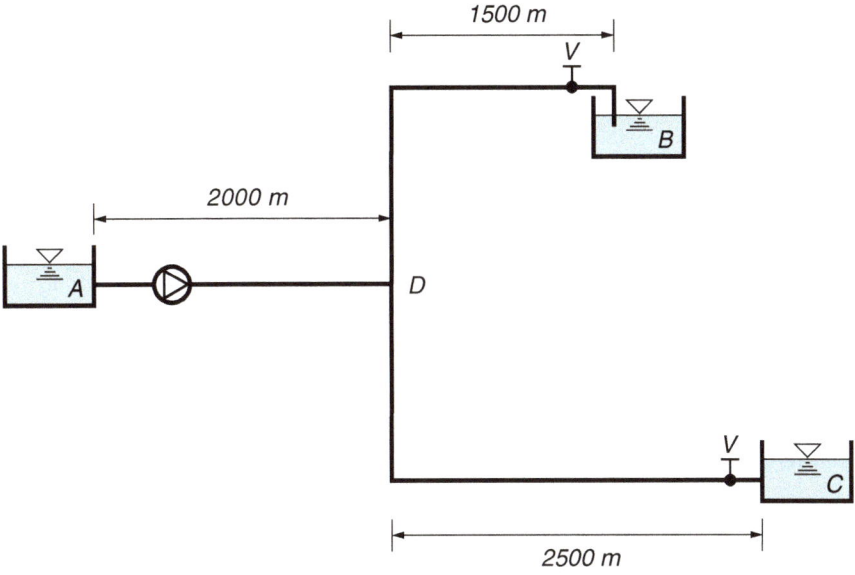

Fig. 11.10 Schematic diagram for the design of a pipeline network

Solution

1. The flowrate Q that feeds the pump is equal to:

$$Q = \frac{w}{\rho} = \frac{120}{1000} = 0.12 \text{ m}^3/\text{s} \quad . \tag{11.131}$$

At node D, this flowrate splits into the two branches D-C and D-B as follows:

$$Q_B = \frac{2}{3}Q = 0.08 \text{ m}^3/\text{s} \quad , \tag{11.132}$$

$$Q_C = \frac{1}{3}Q = 0.04 \text{ m}^3/\text{s} \quad . \tag{11.133}$$

For $D = 348$ mm, the cross-sectional area of the pipe is equal to:

$$A = \frac{\pi D^2}{4} = 0.0951 \text{ m}^2 \quad , \tag{11.134}$$

while the fluid velocity in the branch $A - D$ is equal to:

$$v = \frac{Q}{A} = 1.26 \text{ m/s} \quad . \tag{11.135}$$

The Bernoulli equation, written for sections A-D, D-B and D-C, allows to derive the head of the pump that is required to establish the flow according to the assigned design conditions. For section A-D, the Bernoulli equation gives:

$$p_A + \Delta p_P = p_D + \Delta p_{A-D} \quad , \tag{11.136}$$

and, for the other two sections:

$$p_D = p_B + \Delta p_{D-B} \quad , \tag{11.137}$$

$$p_D = p_C + \Delta p_{D-C} \quad . \tag{11.138}$$

Let us denote the head of the pump by Δp_P and the distributed losses in the different branches by Δp_{A-D}, Δp_{D-B} and Δp_{D-C}, respectively. The distributed losses can be evaluated using the expression:

$$\Delta p = 2f \frac{L}{D} \rho v^2 \quad , \tag{11.139}$$

where the friction factor f, for turbulent flow in commercial pipes, is given by:

$$f = 0.04 Re^{-0.16} \qquad Re = \frac{\rho v D}{\mu} \quad .$$

For section A-D, the following values are calculated:

$$Q = 0.12 \text{ m}^3/\text{s} \qquad v = 1.26 \text{ m/s} \qquad Re = 4.39 \cdot 10^5 \quad ,$$

and:

$$\Delta p_{A-D} = 91.2 \cdot 10^3 \text{ Pa} \quad .$$

For section D-B:

$$Q = 0.08 \text{ m}^3/\text{s} \qquad v = 0.84 \text{ m/s} \qquad Re = 2.92 \cdot 10^5 \quad ,$$

and:

$$\Delta p_{D-B} = 30.4 \cdot 10^3 \text{Pa} \quad .$$

Finally, for section D-C:

$$Q = 0.04 \text{ m}^3/\text{s} \qquad v = 0.42 \text{ m/s} \qquad Re = 1.46 \cdot 10^5 \quad ,$$

and:

$$\Delta p_{D-C} = 12.7 \cdot 10^3 \text{ Pa} \quad .$$

By summing up the Bernoulli equations written for the different branches of the pipeline, the conditions that the head of the pump must satisfy can be derived. To establish a flow between A and B, it must be:

$$\Delta p_P = \Delta p_{A-D} + \Delta p_{D-B} \quad , \tag{11.140}$$

and, similarly, to establish a flow between A and C:

$$\Delta p_P = \Delta p_{A-D} + \Delta p_{D-C} \quad . \tag{11.141}$$

Both conditions are satisfied by calculating the head of the pump as:

$$\Delta p_P = \Delta p_{A-D} + \max(\Delta p_{D-B}, \Delta p_{D-C}) \quad , \tag{11.142}$$

and using the valve in the branch with the lower head loss to regulate the flow until the desired conditions are achieved. For the present case, it is found that:

$$\Delta p_P = \Delta p_{A-D} + \Delta p_{D-B} = 1.22 \cdot 10^5 \text{ Pa} \quad , \tag{11.143}$$

so the valve belonging to the branch D-B will be kept open while the valve belonging to the branch D-C will be throttled until the desired repartition of flowrates is achieved. Finally, it is possible to calculate the pumping power as:

$$P = Q\Delta p_P = 0.12 \cdot 121.6 \cdot 10^3 = 14.6 \text{ kW} \quad . \tag{11.144}$$

2. To choose the commercial pipe diameter that minimizes the total cost, it is first necessary to express the costs as a function of the pipe diameter. The costs are:

$$C_T = 8 \cdot 10^2 D(L_{A-D} + L_{D-B} + L_{D-C}) = 4.8 \cdot 10^6 D \quad , \tag{11.145}$$

$$C_P = 3 \cdot 10^3 (P/10^3) = 3P \quad , \tag{11.146}$$

$$C_E = 0.2 \cdot (P/10^3) \cdot 6000 \cdot 10 = 12P \quad , \tag{11.147}$$

for the given payback period of 10 years and 6000 h of operation per year. The total cost is, therefore:

$$C_{Tot} = 4.8 \cdot 10^6 D + 15P \quad . \tag{11.148}$$

The pumping power is calculated assuming that the highest head losses occur in the branch D-B (assuming a repartition of the flowrates as in the previous point):

$$P = Q[\Delta p_{A-D} + \Delta p_{D-B}] =$$

$$= Q\left[2f\rho\frac{L_{A-D}}{D}\frac{16Q^2}{\pi^2 D^4} + 2f\rho\frac{L_{D-B}}{D}\frac{16\left(\frac{2}{3}Q\right)^2}{\pi^2 D^4}\right] =$$

$$= \frac{32\,Q^3}{\pi^2 D^5}\rho\left[f_{A-D}L_{A-D} + \frac{4}{9}f_{D-B}L_{D-B}\right] \quad . \tag{11.149}$$

Expressing the friction factors as a function of the Reynolds number yields:

$$P = 5.6D^{-5}\left[0.04\left(\frac{4\rho Q}{\mu\pi D}\right)^{-0.16}L_{A-D} + \frac{4}{9}0.04\left(\frac{4\rho\frac{2}{3}Q}{\mu\pi D}\right)^{-0.16}L_{D-B}\right] =$$

$$= 5.6D^{-4.84}16.06 = 89.94D^{-4.84} \quad . \tag{11.150}$$

Substituting this equation into the equation of the total cost, given by Eq. (11.148), yields:

$$C_{Tot} = 4.8 \cdot 10^6 D + 15 \cdot 89.94\, D^{-4.84} \quad , \tag{11.151}$$

and the derivative of C_{Tot} with respect to the diameter is:

$$\frac{dC_{Tot}}{dD} = 4.8 \cdot 10^6 - 4.84 \cdot 15 \cdot 89.94\, D^{-5.84} \quad . \tag{11.152}$$

By setting this derivative equal to zero, the following value of the economic diameter is obtained: $D = 0.323$ m. Considering the available pipe diameters, the closest upper value $D = 348$ mm is chosen.

3. When both valves are open, the repartition of flowrates at node D is different from that considered in the previous points but can be calculated using the continuity equation and imposing the same pressure drops for both sections, D-B and D-C. The following system of equations is obtained:

$$2f_{D-B}\frac{L_{D-B}}{D}\rho v_{D-B}^2 = 2f_{D-C}\frac{L_{D-C}}{D}\rho v_{D-C}^2 \quad , \tag{11.153}$$

$$Q = Av_{D-B} + Av_{D-C} \quad , \tag{11.154}$$

which, after expressing the friction factors as a function of the Reynolds number and after some simplifications, can be recast as:

$$\frac{v_{D-B}}{v_{D-C}} = 1.32, \quad v_{D-B} + v_{D-C} = 1.53 \quad , \tag{11.155}$$

Finally, the new velocities and flowrates are obtained as:

$$v_{D-B} = 0.87 \text{ m/s} \longrightarrow Q_{D-B} = 0.068 \text{ m}^3/\text{s} , \qquad (11.156)$$

$$v_{D-C} = 0.66 \text{ m/s} \longrightarrow Q_{D-C} = 0.052 \text{ m}^3/\text{s} . \qquad (11.157)$$

5   The *TransAlaska Pipeline* is designed to transport a flowrate $Q = 1 \text{ m}^3/\text{s}$ of crude oil ($\rho = 800 \text{ kg/m}^3$, $\nu = 2 \cdot 10^{-4} \text{ m}^2/\text{s}$) over a distance $L = 2000$ km.

1. Knowing that costs of the pipeline are $K_T = 250 \cdot D$ €/m, $K_P = 1550$ €/kW and $K_E = 0.05$ €/kWh and that the pipeline should operate for 8400 h/year with a payback period of 25 years, calculate the economic diameter assuming smooth pipes.
2. To minimize the pressure drop, the crude oil is mixed with *Fene-C* (a polymer) that leads to a drag reduction along the pipeline by changing the friction factor according to the relation $f = 0.1 \, \Gamma^\alpha \, Re^{-0.5}$, where Γ is the mass concentration of *Fene-C* in the mixture (which can never exceed 5%) and $\alpha = -0.005$. Knowing that the cost of *Fene-C* (which cannot be recovered once it is mixed with the crude oil) is $K_{Pol} = 17$ €/kg, determine the new economic pipe diameter considering the costs given in the previous point.
3. Calculate the percent reduction in the friction factor produced by the addition of the polymer as compared with the value calculated in 1.

Solution

1. To calculate the economic diameter for the case without *Fene-C*, it is necessary to derive the total cost function, which has the following general expression:

$$C_{Tot}(D, P) = K_T \, L + \frac{K_P + K_E \, N_h \, N_y}{10^3} P =$$

$$= K_T \, L + K_I \, P , \qquad (11.158)$$

with $K_I = (K_P + K_E \, N_h \, N_y)/10^3$. It is now necessary to express the pumping power, P, as a function of the diameter D, so that the cost function may be rewritten as a function of a single variable (D) and the economic diameter can be calculated by imposing:

$$\frac{dC_{Tot}}{dD} = 0 . \qquad (11.159)$$

The pumping power can be calculated as:

$$P = \frac{\Delta p}{\rho} \rho Q , \qquad (11.160)$$

where the term $\Delta p/\rho$ represents the head losses along the pipeline. Assuming negligible changes in the elevation of the pipeline and assuming equal velocity and pressure at the two ends of the pipeline, the head losses are given by the following simplified expression of the Bernoulli equation

$$\frac{\Delta p}{\rho} = 2f\frac{L}{D}v^2 \quad . \tag{11.161}$$

The velocity v can be expressed as a function of the (known) volumetric flowrate as:

$$v = \frac{4Q}{\pi D^2} \quad . \tag{11.162}$$

The friction factor, f, can be determined using the Blasius equation for smooth pipes (assuming turbulent flow):

$$f = 0.079 Re^{-0.25} \quad , \tag{11.163}$$

where the Reynolds number can be written as a function of the volumetric flowrate:

$$Re = \frac{\rho v D}{\mu} = \frac{4\rho Q}{\pi D \mu} \quad . \tag{11.164}$$

Therefore, the pumping power is obtained as:

$$P = \frac{\Delta p}{\rho}\rho Q = \left[2 \cdot 0.079 \left(\frac{4\rho Q}{\pi \mu}\right)^{-0.25} L \left(\frac{4Q}{\pi}\right)^2 Q\, D^{-4.75} \right] \rho Q =$$

$$= K_{II}\, D^{-4.75} = 4.588 \cdot 10^7\, D^{-4.75} \quad , \tag{11.165}$$

where K_{II} is a constant defined by collecting all known terms on the right-hand side of Eq. (11.165). Substituting Eq. (11.165) into Eq. (11.159) yields:

$$\frac{dC_{Tot}}{dD} = 0 \;\rightarrow\; D_{econ} = \left(\frac{K_T\, L}{4.75\, K_I\, K_{II}}\right)^{-\frac{1}{5.75}} = 1.334\, m \quad . \tag{11.166}$$

This value of the economic diameter gives $v = 0.715$ m/s and $Re = 4770$, which implies that the turbulent flow assumption previously made is indeed correct. The friction factor is $f = 9.5 \cdot 10^{-3}$.

2. When the polymer is added, the friction factor is expressed as:

$$f = 0.1\, \Gamma^\alpha\, Re^{-0.5} \quad . \tag{11.167}$$

Therefore, the pumping power is a function not only of the pipe diameter, but also of the mass concentration of the polymer:

$$P(D, \Gamma) = 2f\frac{L}{D}v^2 \cdot \rho Q = K' D^{-4.5} \Gamma^\alpha \quad . \tag{11.168}$$

The total costs in this case are given by the costs of the installed pipes, by the costs of installing and operating the pump, but also by the cost of the polymer (which is proportional to the mass of polymer employed):

$$C_{Tot}(D, \Gamma) = K_T L + \frac{K_P + K_E N_h N_y}{10^3} P + K_{pol} m_{pol} \quad . \tag{11.169}$$

The mass of polymer, m_{pol}, can be expressed as:

$$m_{pol} = m_{crude} \Gamma = (3600 \, Q \, N_h \, N_y \, \rho) \, \Gamma \quad . \tag{11.170}$$

To minimize the cost function, it is necessary that the derivatives with respect to the two independent variables D and Γ are both zero:

$$\frac{\partial C_{Tot}}{\partial \Gamma} = 0 \quad , \tag{11.171}$$

$$\frac{\partial C_{Tot}}{\partial D} = 0 \quad . \tag{11.172}$$

Equation (11.171) yields:

$$\frac{\partial C_{Tot}}{\partial \Gamma} = \frac{K_P + K_E N_h N_y}{10^3} \frac{\partial P}{\partial \Gamma} + K_{Pol} \frac{\partial m_{pol}}{\partial \Gamma} = 0 \quad , \tag{11.173}$$

with:

$$\frac{\partial P}{\partial \Gamma} = \alpha K' \Gamma^{\alpha-1} D^{-4.5} \quad , \tag{11.174}$$

and:

$$\frac{\partial m_{pol}}{\partial \Gamma} = m_{crude} \quad . \tag{11.175}$$

Substituting Eqs. (11.174) and (11.175) into Eq. (11.173) yields:

$$\Gamma^{\alpha-1} D^{-4.5} = -\frac{K_{pol} \, m_{crude}}{\alpha K' \left(\dfrac{K_P + K_E N_h N_y}{10^3}\right)} \quad . \tag{11.176}$$

Equation (11.172) yields:

$$\frac{\partial C_{Tot}}{\partial D} = K_T \, L + \left(\frac{K_P + K_E \, N_h \, N_y}{10^3} \right) \frac{\partial P}{\partial D} = 0 \quad , \qquad (11.177)$$

with:

$$\frac{\partial P}{\partial D} = -4.5 \, K' \, \Gamma^\alpha \, D^{-5.5} \quad . \qquad (11.178)$$

Substituting Eq. (11.178) into Eq. (11.177) yields:

$$\Gamma^\alpha \, D^{-5.5} = \frac{K_T \, L}{4.5 K' \left(\dfrac{K_P + K_E \, N_h \, N_y}{10^3} \right)} \quad . \qquad (11.179)$$

Equations (11.176) and (11.179) constitute a system that can be solved by change of variables. The ratio between Eqs. (11.179) and (11.176) gives:

$$\frac{\Gamma}{D} = - \underbrace{\frac{\alpha \, K_T \, L}{4.5 \, K_{pol} \, m_{crude}}}_{K_{III}} \quad , \qquad (11.180)$$

and, in turn:

$$\Gamma = K_{III} \, D \quad . \qquad (11.181)$$

Substituting into Eq. (11.176) gives:

$$D_{econ} = \left[\frac{K_T \, L}{4.5 \left(\dfrac{K_P + K_E \, N_h \, N_y}{10^3} \right) K' \, K_{III}^\alpha} \right]^{-1/(5.5+\alpha)} \quad . \qquad (11.182)$$

Substituting the numerical values of the constants, namely:

$$K' = \left(\frac{\rho}{\mu} \right)^{-0.5} \left(\frac{4Q}{\pi} \right)^{1.5} \rho Q L = 3.251 \cdot 10^7 \quad , \qquad (11.183)$$

and

$$K_{III} = \frac{\alpha \, K_T \, L}{4.5 \, K_{pol} \, m_{crude}} = 5.4 \cdot 10^{-8} \quad , \qquad (11.184)$$

gives:

$$D_{econ} = 0.261^{-0.182} \simeq 1.276 \, \text{m} \quad . \tag{11.185}$$

The corresponding mass concentration of the polymer is $\Gamma = K_{III} \, D \simeq 6.9 \cdot 10^{-8}$.

3. To calculate the percent reduction in the friction factor, it must be considered that, when the polymer is added to the crude oil, the economic diameter is reduced with respect to the value calculated at point 1. This implies that, for the same total flowrate, the average velocity of the crude oil in the pipeline will increase. Indeed, it is found that $v = 0.782 \, \text{m/s}$ and that $Re = 4989$ (the flow is still turbulent, of course).

The value of the friction factor becomes $f = 0.1 \, \Gamma^a \, Re^{-0.5} = 1.54 \cdot 10^{-3}$ and the percent reduction produced by the addition of the polymer is:

$$f_\% = \left(1 - \frac{1.54 \cdot 10^{-3}}{9.5 \cdot 10^{-3}}\right) \cdot 100 = 83.82\% \quad . \tag{11.186}$$

Problems

$\boxed{a}$ In the piping system shown in Fig. 11.11, a constant flowrate of 28 kg/s of fluid ($\rho = 820 \, \text{kg/m}^3$, $\mu = 5 \cdot 10^{-3} \text{Pa} \cdot \text{s}$) flows through the pump P. The pipe diameter is equal to 0.12 m and the roughness is $5 \cdot 10^{-5}$ m.

1. Calculate the pumping power when the valve V is closed.
2. Calculate the flowrate transported between the two tanks when the valve is partially open and induces a head loss corresponding to equivalent pipe lengths of: (a) 1000 m, and (b) 100 m, respectively.

$\boxed{b}$ The pipeline network shown in Fig. 11.12 is used to transfer a flowrate of 50 kg/s of a fluid ($\rho = 880 \, \text{kg/m}^3$, $\mu = 10^{-3} \text{Pa} \cdot \text{s}$) from tank A to tanks B and C, in equal parts. The layout of the pipeline is horizontal while tank C is placed at an height equal to 30 m and tank B is placed at an height equal to 40 m.

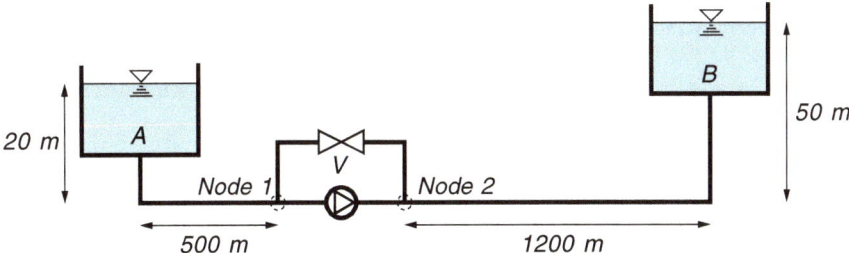

Fig. 11.11 Schematic diagram of a pipeline network designed for the transport of fluid between two tanks

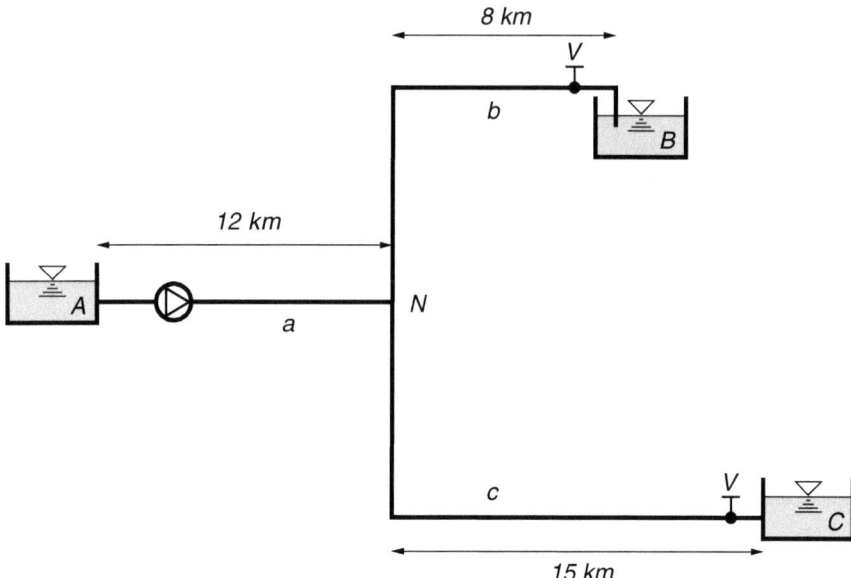

Fig. 11.12 Schematic diagram of a pipeline network

1. Determine the minimum theoretical pumping power considering pipes with a diameter equal to 0.18 m in all sections of the pipeline, assuming that all pipes are smooth and that one of the two valves is fully open while the other is closed as much as necessary to ensure equal flow in the two branches b and c.

2. Determine the economic pipe diameter considering that the diameter of the pipeline in branch a can be different from that in branches b and c, that the capital cost of the installed pipeline is $K_T = 1.2 \cdot 10^3 \, D \, €/m$, the capital cost of the pumping station is $K_P = 2 \cdot 10^3 \, €/kW$ and the operational cost of the pumping station is $K_E = 0.15 \, €/kWh$. Assume that the pipeline operates for 5000 h per year with a payback period of 12 years.

$\boxed{c}$ A cylindrical tank is filled with 2 m³ of a fluid ($\rho = 800 \, kg/m^3$, $\mu = 0.1 \, Pa \cdot s$) that must be periodically transferred to a reactor by means of a horizontal pipeline of length $L = 150$ m. The reactor charging operation must be completed in a maximum time of 1 hr and the volume of fluid that has to be transferred is equal to 1.5 m³.

1. If the initial height of fluid (2 m³) in the tank is equal to 2 m and does not change over time, what is the pipe diameter that allows the charging operation to be completed as requested?

2. Assuming that the tank empties gradually, and considering the diameter calculated at point 1., how long does it take to transfer the required amount of fluid to the reactor?

3. If the purchase cost of the tank in € is $C_t = 250\,S$ where S is the total surface area of the tank in m^2 and the purchase cost of the pipe in € is $G = 600\,D\,L$, with $L = 150$ m and D the pipe diameter in m, how to design the tank-pipe system?

$\boxed{d}$ A flowrate $w = 200$ kg/s of heavy oil (density $\rho = 900$ kg/m^3, viscosity $\mu = 1.5$ Pa $\cdot$ s) must be transported over a distance $L = 25$ km.

1. Determine the economic pipe diameter, D_{econ}, considering the following data:

   a. Capital cost of the installed pipeline: $K_T = 1.6 \cdot 10^3\,D$ €/m;
   b. Capital cost of the pumping station: $K_P = 3.5 \cdot 10^3$ €/kW;
   c. Operational cost of the pumping station: $K_E = 0.2$ €/kWh.
   d. Total hours of operation: $8 \cdot 10^4$.

2. If the oil is mixed with water, the density and viscosity of the mixture change according to the following relations:

$$\rho = (1 - \alpha) \cdot 900 + \alpha \cdot 1000,$$
$$\mu = 1.5 \cdot (1 - \alpha),$$

where α is the volume fraction of water in oil. Determine for the oil flowrate and, for the diameter calculated in point 1., the percentage of water that minimizes the pumping costs.

3. Define and, if possible, solve the economic optimization problem with respect to the two variables D and α.

Appendix A
Suggested Readings

The following list includes texts that may be useful for examining the topics covered in this book from a different perspective and according to different formulations, but also for deepening the study of the subject.

Books Providing an Introduction to Fluid Mechanics

1. M. M. DENN, *Process Fluid Mechanics* Prentice-Hall, Englewood Cliff (NJ), 1980.
2. R. B. BIRD, W. E. STEWART, E. N. LIGHTFOOT, *Transport Phenomena* John Wiley & Sons, New York (NY), 1960.
3. H. ROUSE, *Elementary Mechanics of Fluids* Dover, New York (NY), 1978.
4. I. G. CURRIE, *Fundamental Mechanics of Fluids* McGraw Hill, Singapore, 1993.
5. R. L. PANTON, *Incompressible Flow* John Wiley & Sons, New York (NY), 1960.
6. G. K. BATCHELOR, *An Introduction to Fluid Dynamics* Cambridge University Press, Cambridge (UK), 1981.
7. J. A. FAY, *Introduction to Fluid Dynamics* MIT Press, Cambridge (Mss), 1994.
8. D. J. TRITTON, *Physical Fluids Dynamics* Oxford University Press, New York (NY), 1988.
9. R. ARIS, *Vectors, Tensors and the Basic Equations of Fluid Mechanics* Dover, New York (NY), 1989.
10. S. WHITAKER, *Introduction to Fluid Mechanics* Krieger Publishing Co., Malabar (Fla), 1976.

© The Editor(s) (if applicable) and The Author(s), under exclusive license to Springer Nature Switzerland AG 2024
A. Soldati and C. Marchioli, *Fluid Mechanics for Mechanical Engineers*,
https://doi.org/10.1007/978-3-031-53950-3

Advanced Texbooks

1. H. AREF AND S. BALACHANDAR, *A First Course in Computational Fluid Dynamics* Cambridge University Press, Cambridge (UK), 2017.
2. J. D. ANDERSON, *Fundamentals of Aerodynamics* McGraw-Hill Education, 6^a edizione, New York (US), 2016.
3. A. PROSPERETTI, *Advanced Mathematics for Applications* Cambridge University Press, Cambridge (UK), 2011.
4. A. PROSPERETTI AND G. TRYGGVASON, Computational Methods for Multiphase Flow Cambridge University Press, Cambridge (UK), 2007.
5. S. B. POPE, *Turbulent Flows* Cambridge University Press, Cambridge (UK), 2001.
6. H. TENNEKES AND J. L. LUMLEY, *A First Course in Turbulence* MIT Press, Cambridge (Mss), 1972.
7. J. O. HINZE, *Turbulence* McGraw-Hill, New York (US), 1987.
8. H. SCHLICHTING, *Boundary Layer Theory* McGraw-Hill, New York (US), 1968.
9. L. D. LANDAU AND E. M. LIFSCHIZ, *Fluid Mechanics* Pergamon Press, Oxford (UK), 1989.

Appendix B
Equations in Cartesian, Cylindrical and Spherical Coordinates

B.1 Continuity Equation

Cartesian Coordinates:

$$\frac{\partial \rho}{\partial t} + \frac{\partial}{\partial x}(\rho u_x) + \frac{\partial}{\partial y}(\rho u_y) + \frac{\partial}{\partial z}(\rho u_z) = 0$$

Cylindrical Coordinates:

$$\frac{\partial \rho}{\partial t} + \frac{1}{r}\frac{\partial}{\partial r}(\rho r u_r) + \frac{1}{r}\frac{\partial}{\partial \theta}(\rho u_\theta) + \frac{\partial}{\partial z}(\rho u_z) = 0$$

Spherical Coordinate:

$$\frac{\partial \rho}{\partial t} + \frac{1}{r^2}\frac{\partial}{\partial r}(\rho r^2 u_r) + \frac{1}{r \sin \theta}\frac{\partial}{\partial \theta}(\rho u_\theta \sin \theta) + \frac{1}{r \sin \theta}\frac{\partial}{\partial \varphi}(\rho u_\varphi) = 0$$

B.2 Cauchy Equations

B.2.1 *Cauchy Equations in Cartesian Coordinates*

x component:

$$\rho \left(\frac{\partial u_x}{\partial t} + u_x \frac{\partial u_x}{\partial x} + u_y \frac{\partial u_x}{\partial y} + u_z \frac{\partial u_x}{\partial z} \right) = -\frac{\partial p}{\partial x} + \frac{\partial \tau_{xx}}{\partial x} + \frac{\partial \tau_{yx}}{\partial y} + \frac{\partial \tau_{zx}}{\partial z} + \rho g_x$$

y component:

© The Editor(s) (if applicable) and The Author(s), under exclusive license to Springer
Nature Switzerland AG 2024
A. Soldati and C. Marchioli, *Fluid Mechanics for Mechanical Engineers*,
https://doi.org/10.1007/978-3-031-53950-3

$$\rho\left(\frac{\partial u_y}{\partial t} + u_x\frac{\partial u_y}{\partial x} + u_y\frac{\partial u_y}{\partial y} + u_z\frac{\partial u_y}{\partial z}\right) = -\frac{\partial p}{\partial y} + \frac{\partial \tau_{xy}}{\partial x} + \frac{\partial \tau_{yy}}{\partial y} + \frac{\partial \tau_{zy}}{\partial z} + \rho g_y$$

z component:

$$\rho\left(\frac{\partial u_z}{\partial t} + u_x\frac{\partial u_z}{\partial x} + u_y\frac{\partial u_z}{\partial y} + u_z\frac{\partial u_z}{\partial z}\right) = -\frac{\partial p}{\partial z} + \frac{\partial \tau_{xz}}{\partial x} + \frac{\partial \tau_{yz}}{\partial y} + \frac{\partial \tau_{zz}}{\partial z} + \rho g_z$$

B.2.2　Cauchy Equations in Cylindrical Coordinates

r component:

$$\rho\left(\frac{\partial u_r}{\partial t} + u_r\frac{\partial u_r}{\partial r} + \frac{u_\theta}{r}\frac{\partial u_r}{\partial \theta} - \frac{u_\theta^2}{r} + u_z\frac{\partial u_r}{\partial z}\right) =$$

$$-\frac{\partial p}{\partial r} + \frac{1}{r}\frac{\partial}{\partial r}(r\tau_{rr}) + \frac{1}{r}\frac{\partial \tau_{r\theta}}{\partial \theta} - \frac{\tau_{\theta\theta}}{r}\frac{\partial \tau_{rz}}{\partial z} + \rho g_r$$

θ component:

$$\rho\left(\frac{\partial u_\theta}{\partial t} + u_r\frac{\partial u_\theta}{\partial r} + \frac{u_\theta}{r}\frac{\partial u_\theta}{\partial \theta} + \frac{u_\theta u_r}{r} + u_z\frac{\partial u_\theta}{\partial z}\right) =$$

$$-\frac{1}{r}\frac{\partial p}{\partial \theta} + \frac{1}{r^2}\frac{\partial}{\partial r}(r^2\tau_{r\theta}) + \frac{1}{r}\frac{\partial \tau_{\theta\theta}}{\partial \theta} + \frac{\partial \tau_{\theta z}}{\partial z} + \rho g_\theta$$

z component:

$$\rho\left(\frac{\partial u_z}{\partial t} + u_r\frac{\partial u_z}{\partial r} + \frac{u_\theta}{r}\frac{\partial u_z}{\partial \theta} + u_z\frac{\partial u_z}{\partial z}\right) =$$

$$-\frac{\partial p}{\partial z} + \frac{1}{r}\frac{\partial}{\partial r}(r\tau_{rz}) + \frac{1}{r}\frac{\partial \tau_{\theta z}}{\partial \theta} + \frac{\partial \tau_{zz}}{\partial z} + \rho g_z$$

B.2.3 Cauchy Equations in Spherical Coordinates

r component:

$$\rho \left(\frac{\partial u_r}{\partial t} + u_r \frac{\partial u_r}{\partial r} + \frac{u_\theta}{r} \frac{\partial u_r}{\partial \theta} + \frac{u_\varphi}{r \sin \theta} \frac{\partial u_r}{\partial \varphi} - \frac{u_\theta^2 + u_\varphi^2}{r} \right) =$$

$$-\frac{\partial p}{\partial r} + \frac{1}{r^2} \frac{\partial}{\partial r} (r^2 \tau_{rr}) + \frac{1}{r \sin \theta} \frac{\partial}{\partial \theta} (\tau_{r\theta} \sin \theta) + \frac{1}{r \sin \theta} \frac{\partial \tau_{r\varphi}}{\partial \varphi} - \frac{\tau_{\theta\theta} + \tau_{\varphi\varphi}}{r} + \rho g_r$$

θ component:

$$\rho \left(\frac{\partial u_\theta}{\partial t} + u_r \frac{\partial u_\theta}{\partial r} + \frac{u_\theta}{r} \frac{\partial u_\theta}{\partial \theta} + \frac{u_\varphi}{r \sin \theta} \frac{\partial u_\theta}{\partial \varphi} + \frac{u_r u_\theta}{r} - \frac{u_\varphi^2 \cot \theta}{r} \right) =$$

$$-\frac{1}{r} \frac{\partial p}{\partial \theta} + \frac{1}{r^2} \frac{\partial}{\partial r} (r^2 \tau_{r\theta}) + \frac{1}{r \sin \theta} \frac{\partial}{\partial \theta} (\tau_{\theta\theta} \sin \theta) + \frac{1}{r \sin \theta} \frac{\partial \tau_{\theta\varphi}}{\partial \varphi} + \frac{\tau_{r\theta}}{r} - \frac{\cot \theta}{r} \tau_{\varphi\varphi} + \rho g_\theta$$

φ component:

$$\rho \left(\frac{\partial u_\varphi}{\partial t} + u_r \frac{\partial u_\varphi}{\partial r} + \frac{u_\theta}{r} \frac{\partial u_\varphi}{\partial \theta} + \frac{u_\varphi}{r \sin \theta} \frac{\partial u_\varphi}{\partial \varphi} + \frac{u_r u_\varphi}{r} + \frac{u_\theta u_\varphi}{r} \cot \theta \right) =$$

$$-\frac{1}{r \sin \theta} \frac{\partial p}{\partial \varphi} + \frac{1}{r^2} \frac{\partial}{\partial r} (r^2 \tau_{r\varphi}) + \frac{1}{r} \frac{\partial \tau_{\theta\varphi}}{\partial \theta} + \frac{1}{r \sin \theta} \frac{\partial \tau_{\varphi\varphi}}{\partial \varphi} + \frac{\tau_{r\varphi}}{r} + \frac{2 \cot \theta}{r} \tau_{\theta\varphi} + \rho g_\varphi$$

B.3 Components of the Stress Tensor

B.3.1 Cartesian Coordinates

$$\tau_{xx} = \mu \left[2 \frac{\partial u_x}{\partial x} - \frac{2}{3} (\nabla \cdot \mathbf{v}) \right]$$

$$\tau_{yx} = \tau_{xy} = \mu \left(\frac{\partial u_x}{\partial y} + \frac{\partial u_y}{\partial x} \right)$$

$$\tau_{yy} = \mu \left[2 \frac{\partial u_y}{\partial y} - \frac{2}{3} (\nabla \cdot \mathbf{v}) \right]$$

$$\tau_{zx} = \tau_{xz} = \mu \left(\frac{\partial u_x}{\partial z} + \frac{\partial u_z}{\partial x} \right)$$

$$\tau_{zz} = \mu \left[\frac{\partial u_z}{\partial z} - \frac{2}{3} (\nabla \cdot \mathbf{v}) \right]$$

$$\tau_{zy} = \tau_{yz} = \mu \left(\frac{\partial u_y}{\partial z} + \frac{\partial u_z}{\partial y} \right)$$

B.3.2   Cylindrical Coordinates

$$\tau_{zz} = \mu \left[2\frac{\partial u_z}{\partial z} - \frac{2}{3}(\nabla \cdot \mathbf{v}) \right]$$

$$\tau_{zr} = \tau_{rz} = \mu \left(\frac{\partial u_z}{\partial r} + \frac{\partial u_r}{\partial z} \right)$$

$$\tau_{rr} = \mu \left[2\frac{\partial u_r}{\partial r} - \frac{2}{3}(\nabla \cdot \mathbf{v}) \right]$$

$$\tau_{r\theta} = \tau_{\theta r} = \mu \left[r\frac{\partial}{\partial r}\left(\frac{u_\theta}{r} \right) + \frac{1}{r}\frac{\partial u_r}{\partial \theta} \right]$$

$$\tau_{\theta\theta} = \mu \left[2\left(\frac{u_r}{r} + \frac{1}{r}\frac{\partial u_\theta}{\partial \theta} \right) - \frac{2}{3}(\nabla \cdot \mathbf{v}) \right]$$

$$\tau_{\theta z} = \tau_{z\theta} = \mu \left(\frac{\partial u_\theta}{\partial z} + \frac{1}{r}\frac{\partial u_z}{\partial \theta} \right)$$

B.3.3   Spherical Coordinates

$$\tau_{rr} = \mu \left[2\frac{\partial u_r}{\partial r} - \frac{2}{3}(\nabla \cdot \mathbf{v}) \right]$$

$$\tau_{r\theta} = \tau_{\theta r} = \mu \left[r\frac{\partial}{\partial r}\left(\frac{u_\theta}{r} \right) + \frac{1}{r}\frac{\partial u_r}{\partial \theta} \right]$$

$$\tau_{\theta\theta} = \mu \left[2\left(\frac{1}{r}\frac{\partial u_\theta}{\partial \theta} + \frac{u_r}{r} \right) - \frac{2}{3}(\nabla \cdot \mathbf{v}) \right]$$

$$\tau_{\theta\varphi} = \tau_{\varphi\theta} = \mu \left[\frac{\sin\theta}{r}\frac{\partial}{\partial \theta}\left(\frac{u_\varphi}{\sin\theta} \right) + \frac{1}{r\sin\theta}\frac{\partial u_\theta}{\partial \varphi} \right]$$

$$\tau_{\varphi\varphi} = \mu \left[\left(\frac{u_r}{r} + \frac{u_\theta}{r}\cot\theta + \frac{1}{r\sin\theta}\frac{\partial u_\theta}{\partial \varphi} \right) - \frac{2}{3}(\nabla \cdot \mathbf{v}) \right]$$

$$\tau_{\varphi r} = \tau_{r\varphi} = \mu \left[\frac{1}{r\sin\theta}\frac{\partial u_r}{\partial \varphi} + r\frac{\partial}{\partial r}\left(\frac{u_\varphi}{r} \right) \right]$$

B.4 Navier-Stokes Equations

B.4.1 Navier-Stokes Equations in Cartesian Coordinates

x component:

$$\rho\left(\frac{\partial u_x}{\partial t} + u_x\frac{\partial u_x}{\partial x} + u_y\frac{\partial u_x}{\partial y} + u_z\frac{\partial u_x}{\partial z}\right) = -\frac{\partial P}{\partial x} + \mu\left(\frac{\partial^2 u_x}{\partial x^2} + \frac{\partial^2 u_x}{\partial y^2} + \frac{\partial^2 u_x}{\partial z^2}\right)$$

y component:

$$\rho\left(\frac{\partial u_y}{\partial t} + u_x\frac{\partial u_y}{\partial x} + u_y\frac{\partial u_y}{\partial y} + u_z\frac{\partial u_y}{\partial z}\right) = -\frac{\partial P}{\partial y} + \mu\left(\frac{\partial^2 u_y}{\partial x^2} + \frac{\partial^2 u_y}{\partial y^2} + \frac{\partial^2 u_y}{\partial z^2}\right)$$

z component:

$$\rho\left(\frac{\partial u_z}{\partial t} + u_x\frac{\partial u_z}{\partial x} + u_y\frac{\partial u_z}{\partial y} + u_z\frac{\partial u_z}{\partial z}\right) = -\frac{\partial P}{\partial z} + \mu\left(\frac{\partial^2 u_z}{\partial x^2} + \frac{\partial^2 u_z}{\partial y^2} + \frac{\partial^2 u_z}{\partial z^2}\right)$$

B.4.2 Navier-Stokes Equations in Cylindrical Coordinates

r component:

$$\rho\left(\frac{\partial u_r}{\partial t} + u_r\frac{\partial u_r}{\partial r} + \frac{u_\theta}{r}\frac{\partial u_r}{\partial \theta} - \frac{u_\theta^2}{r} + u_z\frac{\partial u_r}{\partial z}\right) =$$
$$-\frac{\partial P}{\partial r} + \mu\left[\frac{\partial}{\partial r}\left(\frac{1}{r}\frac{\partial}{\partial r}(ru_r)\right) + \frac{1}{r^2}\frac{\partial^2 u_r}{\partial \theta^2} - \frac{2}{r^2}\frac{\partial u_\theta}{\partial \theta} + \frac{\partial^2 u_r}{\partial z^2}\right]$$

θ component:

$$\rho\left(\frac{\partial u_\theta}{\partial t} + u_r\frac{\partial u_\theta}{\partial r} + \frac{u_\theta}{r}\frac{\partial u_\theta}{\partial \theta} + \frac{u_\theta u_r}{r} + u_z\frac{\partial u_\theta}{\partial z}\right) =$$
$$-\frac{1}{r}\frac{\partial P}{\partial \theta} + \mu\left[\frac{\partial}{\partial r}\left(\frac{1}{r}\frac{\partial}{\partial r}(ru_\theta)\right) + \frac{1}{r^2}\frac{\partial^2 u_\theta}{\partial \theta^2} + \frac{2}{r^2}\frac{\partial u_r}{\partial \theta} + \frac{\partial^2 u_\theta}{\partial z^2}\right]$$

z component:

$$\rho\left(\frac{\partial u_z}{\partial t} + u_r\frac{\partial u_z}{\partial r} + \frac{u_\theta}{r}\frac{\partial u_z}{\partial \theta} + u_z\frac{\partial u_z}{\partial z}\right) =$$

$$-\frac{\partial P}{\partial z} + \mu\left[\frac{1}{r}\frac{\partial}{\partial r}\left(r\frac{\partial u_z}{\partial r}\right) + \frac{1}{r^2}\frac{\partial^2 u_z}{\partial \theta^2} + \frac{\partial^2 u_z}{\partial z^2}\right]$$

B.4.3   Navier Stokes Equations in Spherical Coordinates

r component:

$$\rho\left(\frac{\partial u_r}{\partial t} + u_r\frac{\partial u_r}{\partial r} + \frac{u_\theta}{r}\frac{\partial u_r}{\partial \theta} + \frac{u_\varphi}{r\sin\theta}\frac{\partial u_r}{\partial \varphi} - \frac{u_\theta^2 + u_\varphi^2}{r}\right) =$$

$$-\frac{\partial P}{\partial r} + \mu\left[\frac{1}{r^2}\frac{\partial}{\partial r}\left(r^2\frac{\partial u_r}{\partial r}\right) + \frac{1}{r^2\sin\theta}\frac{\partial}{\partial \theta}\left(\sin\theta\frac{\partial u_r}{\partial \theta}\right) + \frac{1}{r^2\sin^2\theta}\frac{\partial^2 u_r}{\partial \varphi^2}\right.$$

$$\left. -\frac{2}{r^2}u_r - \frac{2}{r^2}\frac{\partial u_\theta}{\partial \theta} - \frac{2}{r^2}u_\theta\cot\theta - \frac{2}{r^2\sin\theta}\frac{\partial u_\varphi}{\partial \varphi} -\right]$$

θ component:

$$\rho\left(\frac{\partial u_\theta}{\partial t} + u_r\frac{\partial u_\theta}{\partial r} + \frac{u_\theta}{r}\frac{\partial u_\theta}{\partial \theta} + \frac{u_\varphi}{r\sin\theta}\frac{\partial u_\theta}{\partial \varphi} + \frac{u_r u_\theta}{r} - \frac{u_\varphi^2\cot\theta}{r}\right) =$$

$$-\frac{1}{r}\frac{\partial P}{\partial \theta} + \mu\left[\frac{1}{r^2}\frac{\partial}{\partial r}\left(r^2\frac{\partial u_\theta}{\partial r}\right) + \frac{1}{r^2\sin\theta}\frac{\partial}{\partial \theta}\left(\sin\theta\frac{\partial u_\theta}{\partial \theta}\right) + \frac{1}{r^2\sin^2\theta}\frac{\partial^2 u_\theta}{\partial \varphi^2}\right.$$

$$\left. +\frac{2}{r^2}\frac{\partial u_r}{\partial \theta} - \frac{u_\theta}{r^2\sin^2\theta} - \frac{2\cos\theta}{r^2\sin^2\theta}\frac{\partial u_\varphi}{\partial \varphi}\right]$$

φ component:

$$\rho\left(\frac{\partial u_\varphi}{\partial t} + u_r\frac{\partial u_\varphi}{\partial r} + \frac{u_\theta}{r}\frac{\partial u_\varphi}{\partial \theta} + \frac{u_\varphi}{r\sin\theta}\frac{\partial u_\varphi}{\partial \varphi} + \frac{u_r u_\varphi}{r} + \frac{u_\theta u_\varphi}{r}\cot\theta\right) =$$

$$-\frac{1}{r\sin\theta}\frac{\partial P}{\partial \varphi} + \mu\left[\frac{1}{r^2}\frac{\partial}{\partial r}\left(r^2\frac{\partial u_\varphi}{\partial r}\right) + \frac{1}{r^2\sin\theta}\frac{\partial}{\partial \theta}\left(\sin\theta\frac{\partial u_\varphi}{\partial \theta}\right) + \frac{1}{r^2\sin^2\theta}\frac{\partial^2 u_\varphi}{\partial \varphi^2}\right.$$

$$\left. -\frac{u_\varphi}{r^2\sin^2\theta} + \frac{2}{r^2\sin\theta}\frac{\partial u_r}{\partial \varphi} + \frac{2\cos\theta}{r^2\sin^2\theta}\frac{\partial u_\theta}{\partial \varphi}\right]$$

B.5 Components of the Vorticity Vector

B.5.1 Cartesian Coordinates

$$\omega_x = \frac{\partial u_y}{\partial z} - \frac{\partial u_z}{\partial y} \quad , \quad \omega_y = \frac{\partial u_z}{\partial x} - \frac{\partial u_x}{\partial z} \quad , \quad \omega_z = \frac{\partial u_x}{\partial y} - \frac{\partial u_y}{\partial x}$$

B.5.2 Cylindrical Coordinates

$$\omega_r = \frac{1}{r}\frac{\partial u_z}{\partial \theta} - \frac{\partial u_\theta}{\partial z} \quad , \quad \omega_\theta = \frac{\partial u_r}{\partial z} - \frac{\partial u_z}{\partial r} \quad , \quad \omega_z = \frac{1}{r}\left[\frac{\partial(r u_\theta)}{\partial r} - \frac{\partial u_r}{\partial \theta}\right]$$

B.5.3 Spherical Coordinates

$$\omega_r = \frac{1}{r \sin \theta}\left[\frac{\partial(u_\varphi \sin \theta)}{\partial \theta} - \frac{\partial u_\theta}{\partial \varphi}\right]$$

$$\omega_\theta = \frac{1}{r}\left[\frac{1}{\sin \theta}\frac{\partial u_r}{\partial \varphi} - \frac{\partial(r u_\varphi)}{\partial r}\right]$$

$$\omega_\varphi = \frac{1}{r}\left[\frac{\partial(r u_\theta)}{\partial r} - \frac{\partial u_r}{\partial \theta}\right].$$